BAR INTERNATIONAL SERIES 3036 | 2021

A River in 'Drought'?

Environment and cultural ramifications of Old Kingdom climate change

JOHN W. BURN

BAR PUBLISHING

Published in 2021 by
BAR Publishing, Oxford

BAR International Series 3036

A River in 'Drought'?

ISBN 978 1 4073 5810 9 paperback
ISBN 978 1 4073 5811 6 e-format

DOI https://doi.org/10.30861/9781407358109

A catalogue record for this book is available from the British Library

COVER IMAGE *Author's own photograph, showing the view of the river and desert from Aswan.*

BAR
PUBLISHING

BAR titles are available from:

BAR Publishing
122 Banbury Rd, Oxford, OX2 7BP, UK
EMAIL info@barpublishing.com
PHONE +44 (0)1865 310431
FAX +44 (0)1865 316916
www.barpublishing.com

Of Related Interest

"Yellow" Coffins from Thebes
Recording and decoding complexity in Egyptian funerary arts
Rogério Sousa

Oxford, BAR Publishing, 2020 — BAR International Series **S3010**

Early Bronze IV Village Life in the Jordan Valley
Excavations at Tell Abu en-Ni'aj and Dhahret Umm el-Marar, Jordan
Steven E. Falconer and Patricia L. Fall with contributions by
Ilya Berelov and Steven Porson

Oxford, BAR Publishing, 2019 — BAR International Series **S2922**

Archaeobotanical investigations of plant cultivation and husbandry practices
At the Early Bronze Age settlement Küllüoba in West-Central Turkey:
Considerations on environment, climate and economy
Özgür Çizer

Oxford, BAR Publishing, 2015 — BAR International Series **S2766**

The Management of Estates and their Resources in the Egyptian Old Kingdom
Joyce Swinton

Oxford, BAR Publishing, 2012 — BAR International Series **S2392**

Horus' Eye and Osiris' Efflux: The Egyptian Civilisation of Inundation c. 3000-2000 BCE
Terje Oestigaard

Oxford, BAR Publishing, 2011 — BAR International Series **S2228**

Egyptian Tomb Architecture
The archaeological facts of pharaonic circular symbolism
David I. Lightbody

Oxford, BAR Publishing, 2008 — BAR International Series **S1852**

Acknowledgements

There are many wonderful people that helped me on this journey: initially the lecturers who inspired me to turn this hobby into an obsession: Professor Naguib Kanawati, Associate Professor Boyo Ockinga, Professor E. Christiana Kohler and Doctor Suzanne Binder. I will be eternally grateful to my supervisors, Professor Kanawati and Doctor Linda Evans who allowed me to develop a research path suited to my training, interest, and expertise. Gratitude is directed to my first site supervisors Doctor Sophie Winlaw and Naguib Victor, who helped develop a taste for fieldwork that is still undiminished.

Deep appreciation is directed towards Doctors Mary Hartley and Beth Thompson, who provided never-ending encouragement that there indeed was an end in sight. To the USER crowd for continuing to believe that I would finish, I give thanks. Acknowledgement is also directed towards Associate Professors Yann Tristant and Ronika Powers, and to Leonie Donovan who provided specialist advice when called upon. A special thankyou goes to Doctor Joyce Swinton whose insightful comments made this document infinitely better.

And finally, and most importantly, to Doctor Alexandra Woods. Lecturer, tutor, site supervisor, guide, and inspiration, who sat beside me during one First Intermediate Period tutorial and asked if I had considered applying for fieldwork....... I will be forever grateful.

Dedicated to my partner, Peri, my parents, Jan and Bill, and my whole family who have always supported me and allowed me to follow this dream.

For Beth.

Contents

List of Figures

List of Tables

Foreword and Introductory Remarks

Thus spake Ankhtify:[1]

> *"I gave life to Hefat and Hermer when the sky was clouded and the land in wind, when every man was dying of hunger under the sandbank of Apophis...*
>
> *"I rushed this grain southwards to arrive at Wawat. It travelled downstream, it arriving at Abydos while Upper Egypt in its entirety was dying of hunger, all men eating their children.*
>
> *"Never did I allow death from hunger to occur in this province...*
>
> *"Ankhtify says that this entire land had become like a grasshopper with emptiness as one goes north, the other going south..."*

According to Ankhtify, nomarch of the third Upper Egyptian nome and overlord of the second, a significant famine had struck Egypt during his time in power. The calamity of which he speaks is thought to have come about as a consequence of steadily declining Nile water levels, which had become a serious concern by the end of the Old Kingdom.[2] In an heroic effort, Ankhtify managed to sustain the lives of his own people and due to his remarkable ability, was even able to provide food for some of his neighbours.[3] Conversely, in the same autobiography, Ankhtify identifies the spread of the marshlands as a significant problem when he tried to enforce the authority of the king in the region and to impose his own leadership upon those local rulers who appeared somewhat reluctant to embrace it.[4]

[5] *gmi.n(=i) pr.w ḫww ṯtf(.w) mi grg.t mkḥ3.n* [6] *ir.y=f*

"I found the house of Khui inundated like a marsh, abandoned by him who belonged to it"

Thus, there seemed to have been two competing influences affecting Ankhtify and his contemporaries. The land was in drought while the river had become like a swamp. Are we experiencing narrative hyperbole, or can there exist a situation where both events occur simultaneously?

[1] Author translation (apologies to Nietzsche) from Vandier, *Mo'alla*, Ins. 1–12, 161–174.

[2] Hassan, *Floods & Disasters*, 1–23; Said, *River Nile*, 141–143; Lloyd, *State & Society*, 176–177; Stanley etal, *Nile Flow Failure*, 395–402; Parcak, *Physical Context*, 8–10.

[3] Merer in Lichtheim, *Literature* I, 87, makes similar claims. On the veracity of Ankhtify's claims, see Vandier, *Mo'alla*, 34–44; Coulon, *Véracité et rhétorique*, 117, 129–132.

[4] Grimal, *A History of Ancient Egypt*, 142; Seidlmayer, *FIP*, 129; Manassa, *El-Mo'alla to El-Deir*. http://escholarship.org/uc/item/4pc0w4hg (13/03/2016).

F.1. Egypt is the Nile

Ancient Egypt has been described as a '*Hydraulic Society*',[5] its civilisation developing as a direct consequence of the large-scale manipulation of water.[6] It is difficult to envisage the development of the complex ancient Egyptian civilisation without the adoption of irrigation, something recognised by historians of old: Herodotus, in his *Histories*, recognised the land of Egypt as a "gift from the Nile", while Pliny the Elder claimed it was the river that did the farming.[7] The hydrology of Egypt is unique because almost all of the water that flows through the country comes from external sources, initially falling as snow and rainfall in locales much further south. Each year, the river peaked in an annual flood because of the monsoonal rains in the south, swelling the normally placid flow into a mighty surge. These waters carried masses of rich silt, which were deposited as they flowed over the banks and spread across much of the Nile valley.

The Egyptian calendar year had three seasons, and their names and timings derived from the river and the associated agricultural production. Before the floods arrived every season, the land needed to be repaired and prepared: natural dams were improved and extended, old water channels needed to be repaired and new channels dug.[8] The start of the calendar coincided with the arrival of the floodwater, known as the time of *akhet*, 'the inundation' (of water). When the flood occurred, it discharged nutrients carried by the river onto the surrounding countryside, these nutrients,[9] therefore being lost to the river (see Figure F.1).

Peret, 'the emergence' (of land), is the period when water started to recede, leaving rich and fertile alluvial soils ready for planting, while the season known as *shemu,* 'the deficiency' (of water), was the time when harvests began. This regular cycle underpinned the society that developed in the Nile valley.[10]

The Egyptians developed an agricultural system thought to be one of the most productive in the ancient world. Good quality, fertile land was a prerequisite for successful

[5] Butzer, *Hydraulic Egypt*, 7–10; Alleume, *Systemes Hydraulic*, 301–322.

[6] Wittfogel, *Oriental Despotism*, 19, 311. Also Wörster, *Rivers of Empire*; Butzer, *Hydraulic Egypt*.

[7] Herodotus, *Histories* 2.5; Pliny the Elder, *Natural History*, 18, 47 & 167.

[8] Butzer, *Hydraulic Egypt*, 41–47.

[9] Said, *River Nile*, 131–135.

[10] Hildebrand & Schilling, *Early Nile Agriculture*, 81–95, Alleume, *Systemes Hydraulic*, 301–322.

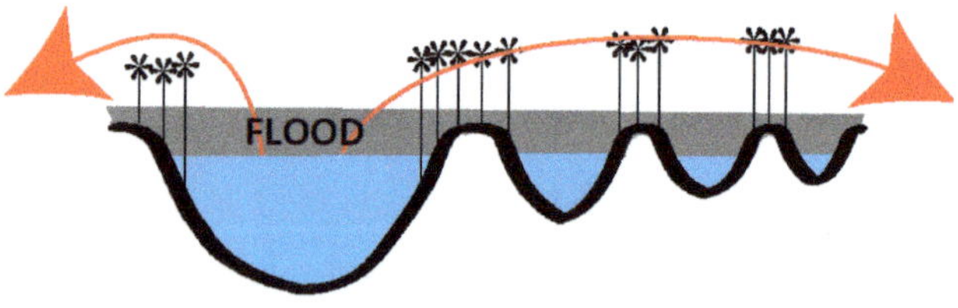

Figure F.1: An inundation deposits resources onto the land.

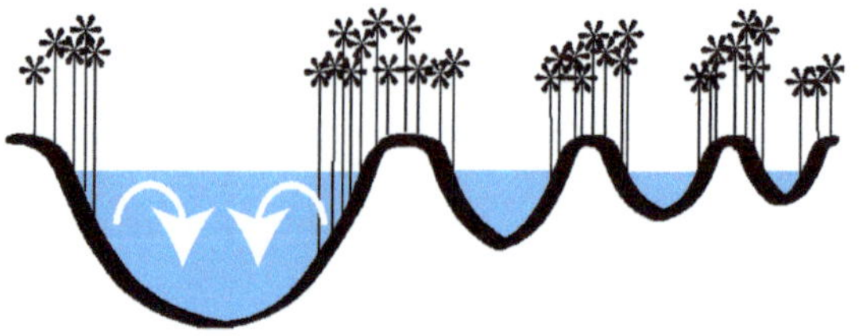

Figure F.2: No inundation means that the nutrients remain in the river.

economic development.[11] Field crops such as wheat, barley, flax, vegetables and fruits provided food, drink and fabric. To ensure a successful harvest, the height of the Nile floods was crucial, and it was measured each year. Nilometers, stone steps leading from the riverbank down into the water, registered the level and, during the Old Kingdom, the heights of 63 annual floods were recorded by 11 different kings. With increased control of water, agricultural surplus could develop.[12] Agricultural success was then followed by rapid population growth. As long as resources were available, overpopulation was not a significant problem.[13] However, though the annual flood was mostly reliable, it could sometimes appear lower than expected.[14] Regular lower-than-average floods led to an onset of lean times with numerous accounts of famine recorded in Egypt's past, with evidence from objects such as the Palermo Stone suggesting short-term fluctuations.[15]

It seems, however, that during the Old Kingdom, the power of the river and the inundation associated with its annual flood was gradually weakening.[16] During 'drought', the river would still have carried these nutrients downstream but did not have the necessary volume to overflow its banks, resulting in nutrient retention within the river channel. As the current slowed, it would have lost the necessary momentum to carry these nutrients, which would have been deposited on the riverbed and river's edges rather than beyond (see Figure F.2), providing excess biological energy to be utilised within the riverine habitat.

F.2. The Effect of 'Drought' upon the River

Since reliable agriculture has been linked to the prosperity and stability of the society[17] a continuous slow decline in water levels[18] and a corresponding increase in aridity should have been noticed at the end of the Old Kingdom.[19] It has been suggested that the drought had a major impact upon the culture of ancient Egypt – in some cases, the fall of kings.[20] Evidence suggesting a continuous drying out of the country from the Predynastic era onwards implies that famine and drought were at least a regular occurrence,[21] one that perhaps should not have had as significant an impact upon the civilisation as many claim.[22] The developing aridity of this time may have encouraged improved production methods.[23] If the current modelling is correct, however, and if ancient Egypt was experiencing climate change, then, in addition to regular lower than usual levels of inundation, the country may also have experienced regular above-average floods as well as unexpected significant rainfall events. Said, in describing the various fluctuations that the River Nile has experienced over its history, comments on the vagaries of its flow.[24]

F.3. The Effect of Excess Water upon the River

While the inundation was relatively predictable, whatever its level, rainfall events in Egypt occurred with irregular and unpredictable timing, thereby interfering with the progression of the 'normal' cultivation cycle. Just as too little flooding meant that irrigation efforts were ineffective, excessive flooding was also problematic.[25] Agricultural success further south, for example, is attributed to lack of large inundations, with excess water flowing downstream.[26] Just as the flood governed the annual outcomes of agriculture, the timing of rainfall events would also have impacted upon the success, or otherwise, of the cultivation cycle.[27]

However, high water levels do not automatically equate to a positive environment.[28] A problem almost as significant as lower than usual flood heights was the occurrence of higher-than-average flood levels. Potential cultural responses to the environmental impact of variable flood levels have been proposed by Butzer and Bárta.[29] Extreme

[11] Soroush & Mordechai, *Short-term Cataclysm*, 1, 8–10.
[12] Said, *River Nile*, 188–208.
[13] Hely, etal., *Climate Water Tradeoff*, 681–686.
[14] Said, *River Nile*, 131–133, 141–142, 149–152.
[15] Macklin etal., *River Dynamics*, fig. 7.
[16] Said, *River Nile*, 134–138; Hamdan etal., *Climate & Collapse*, 89-100.
[17] Bard, *Archaeology of Ancient Egypt*, 51–53; Lawler, *Collapse Revisited*, 907–908. Santoro etal., *1000 Years of Famine*, shows extremes of flood and drought following a semi-regular unpredictable pattern.
[18] Hoffman, *Before the Pharaohs*, 311; Butzer, *Discontinuity*, 106. Park, *Class Stratification*, fig. 3, 103, infers a gradual increase in aridity throughout the Old Kingdom.
[19] Bárta, *Long or Short Term*? disputes the role of climate. Said, *River Nile*, 131–133, identifies Pre- and Early Dynastic droughts.
[20] Schneider, *Existence Now Silenced*, 311.
[21] Brewer, *Predynastic Temperatures*, 299–300.
[22] Said, *River Nile*, 7–91; Phillips etal., *Mid-Holocene Egypt*, 64–74; Butzer, *Collapse, Environment & Society*, 3634; Bárta, *Collapse Hidden in Success*, 18–28; Stocker etal., *Climate Change* 2013, 33–115.
[23] Midant-Reynes, *Prehistory*, 253 – contra Weiskel, *Environmental Lessons*, 99.
[24] Said, *River Nile*, 129, 143–148, 176.
[25] Coulon, *Famine*, 1.
[26] Macklin etal., *Nile Floodwater Farming*, 698.
[27] Morris, *Art of Not Collapsing*, 79–81, Soroush & Mordechai, *Short-term Cataclysm*, 7–8; Nicholson & Shaw, *Materials & Technology*, 514.
[28] In fact, the flood levels during the First Intermediate Period appeard to be above average; see Schneider, *Existence Now Silenced*, 314.
[29] Butzer, *Discontinuity*, 254–257; Bárta, *Collapse Hidden in Success*, 23–25.

water levels could wash away years of careful planning and preparations; canal and boundary markers were lost, field fences washed away and the organised sharing of excess water severely disrupted.[30] The loss of infrastructure would have had significant negative consequences to ineffective governments.[31] Thus, excessive floods and unusual rainfall events may have interfered with the development of agriculture and the maintenance of cultivation in the region.[32] Other scholars have even suggested that excess rainfall events and higher-than-average floods in the Nile valley may have hindered the adoption of agriculture in the region.[33]

At this time in history, high rainfall events causing large scale erosion in wadis was followed by instances of dry sand entering the Nile Valley. Excessive siltation can produce dense, compacted and poorly aerated soils; this saturated soil is less productive than well-worked land.[34] High volumes of water can result in severe weathering and rapid erosion as large amounts of fertile soil, laid down in preceding years, are washed away back into the river,[35] resulting in the riverine habitat receiving extra nutrient resources (see Figure F.3)[36]. Recent evidence points to the likelihood that, at a time when the flood had become less reliable, the Nile valley experienced significant rainfall events.[37] While unexpected, rainfall within the Nile valley has been shown to have always occurred, and mounting evidence suggests that severe rainfall events were becoming more commonplace during the timeframe under investigation.[38]

The physical force of the pluvial cascade impacted upon the integrity of some tombs built during this time, for example the tomb of Merefnebef,[39] while other tombs were structurally re-designed to resist the negative impact of a rain-fuelled deluge.[40] Interestingly, much of the land abandonment that occurred in Mesopotamia at this time has been linked to unreliable or irregular rainfall patterns,[41] suggesting a connection between unusual rain events in

Figure F.3: Rainfall returns nutrients to the river.

Egypt with those which occurred in other places around the region at this time.[42]

F.4. Scientific Evidence of Irregular Water Supply

A large amount of corroborating scientific evidence identifies the timing and significance of the increasing aridity experienced at the end of the Old Kingdom and the consequent decline in agricultural success.[43] According to current scientific opinion, this phenomenon reached its nadir 4200 years ago.[44] Scientific evidence supporting this date includes analyses of ancient rainfall precipitation,[45] Nile delta sediment studies,[46] Mediterranean seabed silt deposit investigations,[47] Isotopic examination of fish bone remains,[48] chemical residue analyses of a variety of lakebed sediments;[49] as well as pollen deposit variances from various Nilotic plants[50] and radiocarbon examination of miscellaneous genetic material found in ancient remains.[51] The conformity of 'agreement' amongst the findings could make a cynical person suspicious, but there does seem to be a vast amount of evidence from many branches of science supporting the assertion of a significant environmental event taking place over the time in question: this time frame identified as approximately 4,200 years before the present. [4200 BP].

Recently, however, it has begun to appear that to some scholars that the idea of a long-term drought seems perhaps a little too simplified to explain all the observations and interpretations that have been made.[52] Very little distinction has been made between the various types of floods and their different impacts upon the society. These disruptive events may be indicators of a climate changing more gradually over the long term. Current climate change modelling suggests that extremes of weather will become

30 Said, *River Nile*, 7–91; Butzer etal., *Urban Geoarchaeology*, 3361–3364, fig. 19; Morris, *Art of Not Collapsing*, 80-81.
31 Butzer, *Pleistocene Nile*, 253–280; Nicholson & Shaw, *Materials & Technology*, 514.
32 Butzer & Hansen, *Desert & River*, 115, 129, 278; Butzer etal., *East African Lake Levels*, 1069–76.
33 Hassan, *Food Security*, 321–334; Butzer, *Hydraulic Egypt*, 9.
34 Butzer, *Hydraulic Egypt*, 17–18; Said, *River Nile*, 143–148.
35 Butzer, *Hydraulic Egypt*, 15–17, 51–53; Sampsell, *Traveller's Geology*, 12.
36 Hassan, *A River Runs Through*, 22–25; Bunbury etal., *Water Historical Perspective*, 52–71; Bunbury etal., *Memphite Floodplain Landscape*, 80–81; Butzer, *Quaternary Nile*, 161–162, 171.
37 Myśliwiec etal., *Saqqara Geoarchaeology & Paleoclimate*, 294–295; Hayes, *Most Ancient Land*, 96; Geriesh etal., *Groundwater*, 605, fig 21; Touzeau etal., *Nile Valley Aridity*, 92–100.
38 Welc & Marks, *OK Climate*, 124–133; Sowada, *Weather Evidence*, 69–74; Welc & Trzciński, *Dry Moat Geology*, 323–343; Trzciński etal., *West Saqqara Geoarchaeology*, 194; Embabi, *Landscapes & Landforms*, 32–35.
39 Myśliwiec etal., *Merefnebef*, 41, pl. 7b.
40 Kuraszkiewicz, *Architectural Innovations*, 32.
41 Myśliwiec etal., *Saqqara Geoarchaeology & Paleoclimate*, 295–298; Fiorentino, *Syrian Climate Change*, 56.

42 Kowalska & Kuraszkiewicz, *End of a World*, 173–176; Walker etal., *Holocene Subdivision*, 653–656.
43 Rzóska, *Nile*, 35–36.
44 See, for example, Krom etal., *Nile Sediment Fluctuations;* Stanley etal., *Nile Flow Failure*; Ducassou etal., *Nile Floods*.
45 Williams & Nottage, *Extreme Rainfall Impacts*; Butzer, *Discontinuity*; Bárta, *In Mud Forgotten*.
46 Arz etal., *Red Sea Dry Event*; Marriner etal., *Nile Delta's Sinking Past*; Shaltout & Azzazi, *Nile Delta Climate Change*.
47 Ducassou etal., *Nile Floods in Sediments;* Krom etal., *Nile Sediment Fluctuations*.
48 Brewer, *Predynastic Temperatures*.
49 El-Wakeel, *Lake Qarun Deposits;* Touzeau etal., *Nile Valley Aridity*.
50 Langutt etal., *Late Bronze Collapse*.
51 Butzer etal., *East African Lake Levels*; Buzon & Simonetti, *Strontium Isotope Variability*; Zanchetta, *Tephrostratigraphy*; Williams etal., *Late Quaternary Floods & Droughts*.
52 Bard, *Archaeology of Ancient Egypt*, 176–180, Müller-Wollermann, *End of the Old Kingdom*, 5.

more 'regular', longer droughts, wetter rainy seasons, cooler cold days and warmer hot days.

Rather than a predictable climate (climate being what you expect over a period of time), the world would seem to be entering a phase where the weather (weather being what you get day to day) experienced appears more unsettled.[53] The same phenomenon may have been experienced in Ancient Egypt with an 'oscillating' pattern of uneven weather displaying more irregular behaviours than previously experienced.[54] Developing geomorphological technologies have allowed scientists to improve the understanding of the nature of the historical Egyptian environment.[55] Kilimanjaro ice core analyses, for example, have identified stages in Egypt's recent past, with the most recent example dating from 5000–3500 BCE, when so much water would have been available for cultivation that irrigation should *not* have been necessary.[56] The annual regeneration of the soil and the subsequent abundant natural yield allowed for the retention of a herder-forager lifestyle to remain as effective cultural behaviours for many prehistoric Egyptians. This notion is at first glance counter-intuitive, but it is feasible to look on the river as a natural provider rather than the source of labour-intensive agriculture-based sustenance.[57] It was possible that the natural growth of pasture/vegetable greens/vines meant the Egyptians did not have to farm. Perhaps the continued practice of a herder-forager lifestyle in many parts of the country was as a consequence of unpredictable river behaviour, with the adoption of agriculture as an option adopted in response (i.e., choosing to) to a changing climate and not as a consequence of it (i.e., being forced to).[58]

F.5. Rationale, Hypotheses and Goals

When a river is expected to flood annually but an inundation does not occur, then the land bordering the river experiences a so-called 'drought'. In the cultural exemplar of many Western countries, a drought, usually defined as "*a prolonged period of abnormally low rainfall, leading to a shortage of water,*"[59] is what happens to the land, and its associated affects are usually applied to the impact that it has upon that land. In discussion of the Old Kingdom, a drought is usually seen as a dramatically low or absent Nile River inundation, and, similarly, most commentators considered the drought's effects on the land and the implications that a low flood had for the society.

It is important to not confuse our modern understanding of the term "drought;" with its use in the Egyptological discourse. In the modern vernacular, drought refers… "*as a period of unusually dry weather that persists long enough to cause problems such as crop damage and water supply shortages.*"[60] Traditionally, Egyptological terminology equates the term 'drought' to the lack of the regular/ reliable annual flood,[61] which may be more precisely identified as an "agricultural drought"; one whereby "*the water needs for crops at various growing stages is not being met*".[62] Examples may include not enough moisture at planting time to germinate seeds or inadequate water while growing, leading to reduced yields. In the terms of this study, a 'drought' refers to a long period of less than normal river levels. This is more precisely identified as a "Hydrological drought" in the modern science literature, referring to "*persistently low water volumes in streams, rivers and reservoirs.*" The problems that will be discussed are based on the understanding that, during the time frame under investigation, the river was in a state of decline and the amount of annual relief that was normally delivered during the inundation was less than anticipated or expected.

"*A society is circumscribed by its environment.*"[63] To help us understand ancient Egyptians, we first need to develop an understanding of their environment.[64] Analysis of the long-term environment of a site or region allows for the investigation of a number of inferences that may arise based upon current scientific considerations,[65] improving the understanding of the cultural development of that site or region, as opposed to it being a centre for redistribution.[66] Hierakonpolis, as an example noted earlier, had developed into an important cultural centre by Predynastic times despite having no resources of note.[67]

Despite the apparent importance of the "drought" in the discourse of the end of the Old Kingdom, a critical aspect of investigations into this era appears to have been left out by the scholarship so far: scientific analyses of the ecological changes that would have taken place within the river itself during times of drought. From an ecological standpoint, when a river floods, it loses sediments and nutrients to the surrounding landscape. By contrast, a river that does NOT flood retains these nutrients. This organic bounty remains

[53] See Stocker & Qin, *Climate Change* 2013.
[54] Bunbury etal., *Memphite Floodplain Landscape*, 89–90; Steward etal, *River Runs Dry*, 202–203.
[55] see Butzer, *Archaeology & Geology*, 1617–1624; Welc & Marks, *OK Climate Change*, 1–10.
[56] Macklin etal., *Nile Floodwater Farming*, 696, fig. 3.
[57] A similar lifestyle to those 'fringe-dwelling' populations of the Mesopotamian marshlands, see http://www.clw.csiro.au/publications/consultancy/2004/ Mesopotamian-marshlands-soil.pdf (22/03/2018).
[58] Midant-Reynes, *Prehistory*, 137–138.
[59] https://en.oxforddictionaries.com/definition/drought (11/11/2016).
[60] https://www.livescience.com/21469-drought-definition.html (12/03/2016).
[61] Baines & Malek, *Encyclopaedia*, 18.
[62] https://www.livescience.com/21469-drought-definition.html (12/03/2016) (note: this site used for all drought based definitions).
[63] Butzer, *Discontinuity*, 102.
[64] Butzer, quoting Keimer, *Ecological Archaeology*, 106–111.
[65] Hassan & Stucki, *Climatic Change*, 37–46; Crumley, *Historic Ecotonal Shifts*, 377–384; Downing etal., *African Climate Change*, 19–44. See also Talling, *History of Nile Research*. Höflmayer, *Dating Catastrophes*, 133, points out that, including himself, many scholars still see no direct evidence.
[66] Butzer, *Hydraulic Egypt* and *Environment & Human Ecology*. Seidlmayer, *OK Elephantine*, 108–111, suggests that not all centres developed along the same lines or for the same reasons.
[67] Hoffman, *Before the Pharaohs*, 160–161; Butzer, *Archaeology & Geology*, 1620–21.

within the river and should significantly alter the normal biological networks of the river. Few investigations into how the Nile's ecology and its related ecosystems changed in response to a changing climate have been undertaken.[68] Changing ecological conditions would have changed the environmental circumstances of the river and, therefore, should have impacted upon the culture that developed along its banks have been undertaken. Just how the river may have been affected by drought and how this may have impacted upon the surrounding culture is the basis for this investigation.

In this examination, the consequences to the river itself of an insufficient or dramatically low inundation will be investigated. In a similar fashion, rainfall events ADD nutrients to the river, another source of nutrient concentration.

The rationale for the investigation is as follows:

1. *A weakened river (one with less ability to deliver the expected and desired flood levels necessary to sustain the annual Egyptian agricultural cycle) presumably carried a lower volume of water. The resultant inundation would have deposited fewer nutrients onto the land.*
2. *Unexpected rainfall events at this time would have resulted in extra nutrient flow into the rivers.*
3. *Consequently, the amount of nutrients remaining in the river should have increased, thereby impacting upon the Nile's ecology and its immediate environment. Because the river was so important to the society, any changes to it would have had an impact upon that society.*

F.6. The A.R.I.D. Hypothesis – an outline

From the above rationale, the following hypothesis was developed:

1. *If Egypt experienced an environmental change at the end of the Old Kingdom, then this must have impacted upon the river, and therefore the society that relied upon it.*
2. *This ought to have left an impression upon the visual culture produced by that society. It should be possible to identify changing influences in the artwork produced by observing the variances in the tomb decoration patterns over the time frame under investigation.*

The overall aim of this investigation, therefore, is to suggest what may have happened to the river Nile when the land was in 'drought'[69] and, the river increasingly nutrient-rich, to attempt to identify signs of a societal response. By attempting to identify the potential physical implications of the phenomenon of a declining river, this investigation will seek to identify how ancient Egyptian society responded to the phenomenon of A River In 'Drought'; hence the A.R.I.D. Hypothesis.

The project has been planned to follow two pathways of consideration, environmental and social: 1– In order to investigate any environmental change within the river, an examination of the ecological processes that could have been occurring is provided; 2– The changing compositions of tomb scenes is traced in order to identify any subsequent cultural change. The report is broken into four parts.

Part A will provide a palaeo-climatic overview, summarising and emphasising the importance that climate played in the developing riverine civilisation that came to be ancient Egypt. It will attempt to ascertain the ways that late Old Kingdom and early First Intermediate Period art may have indicated these environmental changes.

Part B will present a summary of basic ecological terms and will introduce the central principles of riverine ecology that have to be applied to the present investigation. From there, modern ecological examinations of current waterways experiencing similar irregular or unnaturally low water flows will then be applied to the situation that may have existed in ancient Egypt. From these, the potential habitat changes that may have occurred to the Nilotic system at the end of the Old Kingdom will be suggested.

Part C will attempt to identify if any perceived cultural response to this environmental change can be detected through changes to the decoration repertoire as depicted in tombs from the period in question. The progression of tomb scenes produced in the Old Kingdom and into the First Intermediate Period will be analysed for any apparent change in the patterns of representation.

Part D will discuss the results and summarise the findings. It will suggest potential directions for future examination. Finally, a conclusion will be presented.

[68] Beyond three seminal texts: Butzer, Early Hydraulic Civilization in Egypt: A Study in Cultural Ecology (1976), Rzóska, The Nile, Biology of an Ancient River (1976) and Said, The River Nile: Geology, Hydrology and Utilization (1993).

[69] The term 'drought' is used advisedly; as will become apparent.

Part A

Environment, Society and Culture

1

Climate and Society in the Old Kingdom

The surroundings in which the ancient Egyptians found themselves significantly influenced their outlook towards life, their art, and their religion. Many religious festivals were linked to the annual recurrence of natural events: for example, Utterance #581 (*...inundations are upon the land...*) of the Pyramid Texts praises the river for its role in ensuring the continued fertility of the land.[1] To the ancient Egyptians, oases, swamps, and marshlands provided important wetlands, while the eastern and western deserts formed natural borders to their everyday world. The traditional name in ancient Egypt for the fertile valley with its alluvial soil, *Kemet*, meaning 'black land', related directly to the rich sediment 'gifted' by the river. This contrasted significantly to the remainder of the country: *Desheret,* the 'red land' of the desert. The moist and fertile soil of the valley was dedicated to agriculture, animal husbandry and the living; none of it was wasted on the dead. Cemeteries, up and down the Nile, were either constructed on desert escarpments or hewn into mountain rock; this land being agriculturally unproductive. It was, however, very productive in terms of its characteristics being conducive to the preservation of the body: an interesting paradox. From the deserts to the arable lands to the river, from the inundation to the harvest, each component of the surrounding environment had an impact upon the culture of the ancient Egyptians. Any change to these basic characteristics would have had a noticeable impact upon the traditions of the society.

1.1. Egypt and the Nile

As the formation of the Nile valley began only about 25 million years ago; in geological terms, the Nile River may be regarded as relatively young.[2] The river carried water all the way from the Ethiopian highlands and the lakes of Central Africa, over time carving its way through the northeast corner of Africa to reach the Mediterranean Sea. The river branched before reaching the coast in the locality of modern-day Cairo, with several tributaries forming a region now called the Delta. As the sea-level of the Mediterranean fell, about 100,000–120,000 years ago, the speed of the Nile flow increased, as did the rate of erosion, digging ever deeper into the landscape to form the Nile valley.[3]

After the annual Nile floods, natural high points, known as *gaziras*, produced areas that were safe to occupy and close to areas fertilised by a build-up of alluvial soil; these *gaziras* could also serve as natural dam walls.[4] Sometimes, along the length of a river, secondary channels formed.[5] These secondary channels were a natural part of the river system and formed as the river branched as it moved onwards winding its way downstream to the sea.[6]

Along the Nile these secondary channels are known as '*khors*',[7] and they represented economic and ecological benefits to the local population – a potential for exploitation that still exists in the modern age.[8] By the time of the Dynastic era, alluvium and marshlands were present along both flanks of the river and a number of secondary channels probably existed (see Figure 1.1).

1.2. A Developing Riverine Civilisation

About 10,000 years ago, the Sahara plateau was a place that was much more appealing than is generally understood today. As equatorial monsoons reached that far north,[9] grasslands were common, and herds of migratory animals regularly crossed it. Animals thought of as typically southern African today, such as the giraffe, elephant and even rhinoceros, were common sights and recorded in the rock art of the prehistoric inhabitants of the region[10]

Between 7500 and 5500 years ago, as the monsoons retreated southwards,[11] northern Africa began to dry out. As the number of available watering sites decreased[12] and the quantity and quality of available resources diminished,[13] the variety displayed within the distribution and abundance of flora and fauna began to decline.[14] Due to increasing aridity, partly because of human activity,[15]

[1] Allen, *Pyramid Texts*, 110.

[2] For a broad overview, see Said, *River Nile Evolution*; Bunbury, etal., *Memphite Floodplain Landscape*, 74–76, Hassan etal., *Memphite Floodplain Alluvium*.

[3] Williams & Williams, *Nile Basin Evolution*; 207–224; Salama, *River Nile Evolution*, 899–913; Salama, *African Basins,* 105–149; for an overview see Embabi, *Landscapes & Landforms*.

[4] Bunbury, etal., *Memphite Floodplain Landscape*, 74; Ghilardi etal, *Nile Evolution*, 13; Vermeersch & Van Neer, *Late Pleistocene Nile*, 162; Holz, *Man-made Landforms*, 253–269.

[5] Butzer, *Hydraulic Egypt*, 16–18, 33–34.

[6] Hassan etal., *Nile Floodplain Alluvium*, 65–67; Vermeersch & Van Neer, *Late Pleistocene Nile*, 162.

[7] Dumont, *Nile*, 127; Shahin, *Hydrology*, 37; Sallam & El-Barbary, *Secondary Channels*, 489.

[8] Macklin etal., *River Dynamics*, 110–114; Sallam & El-Barbary, *Secondary Channels*, 498.

[9] Kröpelin etal., *Saharan Succession*, 765–768.

[10] Riemer etal., *Desert Animals*, 114; Hendrickx etal., *Western Desert Rock Art*, figs. 6, 17; Köln University, *Project A1*.

[11] White & Mattingly, *Ancient Saharan Lakes*, 58–65.

[12] Said, *River Nile,* 88–91, Midant-Reynes, *Prehistoric Cultures*, 3.

[13] Manning & Timpson, *Saharan Holocene Demographics*, 30.

[14] Pollath, *Prehistoric Gamebag,* 93; Riemer, *Risks & Resources*, 145–147.

[15] Wright, *Humans as Agents,* 1–14; Brierley, etal., *Green Sahara Pastoralism*.

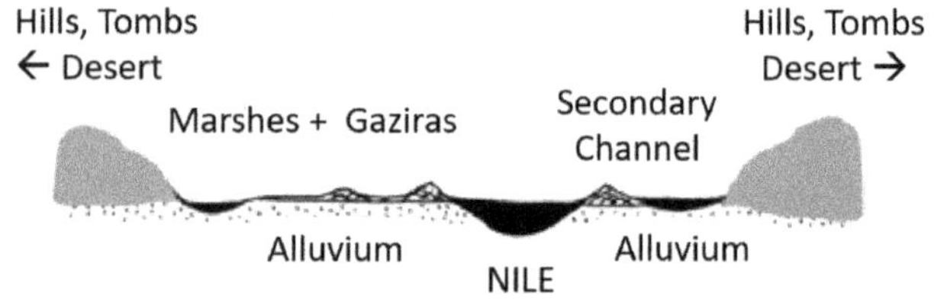

Figure 1.1: Simplified cross-section of the Nile valley.

regions once relatively fertile, began to deteriorate. Hunter-gatherer societies became less successful and were forced to adjust their lifestyle by moving nearer to more predictable and consistent supplies of water.[16] The culture appeared to have transitioned from a community of hunter-gatherers into one resembling more a society of herder-foragers.[17] As large inland water deposits diminished, pastoralists, once able to survive in the marginal lands, in their turn, narrowed the range of their nomadic wanderings and moved closer to more dependable water sources.[18] Populations occupying these areas appear to have become less like herder-foragers and more like agriculturalists, living within a narrower range of the landscape; with the time of most rapid population growth seeming to coincide with that of an increasingly arid environment.[19]

Perhaps a decline in near-to-hand resources stimulated a westward expansion of the Egyptian civilisation onto the oases.[20] Perhaps agriculture was adopted in response to a changing climate that was limiting available resources suggesting that the population did not have to adapt to agriculture but did so out of "*choice rather than necessity*."[21] Cereal production did not replace previous practices but supplemented them,[22] eventually becoming a major force for stability in the valley.[23]

All these groups gravitated towards the major unvarying sources of water in the region: the oases and, more significantly, the River Nile.[24] Villages developed around these areas not affected by the rising floodwaters. The distribution of these villages was not geometrical, because the sites of occupation were simply a consequence of that site's ability to endure the annual flood.[25] Ancient cemeteries and village remnants all along the river suggest a continuous occupation history of more than 8000 years.[26]

Over time, the meandering of the river caused the riverbed to gradually shift position as it migrated from the west to the east.[27] Some settlements became ultimately covered by the inevitable inexorable movement, while others were eventually made uninhabitable due to their ever-increasing remoteness from the river.[28] Buto, for example, appears to have been abandoned, not for political reasons, but primarily due to its increasing remoteness from the nearby waterway.[29]

Some of these villages grew into major towns, developing eventually as centres of administration, trade and religion; with emergent cultures unique to themselves.[30] Spatial efficiency in the distribution of population centres appears to have developed early on, though not all important towns had begun as administrative hubs.[31] The significance of Hierakonpolis, for example, was primarily cultural, not industrial nor agricultural.[32] The uneven distribution of the nomes suggests that this process was haphazard. Upper and Lower Egypt, while exhibiting similar lifestyles, worshipped different gods.[33] The great variety of creation myths and the myriad local gods worshipped by particular regions and towns also suggests that for much of their formative periods, the cultural development that occurred did so in a decentralised, non-directed manner.[34]

Certain centres had other attributes that led them to develop into significant centres, as they progressively (may we say... aggressively?)[35] absorbed politically weaker adjacent centres or amalgamated with centres of similar power and influence. Not all centres developed along the same lines or for the same reasons, but by a certain stage, all had begun to perform similar administrative functions for the Egyptian state.[36] Government became increasingly regional.[37] Eventually, a common culture, based around the exploitation of the resources of the river, as well as encouraged by its natural geographical isolation,[38] led to centralised government control and these localised

[16] Hambrecht & Gielen, *Hunter-gatherer to Sedentary*, 60–61. Connor & Marks, *High Water Adaptations*, 171–199.
[17] Hughes, *Sustainable Agriculture*, 20; Midant-Reynes, *Prehistoric Cultures*, 2. Pennington etal., *Saharan Aridification*, 2, fig. 1, comparing sediment ratios.
[18] Hély & Lézine, *Holocene Vegetation*, 681–686; Pennington etal., *Saharan Aridification*, fig 5b.
[19] Butzer, *Hydraulic Egypt*, 85; Park, *Class Stratification*, 102; Pennington etal., *Saharan Aridification*, 9; Williams & Hill, *Meat & Evolution*, fig. 16.
[20] Ayyad, *Vegetation & Environment* 1, 509–523. It has been argued that sophisticated communities moved eastwards to the Nile valley as the Sahara dried out: Wilkinson, *Political Unification*, 377–395; Midant-Reynes, Prehistory, 137–138. Professor Colin Hope (personal communication July 22, 2011) believes evidence from his work at Dakhleh supports this notion.
[21] Midant-Reynes, *Prehistory*, 137–138.
[22] Williams & Hill, *Meat & Evolution*, 9–11.
[23] Murray, *Cereals*, 505–507.
[24] Riemer, *Risks & Resources*, 152–153; Hassan etal., *Nile Floodplain Alluvium*, 52.
[25] Butzer, *Hydraulic Egypt*, 85; Park, *Class Stratification*, 90–92; Butzer, *Archaeology & Geology*, 621.
[26] Church & Bell, *Settlement Patterns*, 701–714.
[27] Said, *River Nile*, 53–55; Bunbury & Jeffreys, *Ancient Landscapes*, 66–68; Bunbury, etal., *Memphite Floodplain Landscape*, 75–78, figs. 5, 6.
[28] Bunbury & Lutley, *Nile on the move*, 3–5; Said, *River Nile*, 149–152; Wilkinson & Stevens, *Environmental Archaeology*, 277.
[29] Müller-Wollermann, *End of the Old Kingdom*, 5.
[30] Chew, *Harappa to Mesopotamia*, 109–112.
[31] Church & Bell, *Settlement Patterns*, 701–714.
[32] Hoffman, *Before the Pharaohs*, 160–161, identifies Hierakonpolis, despite having no resources of note, as having developed into an important cultural centre by Predynastic times.
[33] Midant-Reynes, *Prehistoric Cultures*, 6–8.
[34] Wilkinson, *Gods & Goddesses*, 6–15.
[35] Bard, *Encyclopaedia*, 30.
[36] Seidlmayer, *OK Elephantine*, 108–111.
[37] Baines & Malek, *Atlas*, 14–19.
[38] Morris, *Art of Not Collapsing*, 63.

cultures blended to form the state that we know as ancient Egypt.

There seems to have been a positive correlation between normal flood levels and the attainment of national prosperity and government stability,[39] since "*without reasonable floods, there were no reasonable harvests*."[40] With the presumed failure of these regular floods, the resultant lack of reliable water interfered with the production of abundant food, the land could not be re-invigorated and so could not deliver the agricultural excess necessary for a highly stratified and well-ordered society supporting artisans and administrators.[41]

1.2.1. Egyptian versus Mesopotamian Irrigation

"*Water drives the world*":[42] whoever controls the water controls the territory.[43] As in other parts of the Middle East, complex human societies developed as the agricultural revolution encouraged people to come together to exploit water as an agricultural resource.[44] While irrigation practices began at roughly the same time, the timing of the flood meant that farmers in Mesopotamia needed to work to make the soil rich, and they needed to store water until the appropriate season for its use arrived.[45] People worked collectively to build canals and storage areas for the water flowing down the Tigris and Euphrates Rivers.[46] The logistical skills needed for this achievement required a complex bureaucracy, which is why organised agriculture developed much earlier in Mesopotamia than Egypt.[47] Large-scale agriculture allowed civilisations to grow and flourish and the city-states of Mesopotamia developed into huge administrative and trade centres.[48]

Egypt's situation was somewhat different to Mesopotamia's, since irrigation and agricultural cooperation appears to have been proceeding well before unification.[49] Small flood basins were occupied first as they were easier to manage with small numbers. Because fewer participants were needed, the logistical demands were not as significant, so the organisation of this fieldwork remained localised, not centrally organised as in Mesopotamia.[50] Technological 'advances' that led to the eventual decline in agricultural productivity within the Fertile Crescent do not seem to have been practised in the Old Kingdom.[51] There appears to be minimal evidence suggesting central government control of irrigation in early Egyptian history,[52] since naturally occurring dams already existed, meaning extensive irrigation works like those undertaken in Mesopotamia were not needed.[53]

No heavily fortified city-states developed along the Nile because there was minimal conflict over water usage and storage.[54] Rather than acting as overlords upon it, Egyptian population centres were an integral part of the surrounding landscape, arising in response to the need to help manage the resources of the land.[55] The spontaneous evolution of significant population centres all along the river explains the simultaneous emergence of multiple major centres of regional administration and helps us understand why the city-state phenomenon as experienced in Mesopotamia did not evolve within the Nile valley.[56]

1.3. Impact of Climate on Old Kingdom Society

It appears that agriculture began in prehistoric Egypt in areas with the narrowest floodplain, those areas of substandard fertility, those areas receiving less of the flood's gifts.[57] In fact it seems that the establishment of dynastic Egypt coincided with a time of increased variability in riverine flows, with increased aridity and lower average flow volumes.[58] This suggests that the less water that was available, the more organisation was required in order to utilise it most effectively.[59] The inference that an increasing aridity should have been an impetus to have driven an update of increasingly complex technology, however, does not seem supported by evidence.[60]

Without the benefits of alluvial deposits from the flood, the capability and productive capacity of the land would be severely diminished. It was difficult to adequately store sufficient surplus food and while the deficits caused by an occasional low flood might be overcome by compensating surpluses in more productive flood years.[61] Increasing aridity is thought to have encouraged the adoption of more efficient production by the Egyptian population.[62] However, as hinted earlier, it may not have been climatic conditions that drove the gradual change from a herder-

[39] Hughes, *Sustainable Agriculture*, 14; O'Connor, *Political Systems*, 15–38.
[40] Said, *River Nile*, 149–152; Bárta, *Collapse Hidden in Success*, 26.
[41] Said, *River Nile*, 88–108; Macklin etal., *River Dynamics*, fig. 1.
[42] Beinart & Hughes, *Environment & Empire*, 130.
[43] Wittfogel, *Oriental Despotism*, 19, 311; Willcocks, *Irrigation*, xvi; Bard, *Archaeology of Ancient Egypt*, 52–53.
[44] Renfrew & Bahn, *Archaeology*, 285–288; Weiss, *Collapse*, 94–95.
[45] Soroush & Mordechai, *Short-term Cataclysm*, 4–7.
[46] ~6000BCE: see Van De Mieroop, *Ancient Near East*, 8–13; Powell, *Sumerian Agriculture*, 291–299; Soroush & Mordechai, *Short-term Cataclysm*, 1–2.
[47] Said, *River Nile*, 175; Jursa & Morenó García, *Near East & Egypt*, 119–120, 144–146.; Soroush & Mordechai, *Short-term Cataclysm*, 7–9.
[48] Leick, *Invention of the City*, 43–53; Midant-Reynes, *Prehistory*, 4–5.
[49] Butzer, *Hydraulic Egypt*, 7. Collon, *Ancient Near Eastern Art*, 13–16; Maisels, *Early Civilisations*, 60–69, 152–169, 336–338; Midant-Reynes, *Prehistory*, 46; Soroush & Mordechai, *Short-term Cataclysm*, 1, 8–9.
[50] Butzer, *Human Ecology*, 261; Peréz Largacha, *Hydraulic Relation*, 82; Soroush & Mordechai, *Short-term Cataclysm*, 1, 8–9.

[51] Weiskel, *Environmental Lessons*, 99; Lawton & Wilke, *Ancient Agriculture*, 1–44; McAdams, *Heartland of Cities*, 19–21, 243–247.
[52] Hassan, *Riverine Civilisations*, 52–54; Butzer, *Human Ecology*, 147.
[53] Soroush & Mordechai, *Adaptation to Cataclysm*, 351–356.
[54] Hassan, *Riverine Civilisations*, 6; Butzer, *Human Ecology*, 261.
[55] Hassan, *Riverine Civilisations*, 69; Midant-Reynes, *Prehistory*, 46.
[56] Van den Brink, *Settlement Patterns*, 301–302; Butzer, *Archaeology as Human Ecology*, 51.
[57] Macklin etal., *River Dynamics*, 118.
[58] Macklin etal., *River Dynamics*, 118–120; Hassan etal., *Nile Floodplain Alluvium*, 65–66.
[59] Said, *River Nile*, 187.
[60] Brewer, *Predynastic Temperatures*, 299–300; Weiskel, *Environmental Lessons*, 99.
[61] Hamdan, *Nile Floodplain Sediments*, 286; Hassan etal., *Nile Floodplain Alluvium*, Figs. 7–8. Macklin etal., *River Dynamics*, fig. 7 and Qianli etal., *Climate-induced Variations*, fig. 7, identify the cycles.
[62] Midant-Reynes, *Prehistory*, 253.

forager society to one with an increasing emphasis upon cultivation.[63]

Since climatic conditions have an important impact upon the fertility of the land, it seems logical that agricultural societies developed a keen awareness of the environmental conditions around them.[64] With the flood arriving at a (mostly) predictable time every year, the most significant technological advance, the calendar, has been applied to agricultural circumstances from well before written history.[65] The passage of the seasons was such a significant event in Egyptian life that it was recorded by several Fifth Dynasty kings on the walls of their temples.[66] From early Dynasty Six, a very rare scene of an important official depicted painting the seasons is preserved in the tombs of two viziers at Saqqara.[67]

As a consequence of the success of agriculture, so too did the environment in ancient Egypt change.[68] Despite the repeated floods continually regenerating the land in an unchanging cycle, the cultural experience did change.[69] Notwithstanding the cultivation being well-managed, in the long term all extensive agricultural systems change their environment in fundamental ways.[70] Despite these regular inundations, land management practices categorised as "*aggravated by collective and cumulative human behaviour*"[71] eventually resulted in increasing salinity of the land around the river. Deforestation, for example, removed the protection afforded to the soil by permanent deeply rooted plants and became a factor in the raising of the water table and an increase in salinity.[72] Overgrazing, for another example, resulted in the removal of valuable nutrients from the soil.[73] Removal of non-agricultural vegetation and its replacement by an annual crop is a significant factor in raising the water table over time.[74] Waterlogged soils are not as productive as well-aerated soil.[75]

1.3.1. The End of the Old Kingdom

The ending of the Holocene Wet phase (from ~ 7500–7000 BCE → ~ 3500–3000 BCE)[76] appeared to coincide with the dislocation of a number of communities around the Aegean.[77] These climatic events are supposed to have brought about the decline of the first urban centres in the Levant,[78] the fall of the Akkadian Empire,[79] as well as the deterioration in the Palatial Culture of the islands in the Mediterranean.[80] It is believed that, during this time, the Old Kingdom declined and 'fell' as Egypt experienced a series of poorer-than-average inundation events, resulting in a number of significant drought-like episodes.[81]

Despite indications of increasing aridity from as early as the 3rd Dynasty,[82] the late Old Kingdom 'drought' was thought to have led to a severe and enduring crisis in food supply causing Egypt to experience a long-term famine, affecting society in a debilitating manner, perhaps initiating a 'dark age' in Egyptian history.[83] Many have attempted to discuss the societal changes that may have occurred and to explain administrative and other governmental reforms in the light of this understanding of a 'drought'. Some suggest that it was weak government that exacerbated the negative impacts of weaker food surpluses.[84] One theory is that the disorder and disruption so weakened the fabric of Old Kingdom administration[85] that the society was rendered paralysed and unable to cope.[86] The calamity of 'drought' would have impacted upon the stability of the society as a whole,[87] and in association with an incapacitated government,[88] has been suggested as a direct cause of the end of the Old Kingdom.[89] Perhaps, it was a combination of all these factors: a "*perfect storm*", perhaps, with each negative aspect magnifying one another's undesirable effects.[90]

It is thought by some Egyptologists that the developing societal unrest that began at the end of the Sixth Dynasty and lasted throughout the First Intermediate Period was a direct consequence of this extended low flood levels.[91] Ankhtify, nomarch of the 2nd and 3rd Upper Egyptian nomes during the latter stages of the period under examination, experienced hardships during his time in charge of his provinces but, according to his testimony,

63 Midant-Reynes, *Prehistory*, 59–63, 137–138. See also Butzer, *Pleistocene Nile*; Connor & Marks, *Terminal Pleistocene*.

64 Erman, *Life in Egypt*, 425; Bard, *Archaeology of Ancient Egypt*, 45–54; Said, *River Nile*, 132–142.

65 Butzer, *Hydraulic Egypt*, 4–11, Tab. 1; Hoffman, *Before the Pharaohs*, 311; Butzer, *Discontinuity*, 106.

66 Edel & Wenig, *Sonnenheiligtum des Ne-User-Re*, 250, 252, pl. 14, 16; Borchardt, *Grabdenkmal des Sahu-re*, 2, pl. 17.

67 Kanawati etal., *Mereruka*, pl. 66a; see James, *Khentika*, pl. 10.

68 Said, *River Nile*, 201; Soroush & Mordechai, *Short-term Cataclysm*, 2.

69 Butzer, *Quaternary Nile*, 162, *Desert in Flood*, 3–5, Butzer etal., *Urban Geoarchaeology*, 3365.

70 Uprety, etal., *Climate Change & Agriculture*, 7–29.

71 Weiskel, *Ecological Lessons*, 98–99.

72 Harpur, *Miscellaneous Scenes*, 40–41, fig. 10. For salinity probles as a result of irrigation, see Pitman & Läuchli, *Salinity & Agriculture*, 3–12; Butzer, *Hydraulic Egypt*, 90.

73 Steward etal., *River Runs Dry*, 207.

74 Vermeersch & Van Neer, *Late Pleistocene Nile*, 162, figs. 1, 11.

75 Butzer, *Hydraulic Egypt*, 19, 48.

76 Bloszies, etal., *Lake Turkana Transition*, 64–76.

77 Said, *River Nile*, 138; Bell, *Dark Age Egypt*, 1–26; Höflmayer's overview in *Dating Catastrophes*.

78 Fiorentino etal., *Climate Change*, 51–58.

79 Höflmayer, *Dating Catastrophes*, 118; Weiss, *Collapse*, 93, 101–104.

80 Drake, *Greek Dark Ages*, 1862–1870; Finné etal., *Late Bronze Age Climate Change*; Weiss, *Collapse*.

81 Weiss, *Collapse*, 100–102; David, *Pyramid Builders*, 25–27.

82 Bell, *Oldest Nile Floods*, 569–573.

83 Hassan, *Floods & Climate Change, Nile Floods & Political Disorder*; Bell, *Dark Age Egypt*. Moeller, *FIP Drought?* 153–168, does not see the absolute link.

84 Lawler, *Collapse Revisited*, 908; Bard, *Archaeology of Ancient Egypt*, 176–177.

85 For an excellent overview, see Priglinger, Elisa. '*Historiographie der Ersten Zwischenzeit einst und heute.*' PhD, university Wien, 2010. Also Müller-Wollermann, *End of the Old Kingdom;* Schneider, *Existence Now Silenced*.

86 Romem, *Unravelling Prolonged Stability* and Lawler, *Collapse Revisited*, 908.

87 Hassan, *Re-reading Ipuwer*.

88 Butzer, *Discontinuity*, 102–112; Malek, *Old Kingdom*, 118.

89 Hassan, *Re-reading Ipuwer*, 357–377.

90 As outlined in Hall, *A Perfect Storm*? 75–86.

91 Bell, *Dark Age Egypt*; Negus, *Great African Drought*.

nobody starved.[92] Ankhtify also experienced disruption in the form of warfare:[93] warfare is never good for an agricultural system, or for society or the environment.[94]

The consequences of this dislocation of culture has excited the imagination of many scholars.[95] This has led to commentary describing the negative consequences a long term water shortage may have produced, such as famine, would have had upon Egyptian culture and the types of changes that may have occurred.[96] However, despite these suggestions of significant impact, Vandier identified some texts related to famines experienced during the Old Kingdom but found very few direct references to great suffering.[97] This seems odd considering the supposed severity of the calamity in question.[98] It is to be expected, therefore, that the negative environmental conditions of that era would have presented a challenging time for local societies.[99] Whether or not the people were actively aware of the wider changes, a population as intimately connected to its surrounding environment would have adjusted its behaviours.[100] It is suspected that, with such a change, enough time would have existed for the society (even if only subconsciously) to adapt to the new circumstances: a form of 'resilience' perhaps?[101]

Many scholars hold the view that they cannot see direct evidence of this in the archaeology they have studied.[102] The debate could be said to have become one where the causes of breakdown have distilled into socio-political versus climatic causes.[103] Perhaps other societal factors were significant in the changes also, with a gradual decline in administrative efficiency observed as a political failure on the part of the government to control the growing regional power of the local elites.[104]

As more evidence arises, it appears more likely that the decline of the Old Kingdom should be recognised as a series of negative factors exacerbating one another.[105] As to whether the decline was as significant as some contend,[106] some areas of Egypt appeared to thrive during the time frame in question.[107] Generally, it does not appear that a 'civilization-ending' drought could be held responsible for the end of the Old Kingdom, but a series of climate-related events occurred during this time.[108] In fact, in may be argued that the end of the Old Kingdom may be seen as a time of innovation and invention, rather than calamity: allowing for adaptation as a matter of fact.[109]

1.3.2. Relating Science and Environment to Art and Culture

As all major civilisations began on or near wetlands, the importance of wetlands to world history can be seen as highly significant.[110] Historically, the marshlands associated with them have been often represented as wastelands,[111] with this attitude perhaps affecting scholars' interpretations concerning the importance of the marshlands of the River Nile.[112] In ancient Egypt, when the river did not flow 'as expected', it did not flood over its banks, and the surrounding land was not submerged. This investigation will attempt to identify environmental changes that may have occurred as a consequence of a drought where the inundation did not occur. Modern investigations of this phenomenon from other places around the globe will be utilised to make some valid assumptions about the nature of 'a river in decline when the Nile did not flow as expected.

1.4. Ecological Principles Applied to History

When an object's environment is discussed, it encompasses that object's natural surroundings.[113] As a consequence of a weak or negligent flood event, the inundation would not have occurred, and nutrients and sediments normally 'lost' to the Nile would have remained within. This would have changed the natural environmental circumstances of the river. Since the primary goal of this study is to identify potential environmental changes that would have developed in the river in times of drought, an ecological overview is needed. Ecology is the science that attempts to explain how nature interacts with itself:[114] it is the study of how organisms interact with one another and with their surroundings (i.e., their environment).

[92] Manassa, *El-Mo'alla to El-Deir*, 5.
[93] Bard, *Archaeology of Ancient Egypt*, 177–178; for the hypothesis of foreign military intervention, see Jansen-Winklin, *Der Untergang*.
[94] Nana-Sinkam, *Degradation*, 1, states: "Chronic localised food shortages caused by war and drought are the most visible aspects of Africa's food crisis." Modern Africa or Ankhtify?
[95] O'Brien, *Boredom with Apocalypse*, 296. Soroush & Mordechai, *Short-term Cataclysm*, 9–10.
[96] Said, *River Nile* 176–178; Hassan, *Re-reading Ipuwer*, 357–377; Büntgen etal., *Climate & Human Susceptibility*, 578–582; Weiss, *Collapse*, 105–106.
[97] Vandier, *La Famine*, 2–3, 100–114; extended by Moreno García, *Études sur l'administration*, 88–92.
[98] Schneider, *Existence Now Silenced*, 315; Gee, *Old Kingdom Collapse*? 68–72.
[99] Nana-Sinkam, *Degradation*, 15–16.
[100] Butzer, *Archaeology & Geology*, 1620–1621.
[101] O'Brien, *Boredom with Apocalypse*, 296–297; Renfrew, *Collapse as Social Transformation*, 481–506.
[102] Moeller, *FIP Drought*? and Willems, *FIP*. However, Höflmayer, *Dating Catastrophes*, 123 and Contardi, *Rhythm of the Nile*, 14, identify it continued acceptance in parts of the scholarship.
[103] Dee, *Dating the Decline*, 323; Gee, *Old Kingdom Collapse*? 64.
[104] Grimal, *History*, 88–93; Malek, *Old Kingdom*, 114–117; Lawler, *Collapse Revisited*, 908; Romem, *Unraveling Prolonged Stability*, 1–3.
[105] Gee, *Old Kingdom Collapse?* 60–61; Müller-Wollermann, *End of the Old Kingdom*, 5; Seidlmayer, *FIP*, 129.

[106] Greenberg, *No Collapse: Transmutation*, 33.
[107] Lawler, *Die - or Fade Away*? 3, 7; Snape, *Of Life and Death*, 106–107.
[108] An excellent synthesis of the current state of understanding of the Old Kingdom climate change can be found in *A Compendium of Recent Evidence from Egypt and Sudan for Climate Change during the Pharaonic Period*, by Pearce Paul Creasman, in 'The Gift of the Nile? Ancient Egypt and the Environment', edited by Thomas Schneider and Christine L. Johnston, Tucson, 2020.
[109] Schneider, *Existence Now Silenced*, 314.
[110] Hendrickx & Vermeersch, *Prehistory*, 33–40; Renfrew & Bahn, *Archaeology*, 239; Wilkinson & Stevens, *Environmental Archaeology*, 66–67.
[111] Allen & Pye, *Saltmarsh Morphodynamics*, 1–18.
[112] Rzóska, *Nile*, 5–9; Malby, *Waterlogged Wealth*, 11.
[113] Riemer, *Risks & Resources*, 124.
[114] In 1869, biologist Ernst Haeckel coined the term "ecology" from the Greek: oikos, meaning 'house' or 'place to live' and logos, meaning 'study'.

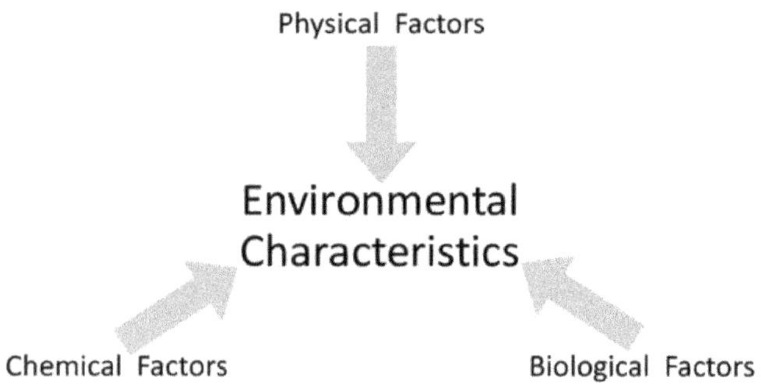

Figure 1.2: Influences on environmental characteristics.

Ecologists investigate a region by working to recognise the various features, aspects, or factors, that characterise an environment. These characteristics can be broken down into measurable and observable factors that, when added together, identify a particular environment with a unique and distinctive character. An environment can then be described and, if need be, classified according to the many characteristics that it possesses. From those attributes, ecologists can apply this awareness to the situation under investigation.[115]

Modern environmental studies apply the principles of ecological science to contemporary situations to make predictions about conceivable hazards or vulnerabilities brought about by a changing or newly developing environmental situation. These principles are well set out and can be applied to many circumstances. The world is currently experiencing rapid climate change with some areas in the modern world experiencing significant shifts in their normal weather patterns. As well as unexpected temperature ranges, many parts of the world are experiencing drought or floods in times of the year where the arrival of (or lack of) these phenomena is atypical, and the intensity of the event is abnormal. It is assumed that these principles would also apply to past situations. By using observations from these modern situations, we may be able to identify similar responses that have occurred in the past. The application of rudimentary ecological principles to the example of an Egyptian drought should enable us to speculate upon the nature of the ecology that developed in response to a gradual drying out of the river.

Egypt itself is in the northeast corner of Africa and consists of a few characteristic ecosystems. The ecosystem types can be broadly divided into Delta, Riverine, Oasis and Desert.[116] Each of these regions displays their own unique characteristics. The major focus of this investigation are the ecosystems that developed along the Nile – with a particular emphasis upon those areas mostly associated with the society and culture of the Old Kingdom, lying north of the cataracts up to the region of the Delta. The time frame in question encompasses the Fourth to the Eighth Dynasties.

[115] Wilkinson & Stevens, *Environmental Archaeology*, 31–33.
[116] Butzer, *Hydraulic Egypt*, 26–38; David, *Pyramid Builders*, 13–19; Zahran & Willis, *Vegetation of Egypt*, 255–256.

1.4.1. Introducing Ecosystems

A.G. Tansley first defined the concept of an ecosystem in 1935:

> *"Simply defined, an ecosystem is the interacting assemblage of living things and their non-living environment."*[117]

Ecosystems display environmental characteristics that are a result of the interacting *abiotic* (non-living) and *biotic* (living) features found within.[118] The abiotic components include various physical and chemical factors, or properties. The biotic components encompass the living things within that environment; these are commonly known as the biological factors. This usually involves the food webs that have arisen, from the producers to the herbivores to the carnivores and their attendant scavengers and decomposer.

The final appearance of an ecosystem is a result of the interaction between these three main factors – physical, chemical, and biological (see Figure 1.2).

1.4.2. Changes to Ecosystems

Thienemann, another pioneer of ecology, elucidated the basic principles of ecosystem response to changing ecological situations.[119] His *Grundprinzip*, or 'Basic Principles', has formed the basis for expressing ecosystem response to ecological change for almost a century.[120] They can be paraphrased as follows:

1. The greater the diversity of conditions in a habitat, the greater the variety of species in that habitat.
2. The more the conditions deviate from the normal, the smaller the variety of species and the greater the numbers of these individual species become.
3. The longer a habitat has been in the same condition, the richer and more varied the species in the habitat, and the more stable that habitat is.

Simply put, as the variability of the habitat increases, so should the biodiversity.[121] Therefore, the converse principle should apply as the circumstances stray from the 'normal' situation, it is expected that some species would flourish while others may flounder.[122] Excess or sparse water supply impacts upon the behaviour of a river. Excess sunlight, in a drought, for example, may speed up decomposition of dead organisms, building a toxic overload or undesirable chemicals in a habitat. Other factors that could change an ecosystem are related to the way that an ecosystem

[117] Tansley, *Terms & Concepts*, 299–302.
[118] Campbell, *Concepts & Connections*, 2–3.
[119] Thienemann, *Grundprinzip*, 421–422.
[120] Perhaps even "dogma"? asks McIntosh, *Ecology*, 141.
[121] Solimen etal., *Nile Island Plant Species*, 81, n. 82.
[122] For example, Al-Hassan & Ofori-Danson, *Plankton Abundance Factors*, 2; Okuku etal., *Plankton Response to Damming*, 114–132; Steward etal, *River Runs Dry*, 205–206.

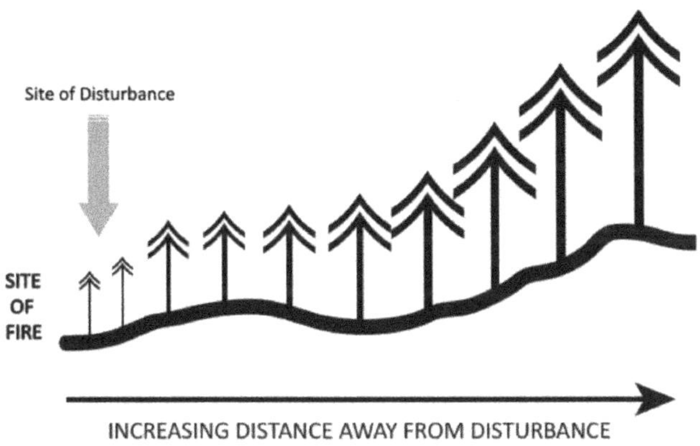

Figure 1.3: A succession in distance.

is utilised or exploited. Changes occur naturally to all ecosystems over time.[123] Seasonal changes producing a regular cycle of change, and ecosystems contain within them organisms that have adapted to this progression.

Among these living things that make up an ecosystem are human beings, and the built environment becomes an important part of the ecosystem.[124] Fundamental to any ecosystem is the human impact upon it.[125] While some ecologists choose in their study of ecosystems to exclude the human species, and some humans think of themselves as somehow separate from ecosystems,[126] this is inappropriate for our study, since humans are an interacting and integral component of ecosystems at many levels, and both contribute to, *and* impact upon, the system of environmental balance.[127]

Ever since the application of fire as a tool, humans have been impacting upon the environment, changing the landscape; altering the configuration of 'space' within their environments.[128] The most significant impact upon the planetary environment has been the adoption of agriculture as the primary source of food supply.[129] Since then, vast swathes of woodland have been cleared; waterways have been dammed, diverted and polluted; and human-induced weathering and erosion has occurred, all leading to rapid changes in the landscape.[130] Compared to the rapid changes experienced in the current climate, however, in historical times, the rate or speed of impact was presumably less.

1.4.3. Succession: Tracing Ecosystem Change

In ecological terms, a 'succession' may be defined as "*the observed process of change in the species structure of an ecological community over time*."[131] Within any community, some species may become less abundant over some interval, or they may even vanish from the ecosystem altogether.[132] Usually, changes occur as a result of some form of disturbance to the natural pattern within the environment.[133] In an ecological context, changes to the distribution and abundance of organisms within an environment usually suggest the influence of some new stimulus upon that environment. In a wildfire, for example, the further from the source of fire, the less damage to the trees can be observed (see Figure 1.3).

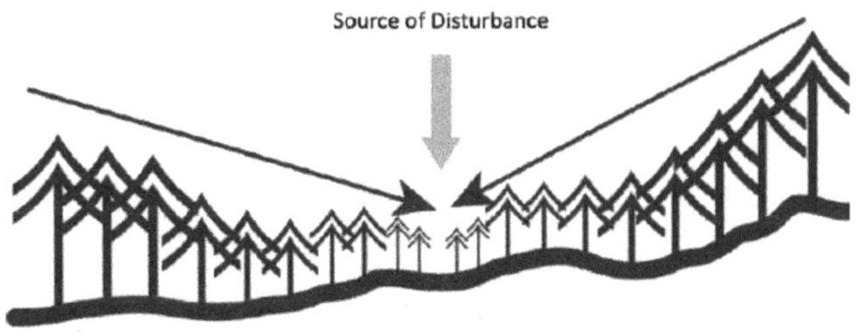

Figure 1.4: Identifying the source of the disturbance.

The further from the fire the greater the proportion of mature trees that should be present. Identifying these changes allows the disturbance to be traced backwards, both in time and in space. This is a 'succession in time' and a well-known ecological principle: the further from a disruption in time and distance, the longer this disturbance takes to manifest itself, and the likelihood that it is observable decreases.[134] The further away from a site in time and distance, the less the noticeable effect. If this disturbance was measured in reverse, then, as well as where the disturbance took place, it is also possible to identify when it occurred, a common enough ecological practice. Measuring the abundance and distribution of a tree species and adding the knowledge of tree growth rates can allow for the time frame of the disturbance to be calculated (see Figure 1.4).

Within the ecological field, potential errors and ambiguities can develop due to a misinterpretation of the data or when particular data is looked for to the exclusion of other evidence. Similarly, when looking for patterns within the decorations that humans produce, the cultural context is vital. Too often, ecology and environmental studies take place independent of the human species and its cultural impact. During this investigation, interpretations or suggestions that arise from interpretations of the data will be tested against the archaeological record and other scientific evaluations that would aid in our understanding.

1.4.4. Applying Ecological Succession to Art

In earlier studies, the author has applied ecologically based succession-style analyses to tomb decoration programmes

[123] Bunbury, *Geographical Development,* 1–12.
[124] Lyle, *Human Ecosystem Design*, 656, recognising that humans continually interact with and are integral to ecosystems, and 657, suggesting humans see themselves as separate from ecosystems.
[125] Wilkinson & Stevens, *Environmental Archaeology*, 253–262.
[126] Lyle, *Human Ecosystem Design*, 657.
[127] Lyle, *Human Ecosystem Design*, 657–660.
[128] Riemer, *Risks & Resources*, 124.
[129] Wright, *Humans as Agents,* 1–14.
[130] Reynolds etal, *Do humans cause deserts*? 1–21.
[131] Tansley, *Terms & Concepts,* 286–287.
[132] Tansley, *Terms & Concepts,* 288–299.
[133] Krebs, *Ecology*, 32–39.
[134] Krebs, *Ecology*, 485–510.

to see if changing patterns within the progression of the decorations could be detected.[135] The assumption made in these studies was that societal disturbances, or influences, would have had a similarly observable impact upon the cultural representations of a society. In these studies, the distribution and abundance of certain tomb scenes over particular periods of time have been analysed to seek to identify developmental changes. Particular changes to the tomb decoration programme have been identified by observing when the changes occurred and linking this time to a cultural disturbance.

Butzer identified how the study of changes to Egypt's early ecology should be used to help develop understanding of how a changing climate may have impacted upon the developing Egyptian culture.[136] Our interpretation of their society may be influenced by the way they responded to the calamity, or how they described its impact.[137] An understanding of the cultural development at any one site or in one region is necessary, since a culture's environmental interaction was, and is, a significant component charting the development of human societies.[138]

If changes to a society can be related to the geo-archaeological history of the region,[139] then it should be possible to relate the cultural history of a region to its environmental history in a similar manner. It should be possible to trace the development of the cultural progression at a particular site or within a particular region.[140] If the ecological analogy outlined above were to be applied to tomb decorations toward the end of the Old Kingdom and into the First Intermediate Period, then it would be logical to assume that the distribution of decorations within tombs would be different before and after the occurrence of the drought, and to pinpoint with some precision when this change occurred.[141] Without a significant merging of the historical and scientific data, key factors that influenced the development of the civilisation may be overlooked.[142]

1.5. Summation: A Changing Environment Changes Society?

The basic points of this chapter include:

- The enduring success of the Egyptian culture indicates its resilience.
- The geo-archaeological history of the region should enable the relationship between the environmental history and the cultural history of a society.
- The culture that developed was influenced by the climate in which it developed.
- A culture's environmental interaction was a significant factor in the charting of the development of that society.

Considering the pivotal role of the river to the long-term success of the society, the exploitation of the riverine environment must have been a significant factor influencing that success. Similarly, the river's ecological health must also have had a substantial impact upon the society's ability to exploit that particular resource. A changing river would have had some impact upon the surrounding environments and that would have impacted upon the local population. While it appears unlikely that a debilitating drought can be held responsible for the decline of the Old Kingdom; it has been identified that over the time frame in question, the water levels in the river were more irregular than in the past.

---

135 Burn, *Mehu* and *Pyramid Texts*.
136 Butzer, *Hydraulic Egypt*, xiii–xv.
137 Pierssené, *Explaining Our World*, 20–22; O'Brien, *Boredom with Apocalypse*, 296–297.
138 Butzer, *Ecological Archaeology*, 106–11; Butzer, *Context*, 417–422.
139 Midant-Reynes, *Prehistory*, 16–17.
140 Church & Bell, *Settlement Patterns*, 701–714. note.
141 see Burn, *Changing Riverine Ecology*?
142 Butzer, *Ecological Archaeology*, 106–11; Butzer, *Context*, 417–422.

2

Art, Society and the Environment

The secondary goal of this study is to identify if any correlation exists between the hypothetical ecological conditions at the end of the Old Kingdom, as suggested by the scientific analysis of the riverine ecology; and the tomb decorations that were produced at the same time. As changing climate impacted upon the daily life of the ancient Egyptian, it could be possible that it influenced the culture as well. While it may seem to the cursory visitor or the casual observer that the culture of ancient Egypt remained stagnant, the society existed in an almost continuous state of change and adaptation. Changes in the way that ancient Egyptians saw themselves and their society should be recognisable in their art.

In this chapter, examples of how a society's environment influences its art will be identified and discussed. The chapter will outline how art, especially tomb decorations, may be used to make inferences about the impact of a changing environment upon the culture. The methodology employed in this study to identify how changing environmental river conditions could have affected the society and the visual culture of ancient Egypt at the end of the Old Kingdom will be presented. A case for the use of tomb wall scenes as indicators of Egyptian society will be made and, finally, a strategy for the utilisation of these tomb decorations as environmental markers will be outlined.

2.1. Egyptian Representations of the Environment

The long-term success of the ancient Egyptian culture suggests an ability to adapt to their immediate surroundings. A culture and society so attuned to the river flowing through its heartland, one so wedded to the regular cycle of inundation, and one so reliant upon its bounty, must have become aware of changes that interfered with this rhythm. The art of the Old Kingdom reveals a deep understanding and detailed knowledge of the environment and illustrates the influence of the physical world on the evolution of Egyptian culture and on the lives and activities of its inhabitants.[1] The clarity, detail and precision of the wall decorations and their various themes suggest that the Egyptians had a great awareness of and interest in their natural surroundings.

Artists took great care to keenly observe and to accurately represent the natural world in which they lived. Water, for example, was represented by a series of zig-zag lines portraying its liquid nature and was often filled with fish, flora, and aquatic animals.[2] The bountiful marshlands of the Nile on which the Egyptian depended for transport, fish, and fowl, constitute a significant component of the wall decorations produced during the Old Kingdom. The numerous types of fish depicted in the water are rendered so clearly, and in such detail, that the various species can easily be identified. Representations of dense papyrus marshes together with their attendant waterbirds, fauna and insects were illustrated in great detail.[3] As well as fish, the water holds dangerous aquatic mammals and reptiles, such as the hippopotamus and crocodile, and even the odd turtle.[4] Frequently, very small animals are also depicted in the marsh scenes, such as frogs, and a wide variety of insects; including a praying mantis, some grasshoppers, butterflies and dragonflies.[5] Illustrated both in and above the papyrus marshlands are many varieties of waterbirds, such as geese, egrets, hoopoes, pied kingfishers, herons, turtle doves and lapwings. Some are shown within vegetation, while others are portrayed flying above the papyrus marshes.

While the river was the lifeblood of the country, the culture also relied on resources from both the desert and the ecotone regions at the edge of the floodplain. The desert landscape was also represented, depicted as wavy lines above the register baseline suggesting the mobility of the sand. In a scene in Niuserra's temple, for example, the desert is represented by an undulating landscape and the fauna and flora of the desert itself is represented by succulents, small shrubs, and stunted trees; as well as many animals, including a few birds and a rare ostrich.[6] The presence of shrubs and succulents suggests an area between the arid desert and the fertile land where water was available and sparse vegetation provided grazing. The great variety of desert animals provided a rich resource and were hunted both for food and for recreational purposes.[7] As such, they were regularly depicted on tomb walls.[8]

The environment, including flora, and fauna such as mammals, birds, reptiles, amphibians, fishes, and invertebrates such as insects and arachnids, was also a source of many Egyptian hieroglyphs. The hieroglyphic determinative for Lower Egypt was a clump of the papyrus

[1] Schäfer, *Principles*, 45, in quoting Goethe, identifying nature as the "eternal spring" of influence.

[2] Kanawati etal., *Mereruka* III, pl. 68.

[3] For example, the Marshland Scenes in Kanawati & Abder-Raziq, etal., *Hesi*.

[4] Kanawati & Abder-Raziq, *Hesi*, pl. 53.

[5] For example, frogs and mantis in the tomb of Hesi: Kanawati & Abder-Raziq, *Hesi*, pl. 54.

[6] Edel & Wenig, *Sonnenheiligtum des Ne-User-Re*, pl. 14.

[7] Herb & Förster, *Desert to Town*, 24–33.

[8] Inumin, depicting hunting dogs attacking ibex: Kanawati, *Inumin*, pl. 47.

that grows in the marshlands, while the determinative for desert lands is composed of two or three adjacent hills, often with red dots to indicate their sandy nature, beyond the fertile Nile.

2.2. Art, Society and Culture

Ancient Egypt, as a society existed in an almost continuous state of gradual cultural evolution,[9] the culture was never in an inert state.[10] Its art also exhibited an almost continuous state of change,[11] whether regarding depicting administration,[12] societal manners,[13] or in religion.[14]

It seems better to see ancient Egypt not as a series of one king following another, in a seemingly never-ending repetitive cycle, but perhaps as in some form of cycle of 'Interrupted Stability,'[15] in which some rulers had the fortunate (or unfortunate?) experience of ruling when significant changes to the society were taking place. Over the time frame in question, changes occurred to both the variety and frequency of decorations displayed in tombs.[16] Within the context of an endlessly evolving cultural milieu, the decorations produced for the elite of the society generally reveal, sub-consciously at least, the current concerns, cultural atmosphere, beliefs, and mores of the society.[17]

After times of disruption, disturbance or upheaval had faded, the artwork presented appeared regularly to have reverted to original, more 'traditional', forms.[18] While not disregarding 'aspirational' themes that influenced many of the decorations,[19] it is hoped that this study will identify some instances where tomb scenes suggest that, as well as symbolic influences, an environment influence was insinuating itself upon the society.[20]

2.2.1. Art as Evidence

Some have asked if we can justify the use of art, in this case tomb wall scenes, as a guide to the society that produced it.[21] Art existed before writing so we should be entitled to see it as some form of symbolic representation,[22] an illustration of the basic concerns of the culture that was producing it.[23] Egyptian "*art was not produced for its own sake,*"[24] but had a role in the culture, and was a significant factor in enforcing the character of the elite within that society.[25] While a "*continuity of culture,*" was evident,[26] art was not static;[27] it being "*no more unchanging than the institutions and beliefs that it served.*"[28] Much of ancient Egyptian culture has been gleaned from the examination and interpretation of its art.[29] Because art contains a multi-layered web of ideas, allusion and symbols,[30] art should not be studied in isolation but in conjunction with other fields.

While it is important to study art in a systematic manner,[31] new thematic and scientific approaches (a new 'lens', perhaps?) may enable the addition of new interpretations to old opinions.[32] While some may question the validity of relying on the artistic evidence as data of its own accord, exactly what is produced allows for the utilisation of the artwork as pieces of archaeological data.[33] A good case in point is the representation of barley and flax harvesting to be found in most provincial tombs. To produce linen, flax needed to be harvested 'green' whereas barley was harvested 'dry' to facilitate the ease of removal of the crop.[34] Most tombs display the harvesting sequence in the correct order, in some form of visual narrative,[35] for example Sheikh Said,[36] Meir,[37] Deir el-Gebrawi[38] and at El-Hawawish.[39]

Since the artists were careful to indicate the correct order of harvesting, then it lends support to the idea that it is

---

9 Vassilika & Bourriau, *Egyptian Art*, 1, suggest that Egyptian art, though evocative, was conservative.

10 For example, Malek, *Old Kingdom*, 23–33; Bard, *Archaeology of Ancient Egypt*, 84–87.

11 Smith & Simpson, *Art & Architecture*, 74; Robins, *Art of Ancient Egypt*, 57, 78; Tiradritti, *Art, Architecture & History*, 20.

12 Smith & Simpson, *Art & Architecture*, 62 and 75, discuss the 'Biography of Weni' in this context. Malek, *Egyptian Art*, 26; Robins, *Art of Ancient Egypt*, 252; Donovan & McCorquodale, Principles & Themes, 1 discuss role of art in reinforcing social structure.

13 Donovan & McCorquodale, Principles & Themes, 1; Smith & Simpson, *Art & Architecture*, 3.

14 Bard, *Archaeology of Ancient Egypt*, 84–87; Grimal, *History*.

15 See Bárta, *Punctuated Equilibrium*.

16 Harpur, *Decorations*, traces the development of Old Kingdom tomb decorations.

17 On the role of symbolism, see Schäfer, *Principles*, 14–18; Malek, *Egyptian Art*, 16–18; Harrington, *Art & Social Theory*, 63–65. The conclusions drawn by Mysliwiec, *Decoration Schemes*, Michalowski, *Art of Ancient Egypt*, Shirai, *Ideal & Reality*, and Hagen & Hagen, *Egyptian Art*, serve as examples of interpretation of what is being represented.

18 Robins, *Art of Ancient Egypt*, 109; Donovan & McCorquodale, *Principles & Themes*, 3; Shirai, *Ideal & Reality*, 333; Hope, *Libya & the Dakhleh Oasis*, 409.

19 Quirke & Spencer, *Ancient Egypt*, 98; Hornung, *Afterlife*, 1–5, 14, 26–27, 30; Allen, *Middle Egyptian*, 315–317; David, *Religion & Magic*, 93; Wilkinson, *Gods & Goddesses*, 42–44.

20 Butzer & Enfield, *Critical Perspectives*, 3630.

21 Seidlmayer, *Beni Hassan People*, 351; Vasiljević, *Hierarchy of Women*, 146.

22 Aldred, *Old Kingdom*, 30–32; Wilkinson, *Reading Egyptian Art*, 9; Donovan & McCorquodale, *Principles & Themes*, 2.

23 Knudsen etal, *Cultural Interpretation*, 303.

24 Aldred, *Days of the Pharaohs*, 11–12.

25 Hendrickx & Eyckerman, *State Development*, 63–64.

26 Smith & Simpson, *Art & Architecture*, 1.

27 Smith & Simpson, *Art & Architecture*, 74; Robins, *Proportion*, 228–248; Binder, *The Physical World*, 29–34.

28 Robins, *Art of Ancient Egypt*, 255.

29 Aldred, *Old Kingdom*, 29; Knudsen etal, *Cultural Interpretation*, 58. Malek, *Egyptian Art*, 3, 21–22.

30 Malek, *Egyptian Art*, 27.

31 Schäfer, *Principles*, xxv.

32 Schäfer, *Principles*, 5–7, Donohue & Fullerton, *Ancient Art & Historiography*, 1–3. Donovan & McCorquodale, *Principles & Themes*, 1.

33 Woods, *Art as Data*, (in preparation).

34 Siebels, *Agriculture*, 118–119.

35 See the commentary in Davies, *Sheikh Saïd*, 19–23: also, Vogelsang-Eastwood, *Textiles*, 270–271.

36 Davies, *Sheikh Saïd*, pl. 16.

37 Blackman & Apted, *Meir* V, pl. 22; Kanawati & Evans, *Meir* II, pl. 83–84; Blackman, *Meir* IV, pl. 14v; Kanawati, *Meir* I, pl. 84.

38 For Ibi: Kanawati, *Deir El-Gebrawi* II, 42–43; Davies, *Deir El Gebrawi* II, pl. 6.

39 For Hesi-Min, see Kanawati, *El-Hawawish* IV, fig. 18.

reasonable to trust that their renditions of other events and phenomena were similarly accurate.[40]

2.2.2. Art as a Branch of Archaeology

Smith and Simpson see art as the "*re-creation of life*",[41] with the artists attempting to reproduce reality as vividly as possible[42] using the resources available, despite some representations that would appear unlikely to have occurred in real life.[43] Interpretations about the whole of society that are made separately from corroborating evidence should be deemed as more likely invalid and flawed.[44] For example, in his attempted construction of a parallel between the pictorial and archaeological records from Beni Hassan, Seidlmayer found it easier to identify parallels for the higher stratum of society than for the lower.[45] While tombs decorated within a royal context may have been produced under stricter, perhaps more religiously motivated, ideals,[46] those tombs produced by individuals beyond the immediate royal family may have been designed with more artistic licence,[47] with some elite adopting royal prerogative during the latter parts of the Old Kingdom.[48]

Art was not wholly derivative;[49] one artist may have been more 'creative' than another in the area,[50] or the scene may have been regarded as an 'informal' decoration, allowing greater scope for freedom and experimentation.[51] Consequently, the tombs of the non-royal elites may be seen as representing a more accurate 'historical' representation of everyday life[52] at that level of society. However, even allowing for conventionalism, in some situations, perhaps in circumstances where the artist was untrained or inexperienced,[53] the artist often drew from what they knew instead of what they saw,[54] so some images may not be realistic.[55]

Schäfer first cautioned against this as some images may be concerned.... "*more with the essential character of objects than with their appearance.*"[56] Most artists drew from what they knew as well as what they saw. A distinction, therefore, needs to be made between 'formal' and 'informal' decorations, with a greater opportunity to witness creativity, variety and complexity from those decorations not directly relating to the god, the king, or the tomb owner.[57]

Within these analyses, it is also important to wary of the problem of applying a modern understanding to decorations that were prepared thousands of years ago for an audience 'knowledgeable' in its meaning. Our modern perceptions need to be restrained to avoid interpretation using our sensibilities,[58] so just as it is important to avoid overly 'emotional' responses to the actual scenes themselves, so is it important to avoid applying overly rational explanations. Nonetheless, the composition of the scenes should enable some inferences to be made; the "*seeking of the message*" [59] of the decorative themes should be possible.

2.2.3. Environmental Influence on Art and Culture

The religious views of the earliest ancient Egyptians indicated they recognised the importance of the river to their society.[60] The land was seen as sacred; the cycle of the river suggested to them the importance of cycles, balance or harmony within the environment rather than domination over it.[61] The goddess of balance, Ma'at, was the personification of harmony with and within nature.[62] One of the gods attested from earliest times was Hapi, the God of the Nile, Lord of the Fishes and Birds.[63] To emphasise the value of the river in providing for the fecundity of the soil, representations of the god were often painted blue,[64] with blue having an especial symbolic meaning as it represented 'precious' things.[65] This male god was depicted with breasts to suggest nurturing and as corpulent to suggest bounty, indicating the especial importance attributed to him.[66] The god Osiris, whom Hapi suckled, was seen as the embodiment of agriculture and vegetation.[67]

Decorations could be unique to an area, where local situations and circumstances had influenced tomb decoration programmes.[68] The provincial art that arose at

[40] Kanawati, *Tomb & Beyond*, 97.
[41] Smith & Simpson, *Art & Architecture*, 1.
[42] Tiradritti, *Art, Architecture & History*, 8.
[43] Robins, *Proportion*, 1.
[44] Donohue & Fullerton, *Ancient Art & Historiography*, 3–6; Knudsen etal, *Cultural Interpretation*, 303.
[45] Seidlmayer, *Beni Hassan People*, 365; Seidlmayer, *OK Elephantine*, 108–127.
[46] Baines, *Schäfer & Egyptian Art*, 10, where not all representations are accurate but more likely 'ideal'.
[47] Davis, *Canonical Tradition*, 199–201; Hartwig, *Painting & Identity*, 49–50; Baines, *Communication & Display*, 476–7 discuss the role of 'prestige' in representations.
[48] Bárta, *Old Kingdom Kingship*, 267–269.
[49] Smith & Simpson, *Art & Architecture*, 75; Lorand, *Four Schools of Art*, 52.
[50] Schäfer, *Principles*, 36–45. Tiradritti, *Art, Architecture & History*, 8, 24 & 32.
[51] Kanawati, *Tomb & Beyond*, 84. Robins, *Proportion*, 23.
[52] Baines, *Schäfer & Egyptian* Art, 8–10.
[53] Michalowski & Guterman, *Art of Ancient Egypt*, 195–6; Robins, *Art of Ancient Egypt*, 80–89, discussing the decline in quality during the First Intermediate Period.
[54] Schäfer, *Principles*, 46; Donovan & McCorquodale, *Principles & Themes*, 1.
[55] Malek, *Egyptian Art*, 21, warns of not relying wholly upon the art, especially out of context, 32. Schäfer, *Principles*, 5–6; recommends that we attempt to avoid emotional responses: a "personal art experience", also Kóthay, *Art & Society*, 19; Knudson etal., *Cultural Interpretations*, 303–304.

[56] Schaefer, *Principles*, viii.
[57] Robins, *Proportion*, 21–23.
[58] Pierssené, *Explaining Our World*, 6. Later, 83–84, Pierssené comments about the 'morality' of an interpretation; contra Kepes, *Arts of the Environment*, 12.
[59] Pierssené, *Explaining Our World*, 71–82.
[60] Brow, *Alluvial Geoarchaeology*, 288.
[61] Hughes, *Pan's Travail*, 48; Malek, *Egyptian Art*, 29.
[62] Málek, *Egyptian Art*, 27.
[63] Wilkinson, *Gods & Goddesses*, 106.
[64] Wilkinson, *Gods & Goddesses*, 107–108.
[65] Williams, *Per-Neb*, 50–52.
[66] Michalowski & Guterman, *Art of Ancient Egypt*, 99.
[67] Hughes, *Sustainable Agriculture*, 14–15.
[68] Vischak, *Elephantine Identity*, 454. Brunner-Traut, *Epilogue*, 434.

the end of the Old Kingdom did so to meet local demand.[69] These factors may result in outcomes that are not 'traditional'.[70] Ideas and contexts change and these have been identified within the art.[71] In the cemetery of Hamra Dom, in tombs 66 and 73, the standard late Old Kingdom provincial repertoire of art is offered, with decorations dominated by scenes of hunting, fishing, fowling, butchering, and the inspection of cattle. However, they are notable for the lack of scenes depicting cultivation. Save-Soderbergh, for example, suggests this may reflect the local environment at the time perhaps not being suitable for agricultural activity.[72] This allows us to contend that the environment must have had some influence (even if only sub-consciously) upon what was seen to be important at that time and place.[73]

2.3. Using Art to Make Inferences about Nature

One of the inferences under exploration is that the changing ecological situation may have led society to a greater awareness of the surrounding environment, influencing how they depicted the 'real' world.[74] If the hypothesis is accurate, then it follows that the tomb wall scenes produced over the time frame in question should allow for the elaboration of suppositions about the ecology in which Egyptians found themselves at the end of the Old Kingdom.[75] Since it was mentioned in writings at the time,[76] it is believed that artistic evidence may also be used to make inferences about the climate.[77] Perhaps an artistic narrative may be developed by tracking changes to the decoration programmes in relation to supposed environmental changes,[78] in addition to the other influences that existed. It may be that this awareness of changes in the surrounding environment was evidenced by an increasing emphasis placed on scenes depicting more authentic surroundings, for example an increase in depictions of fishes able to survive in desiccated pools.[79] As we attempt to identify evidence of an increasing environmental awareness in the depictions that were produced at the time, alterations to the tomb decoration programme may become apparent.[80] Misgivings about the decline in the quantity and quality of available food should be visible upon these same walls.[81] Could it be mere coincidence that iconography relating to the storage of grain seems to become increasingly common during the latter stages of the Old Kingdom[82], or could this greater frequency be related to an increasing importance assigned to this activity in times of scarcity. In a similar manner, the abundance of depictions of the rendering of accounts and punishments of errant estate managers increased over the same time frame,[83] once again suggesting an area of increased concern in Egyptians' daily lives.

2.3.1. Patterns of Change

In attempting to explain the cultural developments that occurred in ancient Egypt at the end of the Old Kingdom, some have suggested a link between the collapse of central government and a deteriorating environment.[84] Ankhtify's autobiography indicates a difficult time for Egypt.[85] Since Egypt had experienced low inundations in the past, [86] the decline in flood levels alone cannot be the significant factor in explaining the fall of the entire Old Kingdom.[87] As agricultural output became less reliable/predictable,[88] it may have become that resources needed to have been sourced from further afield. It may be that the greater awareness of the surrounding environment is evidenced by the increasing emphasis placed on scenes depicting the world as it was, not as it should be.[89] Throughout the Old Kingdom and into the period beyond, changes can be observed to both the variety and frequency of decorations in tombs.[90] It seems likely that the scribes who composed the scenes on the walls of the tombs they were decorating would have been aware of their surrounds.[91]

In his acceptance of Bell's suggestion that a changing environment at the end of the Old Kingdom may have led to a detrimental consequence for the government of the time, Said identified aspects of tomb decoration and literature that may indicate the changing climatic circumstances that were taking place: for example, the increasing use of freshwater fish in artistic representations,[92] and therefore presumably in the diet, suggest that the cereal proportion

[69] Aldred, *Days of the Pharaohs*, 106–112.
[70] Donovan & McCorquodale, *Principles & Themes*, 4, on Mo'alla's unique cattle.
[71] Malek, *Egyptian Art*, 29.
[72] Save-Soderbergh, *Hamra Dom*, 33–35.
[73] Smith & Simpson, *Art & Architecture*, 78, Shirai, *Ideal & Reality*, 329; Siebels, *Agricultural Scenes*, 55–56: are these examples of realism over symbolism?
[74] Church & Bell, *Settlement Patterns*, 713; Kinzig etal., *Assessing Cascading Effects.*
[75] Osborn, *Men of the Stone Age*, 392–434.
[76] Neuberger & Thornes, *Art & Climate*, 95, note references to climate in ancient Egyptian writing. Later, 102, they suggest changes in artistic composition as a result of a changing Mediterranean climate.
[77] Robins, *Art of Ancient Egypt*, 78 and Aldred, *Days of the Pharaohs*, 109.
[78] Kanawati, *Tomb & Beyond*, 97.
[79] Said, *River Nile*, 179.
[80] Pierssené, *Explaining Our World*, 2.
[81] Baines, *Schäfer & Egyptian Art*, 9, on the importance of the role of narrative.
[82] Simpson, *Kawab & Khafkhufu*, 24–25, pl. 36, 37, fig. 47.
[83] Moussa & Altenmüller, *Nianchchnum & Chnumhotep*, 127, Taf. 54 [a, b], Abb. 24; MMA, *Egyptian Art*, 404–407, fig. 126, depicting punishment scenes from the tomb of Tepemankh II.
[84] Bell, *Dark Age Egypt*, 1–3, *Oldest Nile Floods*, 569–573; Büntgen etal., *Climate & Human Susceptibility*, 578–582; Hassan, *Re-reading Ipuwer*, 357–377; Morris, *Nobles Lament, Poor Rejoice*, 60; Cleuziou, *Archaeology & Climate*, 29–36.
[85] Vandier, *Mo'alla*, Inscription #10, 220–222.
[86] Hassan, *Floods & Disasters*, 1–23.
[87] Bárta, *Long Term Short Term*? disputing the role of climate as an absolute cause.
[88] Bard, *Archaeology of Ancient Egypt*, 162–166; Hassan, *Floods & Civilisation*; Dee, *Dating the Decline*.
[89] Smith & Simpson, *Art & Architecture*, 3, 42; Baines, *Schäfer & Egyptian* Art, 8–10; Tiradritti, *Art, Architecture & History*, 24.
[90] Smith & Simpson, *Art & Architecture*, 2.
[91] Baines, *Schäfer & Egyptian Art*, 18–21.
[92] Said, *River Nile*, 138–143, 179.

Table 2.1: OEE Themes

OXFORD EXPEDITION TO EGYPT THEMES					
1	Marsh-related activities	6	Gardening	11	Commerce, including management of workers and possessions
2	Desert and desert-related activities	7	Bird procession and poultry yard scenes	12	Dance, music and games
3	Agricultural pursuits	8	Manufacture and storage of wine and oil	13	Medical procedures, relaxation and bodily care
4	Pastureland and animal husbandry	9	Food preparation and brewing	14	Warfare and war-related activities
5	Orchard scenes	10	Workshop activities	15	Funerary rites and funeral scenes

Table 2.2: OEE Scene types

OXFORD EXPEDITION TO EGYPT 'PASTURELAND' SCENE TYPES					
4.1	Fighting bulls	4.2	Livestock browsing in pasture	4.3	Cattle mating in pasture
4.4	Cow giving birth	4.5	Milking a cow	4.6	Hobbled and tethered calves
4.7	Goats in or near trees	4.8	Feeding domesticated animals by hand	4.9	Force-feeding hyenas
4.10. Overseer seated near animal-tending activities					

of the diet had begun to decline.[93] Perhaps the desert provided a larger proportion of the resources utilised towards the end of the Old Kingdom in response.[94] Desert hunt scenes do not follow the 'normal' conventions of tomb scenes,[95] perhaps indicating their production at times when more authentic depictions were becoming fashionable. As society acquired a greater awareness of the surrounding environment, this increased understanding should have influenced the adoption of decorations depicting a world that the society found more recognisable.[96]

2.4. The Method

In a similar manner to how succession studies identify patterns of change within an environment, this 'succession' approach may be relevant to a study of Egyptian art. In attempting to identify possible cultural influences, the author has, in previous studies,[97] plotted variations in the distribution and abundance of specific tomb decoration themes, with an aim to identify changes in patterns that may indicate variances to the relative importance of an idea or religious belief. The number of times a theme is repeated may be used as an indicator of the relative importance of that theme. The spatial distribution, and the time at which it was produced, were used to suggest the comparative importance of a theme between different locales and time periods.

To investigate changes in tomb decoration programmes over time, it is necessary to identify what types of depictions existed, what they displayed and when they occurred. Harpur's work tracing the development of tomb decorations[98] has been expanded and deposited in digital form in the Oxford Expedition to Egypt (OEE) database. This 'Scene-details Database' is a series of records collating the decoration data for Old Kingdom tombs ("c. 2650–2150 BC") that have so far been recorded.[99] Within this database, the decoration types have been classified into 15 broadly defined 'themes' (see Table 2.1) and each of these themes has been categorised into more precise 'scene types' (see Table 2.2).

While discoveries have been made after the publication of this database, the large number of tombs recorded (495)[100]

[93] Butzer, *Stratigraphy & Climate*, 5–23; Gorny, *Environment, Archaeology & History*, 81–92; Cornevin, *Paléoclimatologie de L'Égypte Ancienne*, 183–203.
[94] Said, *River Nile*, 142–143, 176. Bard, *Archaeology of Ancient Egypt*, 162–166.
[95] Robins, *Proportion*, 6.
[96] Kinzig etal., *Assessing Cascading Effects*. Church & Bell, *Settlement Patterns*, 713; Rzóska, *Nile*, 35, identify changing landscape representations during the Fifth Dynasty.
[97] Burn, *Similarity, Coincidence or Progression*; Pyramid *Texts and Mehu*.

[98] Harpur, *Decoration*.
[99] OEE *Scene-details Database*, doi:10. 5284/1000009.
[100] Total tombs recorded = 495 as at March 18th, 2019. https://archaeologydataservice.ac.uk/archives/view/oee_ahrc_2006/

means that there is a lot of data contained within, with large changes needed before the overall pattern would change in a statistically significant manner. Evidence changing one or two pieces of data would make any overall change to the database inconsequential, in statistical terms. In conjunction with the MastaBase[101] data list; a database containing the iconographic and textual evidence on Old Kingdom elite tombs, using evidence from tombs not included in the database and incorporating more recent discoveries, the statistical analyses that follow utilise the OEE classification terminology.

Regarding the present study, it was thought that it might be possible to identify thematic changes because of the environmental changes that were actually being experienced. Importantly, when judging their significance, this study will look to the *subject* of the scenes and not their *quality* to be the consequential factor. Plotting the distribution and abundance of artistic themes across the latter stages of the Old Kingdom should help identify the changes that the society was undergoing by helping us to identify the changes in ideas and preoccupations of the elite.

Due to the breadth of this study, it was not possible to investigate each tomb and each decoration in order to re-classify it anew, though this would have been the ideal situation. Since the initial investigation was basically a statistical exploration, then it seemed a valid approach to use data already collected. Therefore, it has used the work provided by the OEE and the MastaBase study group as the basis for classification and the formation of the A.R.I.D. spreadsheet. During the latter stages of the study, in the Discussion and Analysis, relevant tomb scenes to the project will be considered. The method adopted here involves plotting the distribution and abundance of decoration themes across the time frame in question – from Dynasty Four to Dynasty Eight. The term 'abundance' is the measure of the number of times a particular decoration theme has been attested. The term 'distribution' describes its appearance within the Dynastic chronology. Where the dating is ambiguous, the dating criteria applied by the MastaBase Study group will be applied, since this is the more recent analysis.

The method applied in this investigation followed these subsequent steps:

1. A tomb was identified as belonging to the appropriate time frame (from Dynasty Four to Dynasty Eight) and assigned to a particular King. (King Teti, for example, was identified as VI.I: as the First King (I) in the Sixth Dynasty (VI).
2. Once a tomb has been categorised according to the above classification system, the different decorative themes were identified and plotted on a spreadsheet.
3. This was performed each tomb within the set time period.
4. A total for each of the themes was determined and the relative proportion of each type compared to all the recorded attestations was calculated.
5. Those calculations were then used to analyse the changing patterns in the tomb decoration programme over the Old Kingdom, from Dynasty Four to Dynasty Eight.

2.4.1. Chronology

A debate that seems to be progressing at an infinitesimally slow pace is the one related to resolving dates of specific stages of Egyptian history. Very little debate exists about the order of Dynasties. The minor discussion regarding orders of succession,[102] or for that matter, who was related to whom,[103] does not interfere with the long-term or overall consequences of this investigation.

Not all the dates suggested within the Scene-details database are accepted within the wider Egyptological community, to help overcome the absence of an absolute chronology, several proposed chronological systems have been put forward.[104] Notwithstanding the fact that an absolute chronology eludes us, it is difficult to discuss historical events without some understanding of the broad chronological sequence. This debate with respect to an absolute time frame is never-ending, and its solution lies beyond the purview of this investigation. Using the database in its entirety from Dynasty Four to Dynasty Eight should allow the internal dating errors to balance one another out; it is to be expected that those dates thought too recent should be balanced out by those dates thought too late. Relative time frames are simple and provide a sense of continuity within the topic. However, when linking context to a potential foreign influence or artefacts originally deriving from an overseas location to a particular place in time and space, a more precise time frame than a relative chronology is needed.[105]

If famine was a significant cause in the decline of good or effective governance, then when this famine occurred in the timeline of dynastic succession is an important consideration. To accurately identify the time frame for the apparent decline in water availability and consequent deterioration of agricultural potential, the chronological framework requires absolute rather than relative figures. With increasingly sophisticated technology available to investigate the history of ancient Egypt and the River Nile itself, a more reliable time frame becomes increasingly

queryTombs.cfm?CFID=55702db0-0133-49de-9715-a7c6b74bd627&CFTOKEN=0.

[101] https://digitalegyptology.org/mastabase/.

[102] Verner, *Who was Shepseskare?*, 581–602; verso Baker, *Encyclopaedia of Pharaohs* I, 427–428.

[103] Krejčí, etal., *Queen Khentkaus* III, 33–35; Verner, *Sons of the Sun*, 24; Callender, *In Hathor's Image*, 149–154.

[104] See von Beckerath, *Chronology*, 174–179; Baines & Malek, *Atlas*, 8; Kitchen, *Chronology*, 206; Shaw, *Oxford History*, 482–483; Strudwick, *Texts*, xxix-xxxi; Dodson & Hilton, *Families*, 287–288; Hornung etal., *Chronology*, 490–491. For comparisons, see Bronk Ramsay etal., *Radiocarbon Chronology*.

[105] Butzer & Enfield, *Critical Perspectives*, 3629.

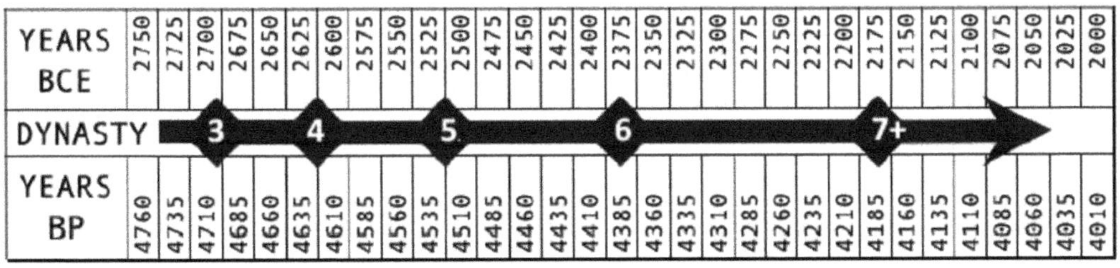

Figure 2.1: Project chronology.

possible. Detailed radio-carbon work by Bronk Ramsay and colleagues has provided a suggested chronology for the end of the Old Kingdom.[106] Large numbers of datable artefacts were investigated in order to produce a reliable date for the era. This systematic investigation represents a significant step forward in the scientific corroboration of Egyptian dynastic chronology.[107]

In their conclusion, Bronk Ramsay and his team suggest that Shaw's 'consensus chronology'[108] is the timeframe that most convincingly fits their data.[109] For this investigation, this chronology as outlined by Shaw[110] and favoured by Bronk Ramsay etal., will be adopted (see Figure 2.1). At the points between the diamond shapes, it seems reasonably certain that the Dynasty in question appears to have been active and in control. The amount of control and effectiveness of administration is yet to be determined. Based upon this outline, the 'drought' that had become severe by 2200 BCE, and has been suggested by many as a major factor in the fall of the Old Kingdom, can be tentatively dated to the latter stages of the reign of Pepy I.

2.5. Summation: Tomb Decoration Change as Artistic Narrative?

The basic points of this chapter include:

- Despite generalisations that ancient Egypt was a stable, though somewhat stagnant, and sometimes inert society, the society experience was always changing.
- Throughout Egyptian history, the culture displayed a continuous evolution in style and form.
- Some of these changes were due to changing religious beliefs, others to regional variations and others to external influences.
- The art produced to decorate the tombs of the elite displayed the major preoccupations and concerns of the society in which the elite orbited.

Disturbances or disruptions to the natural order must have impinged upon the perception of those involved in the organisation and administration of the society, and the society was discerning enough to identify changes to its environment. One of the responses due to this developing perception of environmental change could be in the art produced at this time. In recognising that changes to the decoration programme over time indicate some external stimulation upon the culture, identifying changes to the decoration programme may allow for the identification of current stresses and concerns. Applying succession studies to these artworks should allow for the identification of changes in the patterns of presentations and possible changes in perceptions and emphases over the time frame under investigation.

106 Bronk Ramsay etal, *Radiocarbon Chronology*, fig. 1, table 1.
107 Bruins, *Dating Pharaonic Egypt*, 1490.
108 Shaw, *Outsiders*, 308–323.
109 Bronk Ramsay etal., *Radiocarbon Chronology*, 1556; Höflmayer, *Dating Catastrophes*, 124–125.
110 Shaw, *Oxford History*, 480–481.

Part B

The Ecological Background

3

The Ecology of Rivers

Before it is possible to search for changes to the tomb decoration programme that may have arisen in response to changing environmental circumstances, it is necessary to identify the types of ecological alterations that may have developed in the Old Kingdom in response to changing nutrient levels within the river. As environmental factors alter, the biological profile of the river would also have shifted, the interactions within the environment now take place under baseline conditions fundamentally different to those experienced normally.

This chapter examines the scientific basis upon which a possible decline in Nile flood levels in the late Old Kingdom and subsequent years can be identified. A summary of basic ecological terms will be introduced and a basic outline of the fundamental principles of riverine ecology will be presented. Findings from modern ecological examinations of waterways currently experiencing low water flow and excess nutrient load will be applied to the situation that may have existed in ancient Egypt. From these analyses, suggestions as to the responses made by those organisms reliant on the river's regular patterns can be extracted.

3.1. Aspects of Riverine Ecosystems

Ecosystems are classified according to the physical, chemical and biological properties that they exhibit.[1] Due to variations in geology and geography, as well as the vagaries in flow, each and every riverine ecosystem exhibits characteristics that are unique to themselves.[2] Compared to other bodies of water, the most significant feature of rivers is that they flow in one direction only, leading to varying ecosystems which usually differ along the river's length.[3] The subsequent environmental characteristics can be studied by simplifying them so they can be more readily observed and measured.[4]

3.1.1. Factors within Riverine Ecosystems

The physical factors that have the greatest long-term effect on a river system are those that are related to the geology and geography of the site.[5] Geological factors are usually determined by the underlying rock structure, which impacts upon the nature of the soil and influences the direction that a river may follow.[6] The chemical factors significant to a river include acidity, salinity, and the level of dissolved oxygen, which in turn affect the quantity and quality of nutrients contained within the river.[7] The chemistry of a waterway, therefore, has a significant impact upon the variety and success of its biology.[8] The biological factors that make up an ecosystem develop as a consequence of the interaction between its physical and chemical components. An ecosystem's biological factors include the life cycles of those organisms living within it and are a composite of the interactions of all of the living things within that environment.[9] This interaction usually depends on the flow of energy through the environment and is best investigated by observing the food webs that have developed within it.[10]

3.1.2. Food Chains and Food Webs

All food chains start with a 'producer', an organism that converts sunlight energy into food. Producers are generally plants. Animals are typically 'consumers', an organism that eats another. These animals are preyed upon by increasingly larger animals as the food energy travels up a food chain. All the individual food chains in an area can be represented by a more complex food web.[11] Changes to one biological, physical, or chemical component within a food web can have ramifications to the entire food web mechanism.[12] Changes in nutrient content and composition impact upon the biological nature of waterways.[13] This results in changes to its chemical properties, by altering the acidity, which can influence the solubility of many substances within the waterway. Changes to the acidity within a waterway can subsequently affect those organisms living within it.[14] High acidity in a waterway, for example, can harm the maturation of invertebrate life within, as well as impede

[1] Angelier, *Streams & Rivers*, 215.

[2] Venkatachalapathy & Karthikeyan, *Water Quality Index*, 708.

[3] Bukaveckas, Rivers, 721–727; Allan & Castillo, *Stream Ecology*, 371; Giller & Malmqvist, *Streams & Rivers*, 10–13; Ward etal., *Riverine Landscape Diversity*, 517–539; Fisher & Sponseller, *Rivers as Ecosystems*, 494–496, figs 2, 3.

[4] Rzóska, *Nile*, 84; Allan & Castillo, *Stream Ecology*, 371. Weins, *Riverine Landscapes*, 501–515.

[5] Venkatesan etal., *Fluvial System*, 221–222.

[6] Bukaveckas, *Rivers*, 723–724; Venkatesan etal., *Fluvial System*, 201; Yitayew & Melesse, *Water Resource Issues*, 402–407.

[7] Allan & Castillo, *Stream Ecology*, 68–72; Ramkumar etal., *Land Use Dynamics*, 4; Venkatachalapathy & Karthikeyan, *Water Quality Index*, 708–709.

[8] Dhanakumar etal., *Phosphorous Fractionation*, 478.

[9] Boyd, *Water Quality*, 244; Ramkumar etal., *Land Use Dynamics*, 2–7.

[10] Menon etal., *Species Diversity*, 730–731; Bukaveckas, *Rivers*, 728–731.

[11] Boyd, *Water Quality*, 206–218.

[12] Boyd, *Water Quality*, xiii; El-Sheekh, *Nile River Ecosystems*, 15, Menon etal., *Species Diversity*, 745.

[13] Fisher & Sponseller, *Rivers as Ecosystems*, 498–497, fig. 4; Bukaveckas, *Rivers*, 727.

[14] Giller & Malmqvist, *Streams & Rivers*, 53–54; Mohammad, etal., *Crayfish Development*, 272.

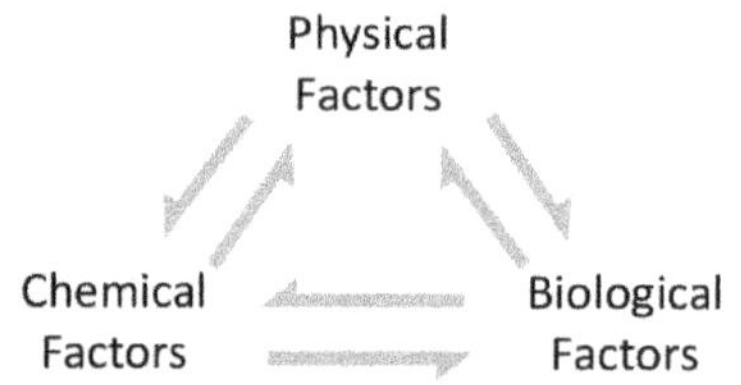

Figure 3.1: Environmental interactions.

particular developmental stages of some fish species.[15] Each of these factors may influence and impact upon the others[16] (see Figure 3.1). For example, the temperature of the water impacts upon the solubility of various chemicals, such as oxygen and carbon dioxide.[17] Excess sunlight in a drought may speed up decomposition of dead organisms, building up a toxic overload or undesirable chemicals in a habitat. The rate of flow changes the dissolved oxygen content in the river, which impacts upon the life cycles of some of its organisms. Other impacts to an ecosystem can arise from the way it is utilised or exploited, for example excessive harvesting of one particular food type or the addition of waste products not naturally associated with that waterway.[18]

3.2. Physical Factors in Riverine Ecosystems

Because this investigation is looking into a relatively small section of geological time, the long-term geological impacts upon the river flow can be ignored. Geographical factors include, for example, latitude, altitude, amount of sunlight exposure, average temperature, annual rainfall and direction of the prevailing winds: and these factors impact upon the value of the surrounding soil, which determines the quality and variety of nutrients washed into the river.[19]

3.2.1. Volume, Flow Rate and Nutrient Transport

The main physical factor impacting riverine ecosystems is the power of the river,[20] usually determined by the volume of water that flows along a river's length.[21] The altitude from which the water travels provides a gravitational force that helps develop the speed.[22] In the case of the Nile, for example, the amount of monsoonal rain that falls over equatorial Africa, the altitude from which it descends and the distance that it has to travel all contribute to its flow rate (see Figure 3.2).

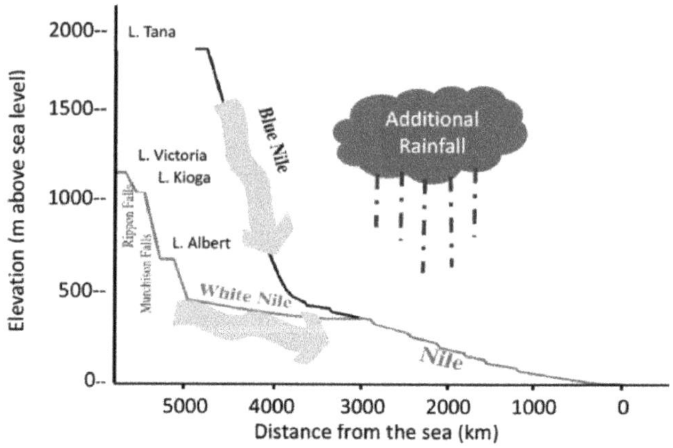

Figure 3.2: Factors affecting Nilotic flow.

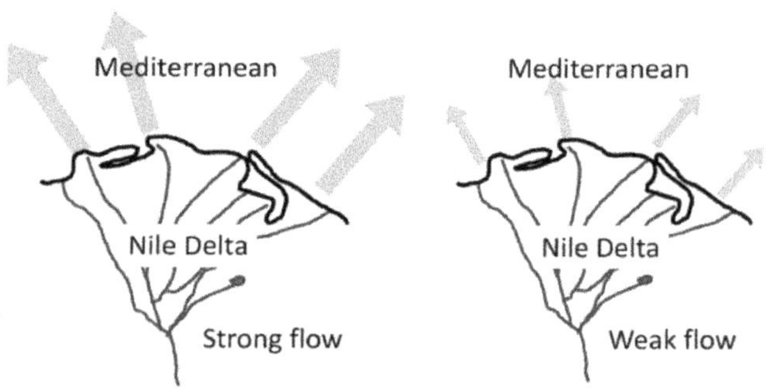

Figure 3.3: Sediment deposition due to river power.

The more rapid the flow, the greater the force of the water.[23] This allows the water to develop a 'momentum', or impetus, which affects sediment movement and nutrient transport. Rapidly flowing – that is, highly energetic – water can carry larger quantities of sediments for a longer time and can therefore transfer more sediment further.[24] The width, breadth and depth of the river is determined by the energy of the water.[25]

3.2.2. Erosion and Secondary Channels

Erosion, the "*moving of sediments from one area to another over a short-term period*",[26] is an important factor in the deposition of sediments.[27] A river's capacity to deposit silt was made up of two factors: speed and volume. The greater its speed, the longer the time and the further the distance the river could carry sediments. The greater its volume, the larger the amount of silt and resources it could deliver to waiting farmers.'[28] The features that allow

[15] Rzóska, *Nile*, 211; Boyd, *Water Quality*, 176–177, table 8.4.
[16] Ramkumar etal., *Land Use Dynamics*, 5; Angelier, *Streams & Rivers*, 215, 709.
[17] Boyd, *Water Quality*, 48–57.
[18] Soliman, etal., Plant Species Nile Islands, 70–92.
[19] Yitayew & Melesse, *Water Resource Issues*, 402–407; Cushing & Allan, *Streams*, 366.
[20] Kotoky & Dutta, *Spatio-temporal Variations*, 258–259; Giller & Malmqvist, *Streams & Rivers*, 55–67.
[21] Melesse etal., *Hydrologic Threats*, 2–3; Bravard & Petit, *Geomorphology*, 387–389.
[22] Bravard & Petit, *Geomorphology*, 392–394; Giller & Malmqvist, *Streams & Rivers*, 31–43.

[23] Venkatesan etal., *Fluvial System*, 221, table 3.
[24] Zahran & Willis, *Vegetation of Egypt*, 254–255; Mekonnen & Melesse, *Soil Erosion Mapping*, 211–213.
[25] Macklin & Lewin, *Rivers of Civilisation*, 235; Thirukumaran & Ramkumar, *Remote Sensing*, 161–167; Venkatesan etal., *Fluvial System*, 206–208, 212–221, figs 4–10. Kotoky etal., *Flood Management*, 240.
[26] Birx (ed.), *Encyclopedia of Time: Science, Philosophy, Theology, & Culture*, http://dx.doi.org.simsrad.net.ocs. mq.edu.au/10.4135/ 9781412963961.n183 (08/04/2016).
[27] Body, *Water Quality*, xi–xiii, 313.
[28] Ramkumar etal., *Land Use Dynamics*, 2–3.

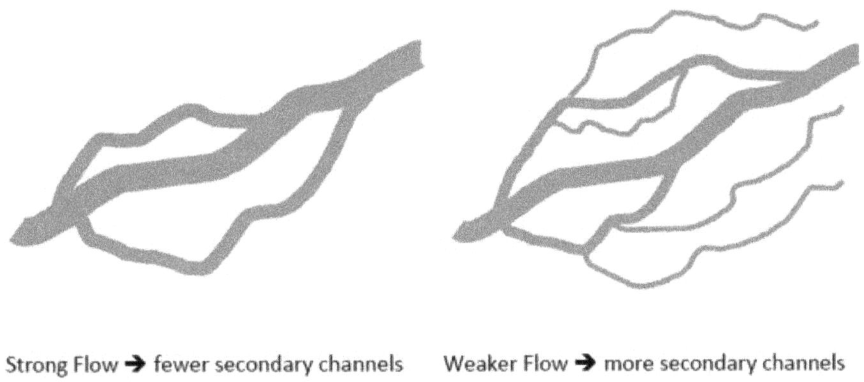

Figure 3.4: Secondary channel formation.

water to develop a 'momentum' also influence the amount of erosion that potentially may occur.[29]

In studies relevant to this investigation, evidence of sapropel (dark-coloured seabed sediments rich in organic matter) sediments from the floor of the Mediterranean, for example, has allowed for the rates of historic erosion depositions of Nile sediments to be determined (see Figure 3.3). The thickness of the layer indicates the amount of sediment carried, and, hence, the volume of the Nile, while the distance from the river mouth indicates the power of its flow at that time. These results have been used to corroborate the idea that the river displayed significant low levels of flow during the time frame under investigations.[30] The rate of erosion impacts upon the width of the marshes that are found alongside river systems.[31] The stability of the soil strata is affected by the energy of the water that is flowing across it.[32] Reduced power leads to a weakening of the river's ability to break up clumps of matter that can encumber the flow and allows for the development of strong stable soil. Strong, stable soil that is well-bound by the roots of plants growing within it acts to resist erosion, allowing plants growing on the riverbanks to develop greater purchase, further promoting marsh survival or expansion.[33]

During times of a weaker river, secondary channels can form where the river does not have enough strength to force its way back to the main flow (see Figure 3.4).[34] It is expected that the number of secondary channels should increase during weaker floods because water with less energy is forced to divert its passage when it encounters a blockage, rather than driving its way through.[35] Because of the less powerful flow, it is also expected that these secondary channels would be shallower, with the added potential to allow more marshland to development.

3.2.3. Turbidity, Light Penetration and Heat Exchange

The physical factor labelled as 'turbidity', describes the 'cloudiness' of a system,[36] and this property affects how readily light passes through it.[37] The depth of penetration of light and heat has been shown to be a function of the flow of the river.[38] Rivers with a weaker-than-average flow are less agitated and are more likely to be less cloudy.[39] The slower a river, the clearer it is, because even the finest particles it carries remain suspended for a shorter amount of time.[40] How deeply light penetrates into the water is influenced by the amount of suspended solid material present,[41] which in turn impacts upon the biological factors within the waterway.[42] Sunlight is needed for photosynthesis,[43] so light is an important ecological factor in determining how deep plants can exist.[44] The deeper light penetrates a waterway, the deeper the 'photosynthetic zone' (see Figure 3.5), allowing for the development of more complex food webs.[45]

While a slower river would typically allow light to penetrate more deeply[46] and improve the potential for an increase in photosynthetic activity,[47] a wider photosynthetic zone can lead to an increase in the growth of algae and other microscopic plants,[48] which could

[29] Kotoky etal., *Flood Management*, 239–241; Cushing & Allan, *Streams*, 366; Kotoky & Dutta, *Spatio-temporal Variations*, 258–259.
[30] Krom, *Nile Sediment Fluctuations*, 71–74.
[31] Steenhuis etal., *Erosion & Discharge*, 146; Allen & Pye, *Saltmarsh Morphodynamics*, 11–12.
[32] Pethick, *Saltmarsh Geomorphology*, 43–47.
[33] Giller & Malmqvist, *Streams & Rivers*, 50–53; Thirukumaran & Ramkumar, *Remote Sensing*, 162–163; Boyd, *Water Quality*, 317–318, 336–337.
[34] Thompson, *Swamp Development*, 177–178; Sallam & El-Barbary, *Secondary Channels*, 489.
[35] Sallam & El-Barbary, *Secondary Channels*, 494–498; Thompson, *Swamp Development*, 181, fig 26.

[36] https://water.usgs.gov/edu/turbidity.html. (12/09/2016).
[37] El-Sheekh, *Nile River Ecosystems*, 6; Allan & Castillo, *Stream Ecology*, 45, 109, 128.
[38] Davies-Colley & Nagels, *Light Penetration*, 5, fig. 3; Smith etal., *Optical Characteristics*, 301–312; Boyd, *Water Quality*, 102, 319.
[39] Entz, *Lakes Nasser & Nubia*, 285–288.
[40] Davies-Colley & Nagels, *Light Penetration*, 1–2.
[41] Abtew, *Land & Water*, 122–123; Boyd, *Water Quality*, 106–107, 112; Ramkumar etal., *Land Use Dynamics*, 4; El-Sheekh, *Nile River Ecosystems*, 6.
[42] Allan & Castillo, *Stream Ecology*, 131–132; Boyd, *Water Quality*, 18, fig. 5.2, 111–112.
[43] El-Sheekh, Nile River Ecosystems, 14–15; Boyd, Water Quality, 108.
[44] Westlake, *Light Climate for Plants*, 99– 119; Boyd, *Water Quality*, 206, fig. 10.7.
[45] El-Sheekh, *Nile River Ecosystems*, 20; Balagurunathan & Shanmugasundaram, *Microbial Biodiversity*, 582–583; Boyd, *Water Quality*, 18, 215, fig. 10.11.
[46] Davies-Colley & Nagels, *Light Penetration*, 3–4.
[47] Boyd, *Water Quality*, 107, fig. 5.2, 213; Bukaveckas, *Rivers*, 731.
[48] Steward etal., *River Runs Dry*, 205.

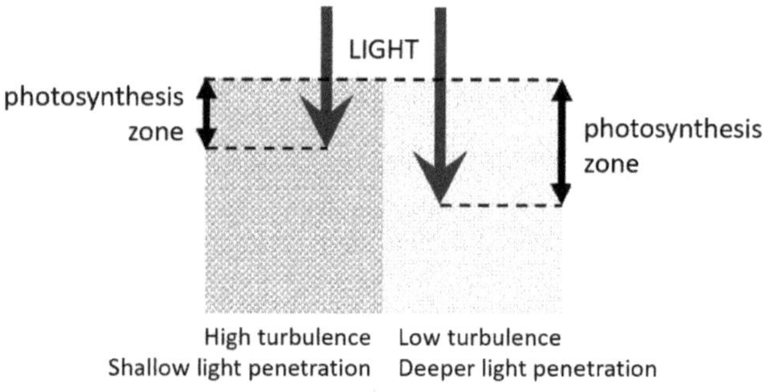

Figure 3.5: Turbidity and zone of photosynthesis.

rapidly make the waterway much cloudier, negating this change, and thereby, further increasing the temperature gradient differences, as the algae absorb heat in the upper layers of the waterway.[49] This growth also impacts upon the wider food web within the habitat.

High turbidity levels can cause problems for an aquatic ecosystem.[50] The flow rate also impacts upon how quickly the surface layer warms because the depth of light penetration also impacts upon the temperature that develops in the upper layers of the waterway.[51] Conversely, while low turbidity levels invite light and heat into the water, high levels of turbidity can also raise the temperature of the top layers because suspended particles near the surface speed up the absorption of heat from sunlight. Distinct layers can be expected to develop due to reduced mixing of materials that are moving at a more sedate pace.[52] Finally, the temperature of the water impacts on the solubility of chemicals within it, which can also influence the type of food webs that may develop. This in turn can impact upon the dissolved oxygen content because less oxygen can be dissolved in warm water than in cold.[53] Also, the different pace might be slower than that to which the organisms within them have become habituated.

3.3. Chemical Factors in Riverine Ecosystems

The major properties that influence the chemical nature of a river include salinity, acidity, and oxygen availability. These properties are significantly transformed by the temperature of the water, since water temperature affects salt and oxygen solubility. These factors may be influenced by one another. Also, as these factors change, the biological capacities of the waterway subsequently exhibit different characteristics.

3.3.1. Salinity

Salinity is a measure of the 'saltiness', or the amount of dissolved salt, present in a waterway.[54] Sunlight causes evaporation which can change the concentration of substances such as salt within the river, typically increasing them.[55] Salinity is an ecological factor of considerable importance, because it can influence the kinds of plants that will grow not just in the water, but in the vicinity, and also influence the types of organisms that live in a body of water.[56] Salinity values can vary across the width, depth and length of the waterway. This factor becomes increasingly significant in areas where the water is shallow or slowly moving, such as in the secondary channels that would have formed along the edges of the Nile.[57] Faster, deeper waterways experience less evaporation, so the situation may arise where chemical concentrations within the river at the same location could be different, depending on speed.[58]

3.3.2. Acidity of the Waterways

The measurement of acidity (pH) is another indicator of the relative health of a waterway,[59] as the acidity of a waterway influences the type and variety of organisms able to inhabit that region.[60] The pH scale ranges from 0 to 14. Substances with pH of less than 7 are acidic; with values ranging from mild (~6.0) to very strong (~1.0). Substances with pH greater than 7 are alkaline; with values ranging from mild (~8.0) to very strong (~13.0).[61] Pure water displays neither acidic nor particularly alkaline properties; it is termed 'neutral', with an ideal pH value of 7.0–7.4,[62] A pH scale with representative examples is shown in the following figure (see Figure 3.6).[63] The acidity of a waterway is influenced by many factors, including changes in nutrient level and salinity,[64] as well as the amount of carbon dioxide absorbed into the water,[65] and the amount of dead and decaying material within the waterway.[66]

[49] de Jonge & Elliott, *Eutrophication*, 318; Boyd, *Water Quality*, 108.
[50] El-Sheekh, *Nile River Ecosystems*, 6–7.
[51] Body, *Water Quality*, 25; Narmada etal., *Groundwater Quality Assessment*, 571.
[52] Entz, *Lakes Nasser & Nubia*, 282, 286; de Jonge & Elliott, *Eutrophication*, 314–315.
[53] El-Sheekh, *Nile River Ecosystems*, 4–5.

[54] http://www.biology-online.org/dictionary/Salinity (15/11/2016); Boyd, *Water Quality*, 8, 71; Abtew, *Land & Water*, 124–125.
[55] Boyd, *Water Quality*, 9–12, 106, table 6.1.
[56] Boyd, *Water Quality*, 99–100.
[57] Narmada etal., *Groundwater Quality Assessment*, 571–572.
[58] Zahran & Willis, *Vegetation of Egypt*, 300–301.
[59] Schindler, *Acid Rain & Freshwater*, 149–157; Boyd, *Water Quality*, 175–176.
[60] Boyd, *Water Quality*, 177–178, table 8.4.
[61] Ramkumar etal., *Land Use Dynamics*, 4; Boyd etal., *Interpretation of pH*, 403–408.
[62] Kemling, *Acid Rain*, 1–2; Boyd, *Water Quality*, 157; Ramachandra etal., *Ecohydrology*, 633.
[63] '*pH – Water Properties*', http://water.usgs.gov/edu/ph.html (19/04/2017).
[64] El-Sheekh, *Nile River Ecosystems*, 4–5.
[65] Boyd, *Water Quality*, 133–136, 157–160.
[66] Boyd, *Water Quality*, 198–199; El-Sheekh, *Nile River Ecosystems*, 15–17.

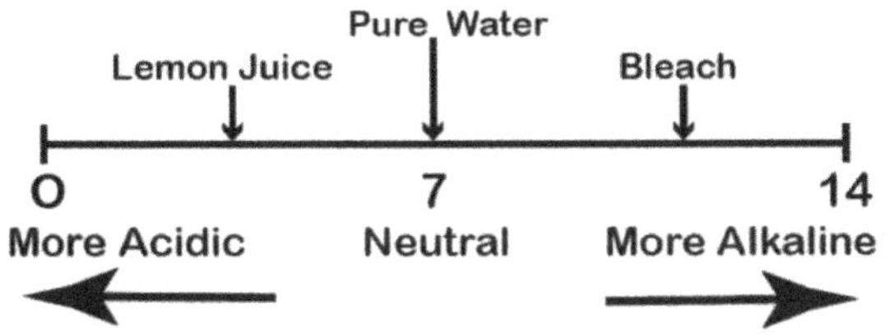

Figure 3.6: pH scale.

3.3.3. Oxygen Availability

Dissolved oxygen is the most important water quality variable in aquatic ecosystems[67] as all aerobic organisms must have an adequate supply of it.[68] Oxygen levels impact upon the type of life that can exist within that area.[69] Oxygen solubility is related to water temperature, and other chemical and biological factors.[70] The slower a waterway is moving, the lesser the absorbance of oxygen from the atmosphere into the water.[71] Higher temperatures mean less oxygen can be dissolved within the water.[72] The concentration of other substances, such as salt, can also interfere with the processes of normal gas exchange.[73] The utilisation of oxygen in the decomposition of nutrients also lowers the oxygen content within the water[74] (see 'eutrophication', below). Similarly, the ability of water to dissolve carbon dioxide is affected by changes to temperature.[75] This is one factor that helps determine the acidity of the waterway.[76] Warmer water absorbs less carbon dioxide, which means that less acid can form. This leads to the waterway departing from its natural level of acidity, one that may have developed over many years of adaptation.

3.4. Biological Factors in Riverine Ecosystems

Historically, inundations delivered vast amounts of silt to the riverbanks alongside the flooding waterway. The number and variety of nutrients carried by a river influence the variety and abundance of food webs that can develop within it.[77]

3.4.1. Nutrient Availability and Changing Habitats

The simplest and fastest-moving organisms take advantage of nutrient delivery most effectively.[78] Photosynthesising algae cover the first few centimetres of the water's surface, competing for light. The nutrients within a river are also utilised by plants as a part of their growth cycle.[79] The chemical by-products of the photosynthetic process make the water more alkaline. Generally, if plant growth burgeons, fragile organisms that cannot survive under these increasingly alkaline conditions will die, while hardier animals that feed on algae will experience population growth.[80] As time passes, the inorganic nutrients upon which the algae rely become depleted, causing the algae to die. These drop to the bottom of the river. Bacteria, which help decompose this dead algal matter, leach oxygen from the water, increasing the acidity of the riverbed, tipping the river pH balance to becoming slightly more acidic.[81] Bottom-feeding animals that cannot handle the low oxygen content and greater acidity die off, potentially decreasing the biodiversity of the environment. This process of decomposition occurs as a natural part of any ecosystem. However, circumstances arise where it unfolds at a pace more rapid than the local habitat can deal with. If a rapid increase in nutrient quantity occurs, for example, then the usual cycles can be interrupted.[82] It is usually a consequence of an over-supply of nutrients leading to a decrease in the amount of oxygen available for normal metabolic processes that causes an imbalance to develop in the local ecosystem.

3.4.2. Food Web Considerations

All food chains rely on the sun as the ultimate source of energy because each chain begins with a plant that utilises the process of photosynthesis to convert sunlight into energy. Plants are eaten by herbivores and they, in their turn, are eaten by carnivores and so on up the food chain. A habitat's food web is built up by combining all the individual food chains together; and this is affected by the availability of nutrients.[83] The usual consequence of an overabundance of nutrients is excess algal growth which can then interfere with the normal biological function of a waterway.[84]

Aquatic plants need two essential minerals to grow: nitrogen and phosphorus. Nutrient build-up, with its commensurate increase in biological pollution, encourages the development of anaerobic conditions within the waterway, so micro-organisms such as blue-green algae begin to thrive.[85] A surfeit of nutrients stimulates the rapid growth of plants and algae, clogging the waterways. Some areas within the waterway can become oxygen-depleted:

[67] Balagurunathan & Shanmugasundaram, *Microbial Biodiversity*, 587–588; Rudolf etal., *Dissolved Oxygen*, 89–91.

[68] Venkatachalapathy & Karthikeyan, *Water Quality Index*, 723; Boyd, *Water Quality*, 133–136, table 6.5.

[69] Saarinen etal., *River Water Acidity*, 360; Mohammed etal., *Crayfish Development*, 281–282; El-Sheekh, *Nile River Ecosystems*, 4, 8, 14.

[70] Boyd, *Water Quality*, 123–126.

[71] Ramachandra etal., *Ecohydrology*, 635.

[72] El-Sheekh, *Nile River Ecosystems*, 6.

[73] Boyd, *Water Quality*, 116–118, fig. 6.1.

[74] Boyd, *Water Quality*, 133, 189, 204–205; James etal., *Environmental Integrity*, 518.

[75] Boyd, *Water Quality*, 168–169, fig. 8.1; Bukaveckas, *Rivers*, 727.

[76] Boyd, *Water Quality*, 201, fig. 10.0.

[77] El-Sheekh, *Nile River Ecosystems*, 3; Boyd, *Water Quality*, 204–207.

[78] Quintana etal., *Predation, Competition & Size*, 52–55; Våge, etal, *Microbial Food Webs*, 1442–1444.

[79] Stevenson, *Algae of Rivers*, 114–122.

[80] Sterner, *Zooplankton*, 678–688.

[81] Zahran & Willis, *Vegetation of Egypt*, 257; El-Sheekh, *Nile River Ecosystems*, 14; Boyd, *Water Quality*, 125–128.

[82] Khan & Ansari, *Eutrophication*, 450–452.

[83] Bukaveckas, *Rivers*, 730–731; Fisher & Sponseller, *Rivers as Ecosystems*, 496–497.

[84] Art, *Dictionary of Ecology*, 196; de Jonge & Elliott, *Eutrophication*, 306.

[85] Rzóska, *Nile*, 209; Khan & Ansari, *Eutrophication*, 473.

this process is known as 'eutrophication'.[86]. Eutrophication commonly develops because of an imbalance within the food web.[87] As the balance of nutrients changes, some life cycles become more difficult and others easier. The number of scavenger and decomposer organisms would increase due to the increased mortality rates among oxygen breathers. In these instances, when the nutrients remain in the river, the organic base should develop more rapidly. An increased nutrient load or significant shift in the food balance may encourage certain plant species to weaken, and others to thrive.[88] In circumstances of environmental change, as the nutrient balance shifts, the proportion of species that made up the initial food web would alter.[89] Consequently, when a waterway becomes nutrient-rich, the ecology of the system changes, with this new situation encouraging overall plant growth,[90] a phenomenon that has also been recorded in the modern river Nile.[91]

3.5. Rivers and 'Drought'

In natural ecosystems, the interactions described above are usually balanced, with the system cycling through regular patterns. In times of disturbance, such as irregular or unnatural water flow, this system is knocked out of equilibrium.[92] The following section will attempt to identify some potential changes that may occur to an ecosystem that is experiencing changes to its normal situation. Since some parts of most rivers are naturally deeper than others,[93] there already exists areas displaying different environmental characteristics within the habitat.[94] As the flow rates diminish or change, these differences become more pronounced, and the subsequent effects become more exaggerated.[95]

3.5.1. Flow Rates Determine Habitat and Biodiversity

Bunn and Arthington have elucidated the basic ecological principles underpinning the consequences of a minimal flow:[96]

1. Flow is a major determinant of physical habitat in streams, which in turn is a major determinant of biotic composition. This physical habitat in turn influences the distribution and abundance of aquatic organisms.
2. Aquatic species have evolved life history strategies primarily in direct response to the natural flow regimes. Alteration to the natural flow regimes can lead to a loss of biodiversity of native species.
3. Maintenance of natural patterns of connectivity is essential to the viability of populations of many riverine species. Species survival diminishes when these populations become isolated.
4. The invasion and success of exotic and introduced species in rivers is facilitated by the alterations to the normal flow regimes.

In presenting a hypothetical response to a restricted flow regime in the Nile River, these principles will be used as the basis for projections.

3.5.2. Sediment Transport and Nutrient Supply

As a river weakens, the distribution of particles to the surrounding environment is also affected:[97] with the largest sediment particles settling out first, and the lightest particles remaining suspended for the longest time.[98] If a river breaks its banks, the alluvial soil deposited is composed of these finer particles. With a weaker flow, heavy particles which could be transported by more energetic water, are not carried as far, so they settle out much more rapidly than normal. Smaller particles, normally suspended in the river, would not remain suspended for as long, and therefore do not travel as far as usual. A river experiencing lower flow or volume than normal, exhibits a different pattern in the distribution of its sediments.[99]

If an ecosystem that is consistently flooded experiences drought, then the overall productivity of the primary producers, the plants, would be lower than normal due to a smaller-than-average water and nutrient supply.[100] Less material could be transported along the river's length and more of it would settle within the environs of the river, closer to its source. Therefore, a series of long-term, non-flood events would impact upon those organisms that have adapted to a more drenched situation. Since less nutrients leave the river, the quality of nutrient supply to the floodplain would deteriorate, explaining why a low flood would result in a downturn in agricultural productivity.[101]

With a weaker flow, the 'cleansing' activity of the river would be lessened; a river that does not 'flush' itself as well as a stronger one experiences organic remains lingering in the waterway for a longer time.[102] A weaker river, therefore, retains the detritus that had accumulated within it, leading to a rapid increase in biological material[103] with the commensurate loss of oxygen and change in

[86] Abtew, *Land & Water*, 126; Boyd, *Water Quality*, 259–261; de Jonge & Elliott, *Eutrophication*, 310, fig. 4.
[87] Khan & Ansari, *Eutrophication*, 450–452.
[88] Khan & Ansari, *Eutrophication*, 461–462.
[89] Thompson, *Swamp Development*, 188–190; Bernhardt etal., *Rapid Pollen Responses*, 736.
[90] Mitsch & Gosselink, *Wetlands*, 150; Tiner, *Wetland Indicators*, 297.
[91] Rzóska, *Upper Nile Swamps*, 5; Sallam & El-Barbary, *Secondary Channels*, 494–498.
[92] Steward etal., *River Runs Dry*, 207–208.
[93] Bravard & Petit, *Geomorphology*, 388–389.
[94] Boyd, *Water Quality*, 17, fig. 1.12.; Fisher & Sponseller, *Rivers as Ecosystems*, 494–496, figs 2, 3.
[95] Bukaveckas, *Rivers*, 728; Fisher & Sponseller, *Rivers as Ecosystems*, 497.
[96] Bunn & Arthington, *Altered Flow Regimes*, 492–493.
[97] Bukaveckas, *Rivers*, 726; Giller & Malmqvist, *Streams & Rivers*, 56–58, fig. 3.7.
[98] Bravard & Petit, *Geomorphology*, 387.
[99] Revel etal., *Nile River Dynamics*, 200–221.
[100] Rzóska, *Upper Nile Swamps*, 5.
[101] Yitayew & Melesse, *Water Resources Issues*, 406–407; Maybeck, *African Waterways*, 303.
[102] de Jonge & Elliott, *Eutrophication*, 310–311; Rzóska, *Nile*, 269.
[103] Giller & Malmqvist, *Streams & Rivers*, 49–51; Sallam & El-Barbary, *Secondary Channels*, 493–496; de Jonge & Elliott, *Eutrophication*, 306, 310–311.

acidity levels.[104] The reported sediment decreases, and subsequent fertility decline in northern Egypt since the construction of the Aswan dam are a modern example of this phenomenon.[105]

3.5.3. Temperature Changes and Evaporation Rates in Weaker Rivers

As a river becomes slower and shallower, its capacity to absorb solar radiation increases,[106] with the top layer heating more readily.[107] Since a weaker river would result in less mixing of water, the waterway would display larger temperature gradients between the surface and depths than would be typical.[108] Slower rivers become warmer rivers because they dissipate less energy in their motion. Warmer rivers experience greater rates of evaporation, which leads to an increase in the salinity level.[109] The shallower areas become warmer than deeper areas, and with more secondary channels developing, the differences in salinity levels between different parts of the river becomes amplified and the consequences more significant.[110] These factors affect the acidity level, altering 'normal' or 'natural' water quality levels.[111] As many species require specific temperature ranges for their metabolic functions,[112] this would lead to variances in local populations.[113] Many riverine organisms respond poorly to increases in the salinity level, especially if the changes are relatively rapid.[114]

Within the secondary channels, the water flow would be slower than that experienced by the rest of the river.[115] Secondary channels would experience the same lowering turbidity and higher algal growth as described previously, only at a faster scale.[116] Warm water has a lower dissolved oxygen content than does cold, so as a river warms the amount of oxygen available for its inhabitants decreases,[117] further exacerbating the differences between secondary channels and the major branches.[118] As the amount of dissolved oxygen drops below normal levels, it impacts upon the normal functioning of the micro-organisms present, altering the food cycles and changing the population dynamics.[119]

[104] Bukaveckas, *Rivers*, 727; de Jonge & Elliott, *Eutrophication*, 312.
[105] Zahran & Willis, *Vegetation of Egypt*, 255; Bard, *Archaeology of Ancient Egypt*, 56.
[106] Boyd, *Water Quality*, 16.
[107] Boyd, *Water Quality*, 215.
[108] Entz, *Lakes Nasser & Nubia*, 282; Boyd, *Water Quality*, 17–18, fig. 1.12.
[109] Boyd, *Water Quality*, 24–25.
[110] Abdo, *Water Quality*, 486–487.
[111] Rzóska, *Nile*, 211.
[112] Boyd, *Water Quality*, 198–199, fig 10.2.
[113] See Suberkropp, *Seasonal Occurrence*, 53–62; Lytle & Poff, *Natural Flow Regimes*, 95–97.
[114] Poff & Zimmerman, *Altered Flow Regimes*, 195–198, 201, figs 2, 3, 4; Fisher & Sponseller, *Rivers as Ecosystems*, 497.
[115] Boyd, *Water Quality*, 18.
[116] Boyd, *Water Quality*, 315; Khan & Ansari, *Eutrophication*, 474.
[117] Thompson, *Swamp Development*, 187; Boyd, *Water Quality*, 48, 131.
[118] Boyd, *Water Quality*, 121, 211, fig. 10.9.
[119] Entz, Lakes *Nasser & Nubia*, 293; Lytle & Poff, *Natural Flow Regimes*, 98–99; Boyd, *Water Quality*, 55–56, fig. 3.4, 201.

3.5.4. Characteristics of Weakened Rivers

When the volume of water becomes too low to initiate an inundation, then the potential for the water to change the landscape is weakened.[120] These physical impacts alter the expression of the chemical and biological characteristics of the ecosystem.[121] In the modern condition, the alteration of flow regimes is often claimed to be the most serious and continuing threat to the ecological sustainability of many rivers and their associated floodplains and wetlands.[122] Most modern-day reductions in river flow are a result of human interaction, such as excessive irrigation or the construction of dams. In riverine terms, if no flooding occurs, then the nutrients that are normally lost during an inundation remain in the river,[123] which results in changes to the biological and chemical nature of the river.[124]

3.6. Summation: Can a River Experience 'Drought'?

The basics points discussed in the chapter include the following:

- A weakening river is a less forceful river and diminishes the potential for the water to change the landscape.
- A weaker river develops more secondary channels
- A weaker river does not cleanse the river as normal as it cannot carry as many sediments as a stronger one, so the deposition of nutrients occurs over a different pattern of sites.
- A weaker river mixes its layers less, leading to more exaggerated differences in the layers' properties and develops wider water temperature and oxygen level variances
- A weaker river warms the surface layers up more readily, increasing surface evaporation. Increased evaporation changes the chemical composition of the waterway and develops greater salinity and acidity ranges.
- A warmer river stores less oxygen than normal, exacerbating the process of eutrophication.
- A weaker river allows light to penetrate more deeply, increasing the photosynthetic zone.
- A weaker river does not flood as fully so retains more nutrients, changing its biology.

Environmental factors interact with one another and impact upon the lifecycles naturally present. Due to an increased nutrient load, and changing physical and chemical factors, the ecological properties of the river and its surroundings change. Changing ecological circumstances affect plant growth and the subsequent food chains that rely on them, resulting in an alteration to 'normal' food webs. It should be possible to apply these considerations to a specific river – the Nile itself.

[120] Macklin & Lewin, *Rivers of Civilization*, 228–244.
[121] Poff & Zimmerman, *Altered Flow Regimes*, 201–204.
[122] Ward etal., *Floodplain Biodiversity*, 125–139; Lundqvist, *Avert Hydrocide*, 428–430; Bravard & Petit, *Geomorphology*, 387; Bukaveckas, *Rivers*, 723.
[123] Rzóska, *Nile*, 269.
[124] Steward etal., *River Runs Dry*, 207–208.

4

The A.R.I.D. Hypothesis: A River In 'Drought'

The ancient Egyptians relied on the inundation to invigorate the soil and provide impetus for a new growing season. In a 'regular' flood, the Nile spilled nutrients onto the nearby land. In riverine terms, nutrients are lost during an inundation; if an inundation did not occur, then those nutrients that were carried in the river remained in the river.[1] In a 'drought' these nutrients remain. If the nutrient load is further enhanced due to rainfall events adding more nutrients rapid ecological change would have resulted in uncharacteristically swift environmental responses.[2] In the terms of this examination, therefore, the context needs to be turned about (inwards?) towards how a situation of lower-than-normal water levels would impact upon the river. If the nutrients were not released by a flood, then their retention in the river would have resulted in changes to the ecology of the river, changing the environmental characteristics of the river itself.[3] Most likely, this would have changed the characteristics of the river and impacted upon the flora and fauna existing within and alongside the waterway.[4]

From current investigations and archaeological evidence, we should be able to propose some modifications to the Nile ecosystem caused by an 'unnatural' or irregular flow. In this chapter, the proposed ecological changes to riverine systems in drought or excess nutrient load identified previously will be applied to the River Nile itself. Modern ecological research into plant species that historically existed in ancient Egypt will provide evidence that we can seek to apply to the situation that may have arisen at the end of the Old Kingdom. Proposed subsequent impacts of changes in the ecology of the Nile upon the society and culture of the Old Kingdom will then be presented.

4.1. Nilotic Ecosystem Response to 'Drought'

The annual flood normally experienced by the Nile maintained a regular cycle to which the living things within the ecosystem had adapted.[5] Riverine species that had evolved to natural seasonal flow would have reacted differently if these patterns became irregular or unreliable. Despite evidence suggesting an increasingly arid countryside from the Predynastic era onwards,[6] the combined environmental effects at the end of the Old Kingdom appeared to have impacted the society in a more significant manner.

4.1.1. Nile Wetland Classification

According to the Ramsar Classification System for Inland Wetland Type,[7] it seems best to categorise the River Nile as an M–Class Inland Wetland (see Table 4.1). These are inland wetlands characterised by permanent plants on the edge of a permanent flowing waterway. While some regions within the river, such as in the delta, or along some of the sluggish secondary channels, may display slightly different characteristics,[8] this classification holds generally true for the major segment of the river. Obviously, this characterisation may be affected by the water use practices of modern Egypt.

For this investigation, it seems reasonable to apply the modern Ramsar classification to that situation existing before the construction of the Aswan Dam as the basis for an analysis of the potential late Old Kingdom changes that have been hypothesised.[9] As the amount of water flowing through the river system declined, then it is arguable that changes would develop between different sections of the river and these changes could become more pronounced or exaggerated if the conditions persisted.[10] Fortunately for our investigation, data was collected from the last 'natural' flood before the closing of the Aswan Dam.[11] We can develop some assumptions based on some of the data that was collected at the time and updated as recently as 2015.[12]

4.1.2. Ecological Impacts of a Low River

The amount of rainfall and glacial melt occurring in central Africa determined the volume of water that flowed along its length. The river carried water all the way from the Ethiopian Highlands and the Lakes of Central Africa,[13] with the maximum altitude for one of the sources of the Nile approximately 2,700 metres above sea level.[14] A riverine drought was occasioned by either declining rainfall

---

1 Rzóska, *Nile*, 269.

2 For example, see Kushlan, *Environmental Stability*, 821–825; Gorman & Karr, *Habitat Structure*, 570–515; Arthington etal., *Habitat Disturbance*, 61–66.

3 Poff & Zimmerman, *Altered Flow Regimes*, 201–204.

4 Arthington etal., *River Ecosystems*, 492; Balagurunathan & Shanmugasundaram, *Microbial Biodiversity*, 576.

5 Sallam & El-Barbary, *Secondary Channels*, 494–498.

6 Bell, *Oldest Nile Floods*, 569–573.

7 Ramsar Convention, https://www.ramsar.org/ (13/03/2017)

8 Thompson, *Swamp Development*, 177.

9 Stanton, *White Nile Marshes*, 375–379.

10 Wilkinson & Stevens, *Environmental Archaeology*, 68.

11 Halim etal., *The Last Normal Nile Flood*, 401–425; also El-Maghraby etal., *Zooplankton Community*, 527–530.

12 See Abd-El Monsef etal., *Aswan Impact*.

13 Abtew & Melesse, *Nile River Basin*, 14–15.

14 Woodward etal., *Nile Evolution*, 261; Wilkinson & Stevens, *Environmental Archaeology*, 66–67; Abtew & Melesse, *Nile River Basin*, 10.

Table 4.1: Ramsar classifications (from Information Sheet Ramsar Wetlands)

INLAND WETLANDS				
Fresh water	Flowing water	Permanent	Rivers, streams, creeks	M
			Deltas	L
			Springs, oases	Y
		Seasonal/Intermittent	Rivers, streams, creeks	N
	Lakes and pools	Permanent	Size > 8 ha	O
			Size < 8 ha	Tp
		Seasonal/Intermittent	Size > 8 ha	P
			Size < 8 ha	Ts
	Marshes on inorganic soils	Permanent	Herb-dominated	Tp
		Permanent/Seasonal/Intermittent	Shrub-dominated	W
			Tree-dominated	Xf
		Seasonal/Intermittent	Herb-dominated	Ts
	Marshes on peat soils	Permanent	Non-forested	U
			Forested	Xp
	Marshes on inorganic/ peat soils	High Altitude (alpine)		Va
		Tundra		Vt
Saline, brackish or alkaline water	Lakes	Permanent		Q
		Seasonal/Intermittent		R
	Marshes & pools	Permanent		Sp
		Seasonal/Intermittent		Ss
Fresh, saline, brackish or alkaline water	Geothermal			Zg
	Subterranean			Zk(b)

or a lack of snow melt. The basic ecological principles associated with the consequences of a minimal flow, as identified previously,[15] can be used to make inferences about the possible consequences to the ecosystem of the Nile. In the modern condition, the reduction of flow regimes is the most significant ongoing threat to rivers and their associated floodplain wetlands;[16] with most causes of reduced flows in modern-day rivers explained as a direct consequence of human intervention – for example, the rapid removal of water from waterways for industry or agriculture.[17] The consequences downstream should allow us to simulate what could have happened to the Nile when it experienced a significant change in its regular flow regime in the Old Kingdom. The consequences to the chemical properties that would have developed have been described in Chapter Three, with increases in salinity in some areas anticipated due to a weaker river developing slightly warmer water leading to increased evaporation. Greater temperature variances would be expected and, as a consequence oxygen solubility would decrease overall, leading to the development of increasingly anaerobic conditions in secondary channels areas. Decreasing aerobic conditions would change the microbial food webs that existed, and this would have increased the probability of significant changes to the biological properties of the organisms existing in these areas.

4.2. Aquatic Plants' Response to 'Drought'

While the historical evidence suggests that the significant Nile River plankton communities have not changed considerably over the past three thousand years,[18] human activity has had a much more considerable impact upon the macroscopic (large enough to see) plants that were important to the ancient Egyptian.[19] From very early on in Egyptology, researchers have been using microscopes to analyse fossil plant remains to reconstruct the history of vegetation in Egypt from geological deposits and archaeological finds.[20] These sediments contain seeds, spores and/or pollen grains deposited in them over historical time. To better understand the biodiversity of plant species present during Dynastic times, sedimentary analyses has allowed for the basic plants to have been identified.[21] It is important to realise that the final resting place of the seeds, spores or grains may not be

[15] Bunn & Arthington, *Altered Flow Regimes*, 492–493, fig. 1.
[16] Ward etal., *Floodplain Biodiversity*, 125–139; Lundqvist, *Avert Hydrocide*, 428–430.
[17] Ramdani etal., *Environmental Influences*, 114–115.

[18] Hamdan etal, *Climate &Collapse*, 44–45: Avnaim-Katav, etal., *Shallow Marine Record*, 144–145.
[19] Taher & Abdel-Motelib, *Microbial Sedimentary Structures*, 341–353.
[20] Wilkinson & Stevens, *Environmental Archaeology*, 23–25, 62–63; Butzer, *Environment & Human Ecology*, 53–78.
[21] Frils etal., *Fossil Water Lilies*, 357.

an indication of their origin.[22] The dispersal properties of pollen, spores and seeds: their size, shape, buoyancy, aerodynamics and density, all impact their dispersal and influence their final resting place.[23] According to the theory of 'superposition', the deeper the sample, the more ancient the time of deposition:[24] using the same concept as applied to archaeological excavations, this allows for an estimation of age of the plant deposits and allows for a succession sequence over time to be developed.[25]

Mudbrick analysis also enables the determination of the mud contents.[26] Studies of mudbricks have yielded well-preserved seeds, spores and pollen grains; all of which combine to serve as an indicator of which cereal type was being harvested at the time, with the rest of the pollen assembly being indicators of the common weeds that were growing within or near the crop.[27] Since the bricks were made on demand, they yield a relatively accurate time frame, serving as "*cultural time-capsules*"[28] preserving a picture of the local environment at that time.[29] Other artefacts also provide information about the plant life present at a particular location, such as wood remains in fires or fossilised insects whose dietary requirements are particularly specific.[30] The combination of this data allows for the reconstruction of past vegetation,[31] and identifies the succession or changes in the biodiversity,[32] as well as identifying and tracing human interaction.[33] Improved technological abilities, such as the scanning electron microscope, have made identification much more reliable.[34]

4.2.1. Fundamental Vegetation Types

Zahran and Willis have presented a classification system categorising the ancient botanical landforms into seven fundamental 'vegetation groups':[35] 1 – Desert; 2 – Salt marsh; 3 – Mountain; 4 – Sand dune; 5 – Reed swamp; 6 – Freshwater and 7 – Saline water vegetation types. For this investigation, the vegetation groups of relevance are those whose profiles relate to the Reed swamp and Freshwater habitats.

Plants identified as historically significant in freshwater and reed swamp habitats include *Phragmites australis* and *Typha domingensis* (cattails/bulrushes) as the most common 'reeds'; and *Cyperus papyrus* and the two lotuses, though, to use proper biological classification suggest these are more precisely termed "lilies",[36] the White (*Nymphaea lotus*) and the Blue (*Nymphaea caerulea*), as the most important waterway plants.[37] More areas along the river, especially in the slower and weaker secondary channels, would have developed properties displaying increasingly similar characteristics to those of mildly saline habitats as the Nile flood levels declined.

4.3. Papyrus and A River In 'Drought'

The simple papyrus plant, *Cyperus papyrus*, is a large, emergent, aquatic perennial, with erect culms that can grow up to five metres tall.[38] Today, it is almost unknown in the modern vegetative profile of Egypt.[39] The remains of papyrus is found commonly in the archaeological record, especially in the marshland regions,[40] and, without bio-archaeological corroboration,[41] the only evidence for its existence at the time frame under investigation would be cultural.[42]

The *Cyperus* genus produces large numbers of seeds and, in 'normal' flood levels, most are lost: they are washed away downstream, too waterlogged to germinate. Some settle too deeply into the mud and cannot germinate due to lack of oxygen; and some sink too deeply into the water so they do not receive the necessary amount of sunlight required to stimulate germination.[43] Their tough roots can extend one metre or more into the soil, and they can also spread through the use of rhizomes: a type of plant stem situated either at the soil surface or underground that contains nodes from which roots and shoots originate capable of producing a new plant.[44] Papyrus displays the capacity to exploit two distinct habitat types: shallow water and the edges where the river meets the land, the land beside the water.[45] This enhanced facility of papyrus to rapidly grow in two different habitats allows it to respond more readily to changes in water level compared to similar plants,[46]

[22] Moore, *Pollen Travel,* 388–389; Caracuta etal., *Amheida Archaeobotanical Analysis*, 45–48.
[23] Wilkinson & Stevens, *Environmental Archaeology*, 96, fig. 39.
[24] https://www.reference.com/science/geology-principle-superposition (23/11/2016).
[25] Bernhardt, *Wetland Responses,* 126–129; Zahran & Willis, *Vegetation of Egypt*, 306, fig. 7.1.
[26] Ayyad etal., *Mudbrick Pollen Information*, 77–96.
[27] Zahran & Willis, *Vegetation of Egypt,* 309–310, fig. 7.4.
[28] Zahran & Willis, *Vegetation of Egypt*, 311.
[29] Ayyad etal., *Mudbrick Pollen Information,* 77–96; Ritchie, *Modern Pollen Spectra*, 1–6; Ritchie, *Dakhleh Oasis Palaeobotany*, 74–80; Schrank & Mahmoud, *Dakhla Oasis Palynology*, 188–189.
[30] Wilkinson & Stevens, *Environmental Archaeology*, 100–103.
[31] Wilkinson & Stevens, *Environmental Archaeology*, 89–95, figs 37, 38; Hamden etal., *Faiyum Sedimentary Record,* 1–3; Ayyad, *Pollen Grain Ecology* (PhD, University of Mansoura, Egypt).
[32] Saad & Sami, *Nile Delta Deposits*, 467–503; Zahran & Willis, *Vegetation of Egypt*, 305.
[33] Kamal etal., *Hydrophyte Nutrient Status,* 1–2.
[34] Zahran & Willis, *Vegetation of Egypt*, 308, fig. 7.3.
[35] Zahran & Willis, *Vegetation of Egypt*, 317.
[36] Conard, *Waterlilies,* 194; Les etal., *Classification of Water Lilies*, 41–44.
[37] Zahran & Willis, *Vegetation of Egypt*, 312, 318.
[38] *'Cyperus papyrus'*, https://www.cabi.org/ISC/datasheet/17503, (03/12/2017).
[39] *'Cyperus papyrus'*, http://data.kew.org/sid/SidServlet?ID=7263&Num=gcm#Salt (23/11/2016).
[40] Bernhardt etal., *Wetland Responses*, 137–138; Butzer, *Environment & Human Ecology*, 53–56.
[41] Butzer, *Environment & Human Ecology,* 53–56; Zahran & Willis, *Vegetation of Egypt*, 311.
[42] Kamrin, *'Papyrus in Ancient Egypt'*, http://www.metmuseum.org/toah/hd/papy/hd_papy.htm and Plantz Africa, *'Cyperus papyrus'*, http://www.plantzafrica.com/plantcd/cyperuspap.htm (21/11/2016).
[43] Hammer, *Freshwater Wetlands*, 263–264; Boar, *Papyrus Water Level Response*, 88–90.
[44] https://biologydictionary.net/rhizome/ (13/11/2017).
[45] Mitsch & Gosselink, *Wetlands*, 390, fig. 12.4.
[46] Jones & Muthuri, *Biomass in Papyrus*, 347.

and, as such, it can rapidly increase its own biomass and density.[47]

4.3.1. Responses to Changing Physical Factors

As noted earlier, the power of the water affects the rate of erosion, thus impacting upon the width of the marshes.[48] A weaker-than-usual river would result in a smaller volume of flowing water, suggesting that a commensurate decrease in the rate of erosion should occur. Lind, in identifying the ability of papyrus to grow on the edge of a riverbank, also notes its capability to 'clump' upon itself, forming islands,[49] and reports that these clumps can also influence the rate of flow of the river. As a consequence of a weaker river, the forces hindering the ability of papyrus to clump would diminish, and so the size of papyrus clumps would be expected to increase.[50] In Egypt, this increased distribution of papyrus plants would effectively produce an obstacle; in Arabic, the term *sudd* relates to such an obstacle or blockage within the river.[51] Over time, these islands further increase in volume as they collect detritus.[52]

Once established, sedge-type plants, such as the papyrus, send out rhizomes to spread along the nearby soil strata,[53] but the spread of these rhizomes is usually limited by the power of the water.[54] The stability of the stratum in which the papyrus is growing is affected by the energy of the water that is flowing across it.[55] Since a weaker river has less energy to carve out new channels, this increases the number of secondary channels as water flows around obstacles not through them. Both a weaker river and a greater supply of habitats delivered by an expanded number of secondary channels should encourage rhizomes to spread more readily. Rhizomes spread better when water availability is lower than the conditions needed for successful seed germination to take place.[56]

In a low flood, erosive conditions become less severe as the momentum of the water reduces – the flow less powerful and the mud becomes less fluid. Additionally, the water is not as deep. In those circumstances, more seeds would have an increased chance of surviving to germination.[57] In times of low water, the incidence of papyrus pollen deposition rates in the Mediterranean decreased,[58] implying that more pollen remained within the river, increasing the potential for germination. A weaker river produces less turbulence; this lowers turbidity and allows light to penetrate deeper into the water, encouraging plant growth.[59]

4.3.2. Response to Changing Chemical Factors

Excess nutrient availability suggests that more marshlands would have developed, but we should never perceive the Nile marshlands as motionless 'swamps.'[60] Since water flow in secondary channels is weaker than in the primary channel of the river, and with the water column becoming increasingly shallow,[61] it has been suggested that the secondary channels of a weaker river would have experienced an increasing level of salinity.[62] Since the development of the Aswan Dam, with the consequent lower flow, agricultural fields in the Nile Valley became saltier and less fertile than earlier records indicated.[63] Due to increased evaporation in shallow water, it is expected that the salinity levels within secondary channels would be greater than those in the main part of the river; however, papyrus has been identified as a plant that appears able to tolerate slight increases in salinity.[64] In a low flood, those nutrients that were normally 'lost' to the land in a 'natural' flood remain, changing acidity levels from those normally existing.[65] Papyrus display an ability to tolerate a wider range of acidity levels than similar plants,[66] suggesting that it is well-suited to survive the acidity changes that could develop during times of lower-than-usual volume.

4.3.3. Responses to Changing Biological Factors

The rate of cycling of nutrients within the ecosystem is affected by the amount of water flowing through it,[67] with the loss of nutrients to the land during an inundation normally inhibiting papyrus growth.[68] Papyrus clumps present considerable absorbing surfaces as they spread out over the water at river and lake edges, and as the floating mats form, large amounts of nutrients would have been trapped within and become incorporated into the plant mass.[69] Combined with the reduction in water current leading to a larger area of sheltered living conditions at the river's edges,[70] the greater nutrient levels would have provided the impetus for a bloom of papyrus. A weaker river also seems to allow an increase in the growth rate of those micro-organisms that 'fix' nitrogen, improving the nutrient quality of the host, in this case the papyrus.[71] Since it has been reported that swift water flow within a river results in a more rapid loss of organic material from

---

47 Lind, *Ugandan Swamps*, 169.
48 Allen & Pye, *Saltmarsh Morphodynamics*, 11–12.
49 Lind, *Ugandan Swamps*, 167.
50 Ellery etal., *Channel Blockage*, 29–31; Thompson, *Swamp Development*, 192.
51 Rzóska, *Nile*, 202; Ellery etal., *Channel Blockage*, 29; Mitsch & Gosselink, *Wetlands*, 39; Ellery etal., *Channel Blockage*, 34–35, fig. 5.
52 Jones & Muthuri, *Biomass in Papyrus*, 355.
53 Jones & Muthuri, *Biomass in Papyrus*, 354.
54 Rzóska, *Nile*, 205.
55 Pethick, *Saltmarsh Geomorphology*, 43–47; Rzóska, *Nile*, 205.
56 Boar, *Papyrus Water Level Response*, 90.
57 Boar, *Papyrus Water Level Response*, 88–89; Mitsch & Gosselink, *Wetlands*, 392.
58 Bernhardt, *Delta Vegetation*, 616–618.
59 Tiner, *Wetland Indicators*, 18, fig. 2.1.A.
60 Tiner, *Wetland Indicators*, 282.
61 Entz, *Lakes Nasser & Nubia*, 282–285.
62 Thompson, *Swamp Development*, 178.
63 Abd-El Monsef etal., *Aswan Dam Impact*, 1882–1883; Halim etal., *Last Normal Nile Flood*, 401–425.
64 Mashaly etal, *Vegetation Analysis*, 1189.
65 Rzóska, *Nile*, 269.
66 Speight & Blackith, *The Animals*, 354.
67 Mitsch & Gosselink, *Wetlands*, 147.
68 Jones, *Papyrus Photosynthesis*, 359.
69 Boar, *Papyrus Water Level Response*, 91; Thompson, *Swamp Development*, 182; Gaudet, *Papyrus Nutrients*, 415–416.
70 Rzóska, *Upper Nile Swamps*, 11.
71 Thompson, *Swamp Development*, 188.

that system,[72] then it seems logical that the reverse – that a slow river retains organic matter – is true.[73]

4.3.4. Ecological Implications for Papyrus in 'Drought'

While excess flooding and rapid drying events hinder germination,[74] slight changes do not affect the setting of seeds. Salinity does not seem to impact upon the germination and growth of papyrus, nor is papyrus affected by slight increases in temperature. It seems, that in response to a new setting presented by a weaker river flow, one with a greater concentration of nutrients, warmer-than-usual water temperatures, and a slightly more saline environment, papyrus would experience a significant boost to its survival chances.

4.4. Phragmites and A River In 'Drought'

Phragmites australis is an indigenous species displaying a long historical association with Egypt. Known commonly as 'reeds', *Phragmites* has been found growing in many diverse habitats.[75] The ability of the huge numbers of seeds produced to be dispersed by wind as well as water assures their rapid scattering.[76] As with papyrus, reeds can reproduce sexually and asexually via rhizomes, but roots are also able to develop from fragments of the rhizome that have settled above ground on moist land.[77] Another form of asexual reproduction employed by phragmites is the ability of any part of the stem that touches the ground to develop a root system, this is known as a 'stolon.'[78] The ability to utilise many reproductive methods results in a widespread distribution within the river.

4.4.1. Responses to Changing Physical Factors

Phragmites is termed an 'invasive' plant, since the reed beds grow into the surrounding environment.[79] These plants can spread at a rapid rate, weakening the soil by retaining moisture, making the waterway more marsh-like;[80] with the 'gaining of ground' by these sedge-type plants is a behaviour characteristic of the genus.[81] Since flooding has been suggested as a control strategy,[82] the inference is that a slower river or a lower water level is conducive to phragmites growth.[83] Along the Nile, with its many secondary channels, it would be expected that the reed populations would be more common in the slower-flowing water, especially in a season of low inundation. Phragmites' internal structure allows it to resist dehydration, enabling it to survive extensive dry periods.[84]

4.4.2. Responses to Changing Chemical and Biological Factors

The reed seems to grow best in areas with slow water,[85] displaying the ability to tolerate slightly more brackish conditions compared to other riverine plants.[86] In the modern era, *Phragmites australis* growth has been promoted by an increase in soil salinity brought about as a consequence of de-icing roads and pathways, in America, for example; suggesting that it responds more favourably to an increase in salinity than similar plants.[87] Due to this tolerance, some reed swamps that exhibit higher than normal levels of salinity have become almost mono-cultures of phragmites.[88] This suggests that an increase in water salt content due to higher rates of evaporation times of excess nutrient load in the river would have no detrimental effect.[89] On the contrary, an advantageous salt tolerance would have improved phragmites survival. Reeds appear to respond to an increase in nutrient supply in a manner similar to papyrus,[90] growing best in areas with slow water and silted substrates.[91] Phragmites' growth, like that of papyrus, does not seem to be as impaired as most plants when their habitat becomes oxygen poor.[92]

4.4.3. Ecological Implications for Phragmites in Drought

Because of the varied reproductive strategies available to phragmites, the reed responds swiftly to habitat disturbance.[93] In the more recent timeframe, various examples of human manipulation, impacts or disturbance to environments appear to promote phragmites growth.[94] Restriction of the tidal inundation of a marsh and sedimentation may favour reeds,[95] suggesting that a lower-than-normal flood event, coupled with extra nutrients from rainfall events, would not have had a detrimental effect, but more likely would have encouraged phragmites response.

[72] Mitsch & Gosselink, *Wetlands*, 150.
[73] Rzóska, *Nile*, 269–270; Hammer, *Freshwater Wetlands*, 74; Krom etal., *Nile Sediment Fluctuations*, 73; Mitsch & Gosselink, *Wetlands*, 107, 145.
[74] Boar, *Papyrus Water Level Response*, 90.
[75] Uchytil, *Phragmites australis*, 1; Zahran & Willis, *Vegetation of Egypt*, 314.
[76] Hosier, *Unwelcome Phragmites*.
[77] Huhta, *Common Reed*, 1–3.
[78] Ailstock, Phragmites Adaptions, 1–2.
[79] Holm etal., *World's Worst Weeds*; Kettenring etal., *Phragmites Invasion*, 1306–1311; Global Invasive Species Database, http://www.iucngisd.org/gisd/speciesname/Phragmites+australis (12/08/2015).
[80] Kettenring etal., *Phragmites Invasion*, 1306–1307.
[81] Hammer, *Freshwater Wetlands*, 156–158, fig. 8.2.
[82] Hellings & Gallagher, *Flooding on Phragmites*, 41–49; Marks etal., *Phragmites*, 7; Zahran & Willis, *Vegetation of Egypt*, 75–76.
[83] Zhao etal., *Nutrient Accumulation*, 19–20, table 1, fig. 2.
[84] Rzóska, *Nile*, 205.
[85] Köbbing etal., *Reed Utilisation*, 1–2.
[86] Chambers etal., *Phragmites Expansion*, 398–406.
[87] McNabb & Batterson, *Phragmites Occurrence*; Marks etal., *Phragmites Australis*, 285–294; Marks etal., *Phragmites Stewardship*, 2–4.
[88] Zahran & Willis, *Vegetation of Egypt*, 56.
[89] Yang et.al, *Phragmites Stress*, 420–426.
[90] Haslam, *Reed*, 1–4; Kettenring etal., *Phragmites Invasion*, 1306.
[91] Zhao etal., *Nutrient Accumulation*, 20–22, table 2; Barkworth etal., *Flora*, 790–799; Meadows & Saltonstall, *Phragmites Distribution*, 99–107; Zahran & Willis, *Vegetation of Egypt*, 66.
[92] Zhao etal., *Nutrient Accumulation*, 24.
[93] Roman etal., *Salt Marsh Vegetation Change*, 141–150; Minchinton & Bertness, *Phragmites Spread*, 1404–1408, tables 1–3.
[94] Marks etal., *Phragmites Stewardship*, 1–3; Kettenring etal., *Phragmites Invasion*, 1307; Minchinton & Bertness, *Phragmites Spread*, 1412–1414.
[95] Hosier, *Unwelcome Phragmites*.

4.5. Cattails and A River In 'Drought'

The primary cattail species, *Typha domingensis*, commonly called a 'bulrush', is usually found in close proximity to phragmites.[96] It is a tall marsh herb able to achieve heights of 2–2.5 metres.[97] Cattails propagate by seeds, which can be broadcast by wind,[98] as well as by stout creeping rhizomes,[99] but cannot spread beyond shallow water.[100] Typha generally do not occupy as wide a range of riverine habitats as that inhabited by phragmites, both due to their biology and physiology.[101] The immature bulrush is thought to be good cattle food.[102]

4.5.1. Responses to Changing Physical Factors

Absence of new water results in a limited growth season.[103] Since erosion has less of an impact in a low river event,[104] plant stability would be increased; however, less powerful water flow results in a weakened ability of the cattail rhizomes to penetrate the subsoil,[105] presumably because the sediment is less disturbed. Experimental investigation of the relationship between water flow and plant growth suggests that low water availability impairs the overall fitness of the species, but excess water also limits growth rate.[106] While higher-than-normal water temperatures inhibit plant growth rates for Typha species,[107] it has been shown that increased soil temperature slows the formation of new rhizomes but encourages rhizomes already present to grow more rapidly.[108]

4.5.2. Responses to Changing Chemical and Biological Factors

Extremely low pH values (highly acidic) are toxic for the growth of typha species.[109] *Typha* are usually less dominant in the more saline parts of the swamps than are the *Phragmites* species.[110] Cattails do not respond well in conditions more saline than normal; as the salinity increases, their overall numbers decline,[111] and the spread of the rhizome is limited. While the mature plant is able to tolerate some minor salinity increase, seed germination is less successful and seedling growth less sturdy.[112] Root structure is modified when subjected to salt stress.[113] However, cattails exhibit different sensitivities to salinity at different stages in their life cycle, displaying a certain resilience to changing environmental situations.[114] In waterways where the sediment is highly nutritious, cattails can predominate, becoming almost mono-cultures.[115] Their deep roots aid them in resisting changes in soil quality, leaving the plant less affected by drying out of its territory.[116] An increase in phosphorus nutrition enables cattails to resist some of the negative effects of salt on their development.[117] An increase in nitrogen-based nutrients, such as rapidly decaying organic matter, does not aid typha growth but it seems to help the plant develop resistance to the toxic effects of cyanobacteria.[118] As the soil becomes increasingly anaerobic, the lack of oxygen limits growth potential for the plant, comparing unfavourably with phragmites' ability in this regard.[119]

4.5.3. Ecological Implications for Cattails in Drought

Typha seems to respond to changing salt levels and variable acidity better than free-floating macrophytes.[120] Typha plants can persist in conditions detrimental to their survival by reverting to a dormant state until conditions improve.[121] This suggests that, while it may not gain biomass in drought-like conditions or in areas inimical to its survival, typha can persist in an environment until the equilibrium is returned. The species appears capable of modifying its morphology in order to adapt to severe conditions.[122] Typha plants respond to a changing environmental situation by adapting their reproductive strategies, emphasising the production of either seeds or rhizomes.[123] Being resistant to cyanobacteria toxins suggests that the species is more likely to persist in conditions where the waterway has become oxygen poor. In studies identifying the distribution of plants in modern times, Zahran and Willis identify Phragmites and typha populations recorded in close proximity to one another,[124] and it seems that these two groups appear to be in competition for similar habitats and resources.[125] When they are in competition with one another in water channels, a community of phragmites is more likely to

[96] Zahran & Willis, *Vegetation of Egypt*, 260–266; Shaltout & Ahmed, *Plant Life*, annex 3.2 and 3.8.
[97] Eid etal., *Typha Growth Dynamics*, 63–72.
[98] Zahran & Willis, *Vegetation of Egypt*, 278.
[99] Shaltout & Ahmed, *Plant Life*, 112.
[100] Zahran & Willis, *Vegetation of Egypt*, 272–275.
[101] Zahran & Willis, *Vegetation of Egypt*, 63.
[102] Shaltout & Ahmed, *Plant Life*, 117; Zahran & Willis, *Vegetation of Egypt*, 298.
[103] Mexicano etal., *Ecological Engineering*, 174.
[104] Grace, *Effects of Water Depth*, 765.
[105] Glenn etal., *Salinity & Growth*, 84; Grace & Wetzel, *Typha Population Dynamics*, 140.
[106] Vivian etal., *Wetland Plant Growth*, 1003–1007, figs 4, 5, 7.
[107] Cary & Weerts, *Growth & Water Temperature*, 115.
[108] Eid etal., *Typha Growth Dynamics*, 68–69; McNaughton, *Typha Ecotype Function*, 323.
[109] Mufarrege etal., *Typha Adaptability*, 462.
[110] Zahran & Willis, *Vegetation of Egypt*, 75–76.
[111] Glenn etal., *Salinity & Growth*, 76, 89; Zahran & Willis, *Vegetation of Egypt*, 70; Neubauer, *Ecosystem Responses*, 500.
[112] Beare & Zedler, *Cattail Invasion & Resistance*, 167–169; Watkins, *Hydrology, Nutrients & Salinity*, 40–41.
[113] Mufarrege etal., *Typha Adaptability*, 462; García-Hernándeza etal., *Typha Salinity Responses*, 201; Mexicano etal., *Ecological Engineering*, 174; Glenn etal., *Salinity & Growth*, 81.
[114] Watkins, Hydrology, *Nutrients & Salinity*, 40–41.
[115] Zahran & Willis, *Vegetation of Egypt*, 50; Grace, *Nutrient Additions*, 87–88.
[116] Goulden etal., *Typha Marsh Evapo-transpiration*, 102.
[117] Macek etal., *Typha domingensis Spread*, 1514, fig. 6.
[118] Macek etal., *Typha domingensis Spread*, 1517.
[119] Pezeshki, *Cattail Photosynthetic Response*, 34; Zahran & Willis, *Vegetation of Egypt*, 70.
[120] Mufarrege etal., *Typha Adaptability*, 463.
[121] Glenn etal., *Salinity & Growth*, 89.
[122] Mufarrege etal., *Typha domingensis Adaptability*, 464.
[123] McNaughton, *Typha Ecotype Function*, 323.
[124] El-Ghani etal., *Environmental Relationships*, 8–11.
[125] Zahran & Willis, *Vegetation of Egypt*, 274.

overcome *Typha domingensis* than vice versa.[126] So, overall, a weaker flood would hinder typha spread. In the modern slow-moving Nile, phragmites are the prevalent reed.[127]

4.6. Lotus and a River in 'Drought'

The plants commonly identified as the blue and white Egyptian lotus are actually members of the water lily family, Nymphaeaceae:[128] however, the term 'lotus' has stuck in the Egyptological lexicon.[129] The plants prefer clear, warm, still and slightly acidic waters,[130] reproducing using both seeds and rhizomes, and in optimum conditions they spread rapidly due to these dual propagation strategies.[131] The growth of the lotus follows an annual cycle and removing the normal water supply can interfere with the normal growth of the plant.[132] However, individual plants take up to four years to grow from seed, suggesting that a rapid response to changing environmental factors would be difficult.

4.6.1. Responses to Changing Physical Factors

As the flow of a river decreases, erosion becomes less problematic for the settling of a new lotus into fine sediment.[133] The lotus can survive briefly as a terrestrial plant, but cannot survive a long-term significant fall in water level.[134] Because its leaves lie on the surface of the water,[135] its growth rate is not affected by changes in turbidity; however, it can shade other plants, lowering their ability for photosynthesis.[136] Changes in water temperature do not seem to hinder plant growth, but temperatures above 15°C accelerate the development of disease.[137] If the depth of the secondary channels decreases, the rate of lotus expansion is slowed and *Typha/ Phragmites* take over, depending on the environmental conditions.[138]

4.6.2. Responses to Changing Chemical and Biological Factors

Lotus growth responds poorly to strongly acidic conditions.[139] Although lotus grows best in acidic soil of pH 4.6, growth is not affected by water with pH values ranging from 5.5 to 8.0.[140] Increased salinity is also a hindrance to growth.[141] Increased nutrients in the sediment increase its acidity, potentially reducing the optimum growing conditions of the plant. An increasing concentration of nutrients can speed up plant and rhizome growth[142] but can hinder seedling germination.[143] While the plants may respond favourably to slight increases in nutrient levels in the water, they struggle to grow when water conditions become more anaerobic.[144] Lower than normal oxygen content in the water slows plant development,[145] and if the oxygen content of the sediment declines, it limits the potential germination of the lotus seed.[146]

4.6.3. Ecological Implications for Lotus in Drought

A slowing river would provide more stretches of water that were shallower than normal, developing higher water temperatures. Increasing water temperature would increase evaporation and hence water salinity, limiting plant growth. Extra nutrients would initially encourage plant growth, but then the anaerobic consequences of excess nutrients in the sediment would slow rhizome growth and limit germination. Shallower, narrower channels intimate a decrease in lotus abundance. Consequently, a weaker river would be less beneficial to lotus growth.

4.7. Low River Habitat and Food Web Changes

Without an inundation, the nutrients carried by the river would have remained within the river and changed the river's environmental characteristics. Extra rainfall events add further nutrient load. As a result, some plant species would thrive while others would experience a downturn in environmental success, resulting in less biodiversity but much greater biomass. Following is a summary table (Table 4.2) comparing the overall responses of the four major Nilotic plants and their potential response to circumstances where the river did not flood.[147]

Papyrus and phragmites plants appear to benefit more from circumstances as expected from the analysis of a

[126] Zahran & Willis, *Vegetation of Egypt*, 20–24, 68, 70, 260–265, 298, 300, 302, 314, 318.
[127] Zahran & Willis, *Vegetation of Egypt*, 318; Boulos & Fahmy, *Grasses*, 507; Sainty, *Weed Control & Utilization*, http://www.fao.org/docrep/field/003/R7236E/R7236E00.htm (27/09/2017).
[128] Hanneder, *Water Lilies & Lotus*, 162; Izham, *Molecular Characterisation*, 8.
[129] Harer, *Properties of the Egyptian Lotus*, 49–51.
[130] Orozco Obando, *Sacred Lotus*, 13; Mohamed & Serag, *Anatomy of Lotus*, 4.
[131] Khedr & Hegazy, *Ecology of Lotus*, 120; Zahran & Willis, *Vegetation of Egypt*, 268.
[132] Ziada, *Economic Potentialities*, 1396–1399, fig. 6.
[133] Khedr & El-Demerdash, *Distribution of Aquatic Plants*, 82; Mahomed & Serag, *Anatomy of Lotus*, 4; Shaltout & El-Sheikh, *Vegetation Environment Relations*, 567; Khedr & Hegazy, *Ecology of Lotus*, 127.
[134] Zahran & Willis, *Vegetation of Egypt* 276; Orozco Obando, *Sacred Lotus*, 14; Khedr & Hegazy, *Ecology of Lotus*, 119.
[135] Zahran & Willis, *Vegetation of Egypt*, 271.
[136] Mohamed & Serag, *Anatomy of Lotus*, 4; Khedr & Hegazy, *Ecology of Lotus*, 119.
[137] Orozco Obando, *Sacred Lotus*, 17.
[138] Khedr & El-Demerdash, *Distribution of Aquatic Plants*, 82.
[139] Khedr & Hegazy, *Ecology of Lotus*, 127; Orozco Obando, *Sacred Lotus*, 13.
[140] Shen-Miller etal, *Long Living Lotus*, 239.
[141] Orozco Obando, *Sacred Lotus*, 13; Khedr & Hegazy, *Ecology of Lotus*, 123, 127, fig.1.
[142] Khedr & El-Demerdash, *Distribution of Aquatic Plants*, 76; Khedr & Hegazy, *Ecology of Lotus*, 127.
[143] Orozco Obando, *Sacred Lotus*, 12; Shen-Miller etal, *Long Living Lotus*, 239.
[144] http://pza.sanbi.org/nymphaea-nouchali-var-caerulea (25/11/2016).
[145] Orozco Obando, *Sacred Lotus*, 12, 25, 117.
[146] Mohamed & Serag, *Anatomy of Lotus*, 11, tables 3 & 5.
[147] This table summarises the work presented in: Burn, J. W., *A River in Drought: Consequences of a Low Nile at the End of the Old Kingdom*, in Environment and Ecology Research, 2018, 6; 5: 446–460.

Table 4.2 Plant response summary

PAPYRUS & PHRAGMITES	BULRUSHES & LOTUSES
Tolerate salinity increase	Salinity increase inhibits growth
Tolerate decreasing oxygen solubility	Narrower oxygen solubility range
Tolerate increasing acidty of waterways	Limited acid tolerance
Asexual (stolon & rhizome) reproduction	Mainly sexual reproduction
Masses collect nutrients and add matter	Do not clump

river in surplus, excess nutrients would not seem to benefit the lotuses and the typha plants (bulrushes/cattails) as it would the other two. As well as the increase in biomass accumulation, papyrus decomposition would have enabled nutrients to be returned to the waterway to be, in their turn, re-absorbed by other components of the riverine food web; especially microbes, with perhaps as much as a third of the nutrients within a papyrus clump being returned to the environment each season.[148]

4.7.1. Food Web Impacts

When papyrus clumps grow denser and taller, as it is hypothesised, they would in a low Nile inundation, the food webs associated with them would have changed as microhabitats increased in number and variety. The considerable mass and height potential of papyrus would have produced a large number of 'microhabitats' within its storeys.[149] In the modern situation, increased vertical habitat zonation of bird species has been observed within the papyrus columns as papyrus clumps grow taller.[150] Whereas papyrus clumps grow denser and taller, phragmites species grow outwards.[151] Biomass accumulation, the forming mats of reeds,[152] would have occurred, but the increase in the number of micro-habitats occurs in a manner different to that provided by papyrus. As well as being an aquatic food source, papyrus and phragmites expansion become a significant attractant to birdlife,[153] with a direct relationship between an increase in marsh size and a corresponding increase in bird abundance.[154] As phragmites masses expand, the centres of these masses would have become safe refuges for a variety of animals to avoid predators.[155]

Modern investigations have recorded the rapid increase in numbers of ducks and geese in developing secondary channels, these channels having become important and secure nesting sites. Among the modern-day avian visitors to the Nile, individuals of historically recorded species have been identified.[156] As well as avian organisms, other terrestrial organisms, invertebrate and vertebrate, such as locusts, rodents and small carnivores are attracted to the increase in biomass in search of food and have been identified as living in, on and around the papyrus.[157] Secondary channels provided an important role in the spawning activities of most Nilotic fish species,[158] whereas the deeper parts of the river provided for a greater variety of the bigger fish species.[159] Migration of fish species can occur along the length of the river and within different parts of the river, providing a variety of resources.[160]

Increased plant material increases areas under which fish species could have sheltered and gathered food.[161] Despite seeming 'monotonous'[162] in appearance, growing papyrus clumps and reed mats attract larger and more varied shoals of fish[163] and higher numbers and varieties of animals and their predators, and this vegetative fringe has been identified as being the most biologically rich habitat of the river.[164] Considering the huge biomass potential along the fringes of the river,[165] and the identification of small fish species as the most important faunal constituent of these vegetative fringes,[166] then it is to be expected that a more determined effort by the ancient Egyptians to exploit this particular resource should have occurred. Therefore, it would be expected that a re-adjustment of piscine habitats may have occurred.[167] A more detailed investigation of the fish species depicted within Egyptian tomb decoration scenes may identify whether these organisms change in abundance over the time frame under investigation. It would be expected that the depiction of smaller fish, with more included varieties should increase. Catfish and lungfish appear less vulnerable to reduced oxygen levels.[168] Similarly, with a possible decline in distribution due to diminishing oxygen levels, the instances of catfish and lungfish depictions would be expected to occur.

Papyrus appears a relatively insignificant plant within the food chains of larger herbivores.[169] Despite its high nutritive value, most large herbivores are poorly adapted to grazing upon such sedge-type marshland flora.[170] However, cattle tend to eat the youngest and freshest shoots, and

[148] Gaudet, *Papyrus Nutrients*, 420.
[149] Britton, *Papyrus Swamps*, 451.
[150] Britton, *Papyrus Swamps*, 459.
[151] Gaudet, *Papyrus Nutrients*, 415; Zahran & Willis, *Vegetation of Egypt*, 298.
[152] Zahran & Willis, *Vegetation of Egypt*, 296.
[153] Mitsch & Gosselink, *Wetlands*, 382.
[154] Maclean etal., *Passerine Habitat Loss*, 349; Gagnon Lupien etal, *Bird Diversity*, 34–38.
[155] Gagnon Lupien etal., *Bird Diversity*, 38–40.
[156] Sallam & El-Barbary, *Secondary Channels*, 496–497.
[157] Rzóska, *Nile*, 213; Speight & Blackith, *The Animals*, 354–359; Rzóska, *Upper Nile Swamps*, 27; Gagnon Lupien etal., *Bird Diversity*, 32–33.
[158] Sallam & El-Barbary, *Secondary Channels*, 498.
[159] Mitsch & Gosselink, *Wetlands*, 400.
[160] Rzóska, *Nile*, 211; Speight & Blackith, *The Animals*, 352–353.
[161] Thompson, *Swamp Development*, 190.
[162] Rzóska, *Nile*, 213.
[163] Thompson, *Swamp Development*, 190–191; Rzóska, *Nile*, 211.
[164] Britton, *Papyrus Swamps*, 450; Rzóska, *Upper Nile Swamps*, 13.
[165] Jones & Muthuri, *Papyrus Canopy Structure*, 482.
[166] Rzóska, *Upper Nile Swamps*, 13–16; Rzóska, *Nile*, 211–212.
[167] Chapman etal., *Papyrus Swamp Variations*, 321.
[168] Chapman etal., *Papyrus Swamp Variations*, 318.
[169] Saunders etal., *Papyrus Carbon Cycles*, 489.
[170] Muthuri & Kinyamario, *Papyrus Nutritive Value*, 23–24.

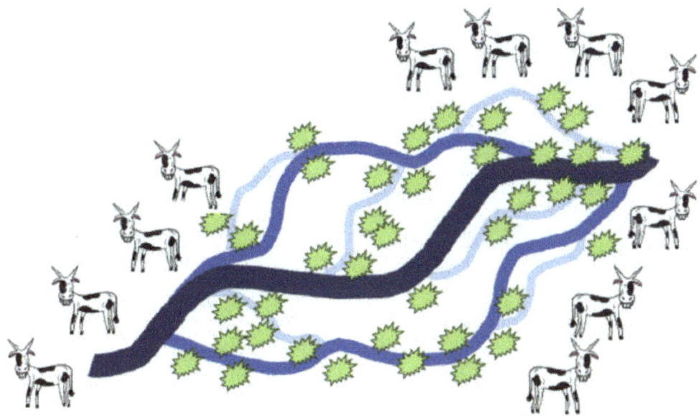

Figure 4.1: More secondary channels provide more grazing opportunities.

these grow quite readily in the secondary channels, (see Figure 4.1) providing extra nutritional resources for cattle. Also, this damage by grazing animals seems to encourage regrowth and therefore does not interfere significantly with the biomass output.[171]

Moreover, the level of grazing has to be quite significant before it appears to have a harmful impact upon the success of sedges, with some species able to accept up to 40 percent grazing and still experience viable reproductive success.[172] So, as the biomass of papyrus and phragmites increased due to changing environmental circumstances, it seems feasible that cattle grazing would have been unable to keep up. Since it is the older parts of the plant and rhizome that store the starch needed for the plant's metabolism, and most of the developing flower heads do not produce flowers but have an important photosynthetic value to the potential of the plant,[173] grazing upon the growing parts of a plant does not significantly impact upon its reproductive ability.[174]

4.8. Nile Valley Rain: Even More Nutrients?

While unexpected, rainfall within the Nile valley has been shown to have always occurred, and mounting evidence suggests that severe rainfall events were becoming more commonplace during the timeframe under investigation.[175] Interestingly, much of the land abandonment that occurred in Mesopotamia at this time has been linked to unreliable or irregular rainfall patterns,[176] suggesting a connection between unusual rain events in Egypt with those which occurred in other places around the region at this time.[177]

[171] van Deursen & Drost, *Cattle Defoliation*, 294–295; Britton, *Papyrus Swamps*, 450.
[172] van Deursen & Drost, *Cattle Defoliation*, 292–293.
[173] Jones & Muthuri, *Papyrus Canopy Structure*, 481.
[174] Muthuri & Kinyamario, *Papyrus Nutritive Value*, 25–27; Baran etal, *Phragmites & Nutrition*, 448.
[175] Welc & Marks, *OK Climate Change*, 124–133; Sowada, *Weather Evidence*, 69–74; Welc & Trzciński, *Dry Moat Geology*, 323–343; Trzciński etal., *West Saqqara Geoarchaeology*, 194.
[176] Myśliwiec etal., *Saqqara Geoarchaeology & Paleoclimate*, 295–298; Fiorentino, *Syrian Climate Change*, 56.
[177] Kowalska & Kuraszkiewicz, *End of a World*, 173–176; Walker etal., *Holocene Subdivision*, 653–656.

4.8.1. Alluvial versus Pluvial Events

The variety, source and movement of sediments within a river are very important, because they influence the way that the living organisms within the river develop and interact.[178] In the 'natural' chain of events, the inundation deposited alluvial sediments; those "*deposited by a flowing waterway*"[179] to the river valley as the water overflowed its banks, a process that appeared to be weakening during the time frame under investigation.[180] Rainfall upon the land drains into the river,[181] producing sediment that becomes a part of the assemblage of the riverbank and its edges.

The vast series of wadis that can be observed on any satellite map indicate that water was flowing towards the river on a semi-regular basis:[182] these wadis have been identified as indications of rainfall.[183] The depth of these wadis suggests that it did so in some considerable volume.[184] With rainfall comes run-off, which flows into the nearest waterway, transporting pluvial sediments: those sediments are "*made of sand and earth that is washed down by rainfall.*"[185] The composition and structure of alluvial sediments is different to pluvial sediments and can be readily identified by scientific means,[186] as can the direction of the pluvial flow.[187]

4.8.2. Rainfall Events and Impacts

While the inundation was relatively predictable, whatever its level, rainfall events in Egypt occurred with irregular and unpredictable timing, thereby interfering with the progression of the 'normal' cultivation cycle. Just as too little flooding meant that irrigation efforts were ineffective, excessive flooding was also problematic.[188] Agricultural success further south, for example, is attributed to lack of large inundations, with excess water flowing downstream.[189] Just as the flood governed the annual outcomes of agriculture, the timing of rainfall events would also have impacted upon the success, or otherwise, of the cultivation cycle.[190]

[178] Allan & Castillo, *Stream Ecology*, 19, 41, 75.
[179] https://www.thesciencedictionary.com/results/alluvial (16/10/2016).
[180] Butzer, *Geological Deposits*, 77; Stanley & Warne, *Nile Destruction Phase*, 801–802.
[181] Woodward etal., *Nile Evolution*, 265.
[182] York & Kennedy, *Google Earth and Archaeology*, 1284–1293; Déodat & Lecoq, *Using Google Earth*, 321–338; and especially, Parcak, *Remote Sensing*.
[183] Butzer & Hansen, *Desert & River*, 121; Butzer, *Hydraulic Egypt*, 13 & 106; Hassan etal., *Nile Floodplain at Saqqara*, 57.
[184] Welc & Marks, *OK Climate Change*, 8; Hassan etal., *Nile Floodplain at Saqqara*, 59–63, figs 5 & 6.
[185] https://www.thesciencedictionary.com/results/pluvial (16/10/2016).
[186] Bárta, *In Mud Forgotten*, 75–82; Welc & Marks, *OK Climate Change*, 1–10.
[187] For example, Myśliwiec etal., *Saqqara Geoarchaeology & Paleoclimate*, 286–289, figs 14–15.
[188] Coulon, *Famine*, 1.
[189] Macklin etal., *Nile Floodwater Farming*, 698.
[190] Morris, *Art of Not Collapsing*, 79–81, Soroush & Mordechai, *Short-term Cataclysm*, 7–8; Nicholson & Shaw, *Materials & Technology*, 514.

The term 'erosivity' is defined as *a measure of the potential ability of material to be eroded by rain, wind, or surface run-off.* [191] The power of the run-off is a function of the volume and speed of water, which influences how much sediment is moved.[192] In recent history, it has been shown that the more variable and irregular the weather patterns, the more readily the soil is eroded and the land degraded.[193] Drainage pathways in ancient Egypt would have been constructed based upon experience of previous inundations. Large amounts of water flowing towards the river must have compromised the design of the irrigation channels and must have also interfered with basic agricultural infrastructure as the channels were designed to take advantage of water moving away from the river not towards it.[194] These events would have developed stresses in the agriculture cycle, something with which an inadequately organised administrative system may have been unable to cope. Poor administration may have meant that the repairs may not have been completed in time to take advantage of the excess water.

4.8.3. Do Rainfall Events Encourage More Marshlands?

During times of rain, the river would have accumulated additional nutrients as the torrent transferred (returned?) pluvial sediments, which would have been built up from a previous alluvial event, from the land back into the river. Organic wastes from previous users of the soil would have been washed into the river in accompaniment. This nutrient surge may have helped precipitate a papyrus and phragmites 'bloom', causing a further expansion of marshlands.

4.9. The Nile in 'Drought': a Re-Interpretation?

Rzóska points out that, in terms of papyrus marshes, a 'non-flood' event due to a low or non-existent inundation, is merely a loss of water (and its accompanying nutrients) to those relying on the excess for success in agriculture.[195] In riverine terms, a non-inundation means that water and its accompanying nutrients is reserved for utilisation within the river.[196] Rainfall events add further nutrients to the surplus.

4.9.1. Does 'Drought' Encourage Marshlands?

One of the consequences of this excess of nutrients may have been promoted growth of the papyrus marshes and reed beds. The processes leading to such rapid growth

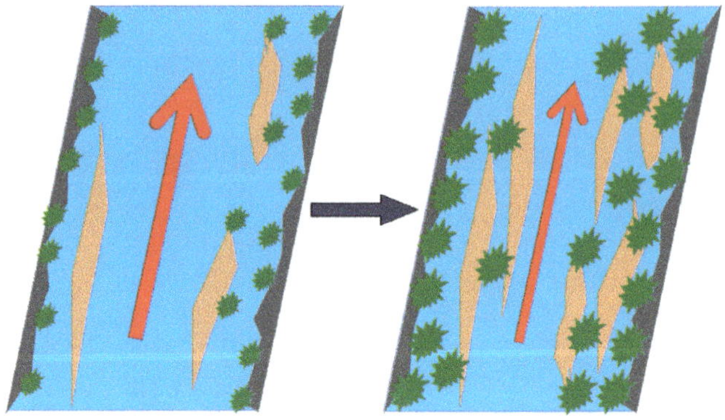

Figure 4.2: Normal versus nutrient-excess.

can be envisaged as follows. In a normal flood, as the river overflowed its banks, the nutrients carried within are deposited on the riverbanks along its length. During 'drought', the river would still have carried these nutrients downstream but did not have the necessary volume to overflow its banks, resulting in nutrient retention within the river channel. As the current slowed, it would have lost the necessary momentum to carry these nutrients, which would have been deposited on the riverbed and river's edges rather than beyond. A long-term decline in water level would have magnified this build-up, providing the biological energy for a 'papyrus bloom' and a phragmites expansion. We should expect the river to have narrowed as the river volume (and therefore flow rate) decreased, 'pulling' the papyrus marshes and reed beds to grow inwards towards the centre of the river (see Figure 4.2).

Ankhtify, the great overlord of the nomes of Edfu and Hierakonpolis, who had described the desperate situation that developed in the south of the country during the famine,[197] bemoaned in his biography the "*the House of Khui; inundated like a marsh (and) neglected by its keeper.*"[198] The 'Marshes of Edfu', those that interfered with the grain distribution (and conquest?) plans of Ankhtify,[199] may have developed as the consequence of a bloom in papyrus and phragmites due to the deposition of nutrients within the river. The rapid growth of papyrus clumps within the river and the increased development of *khors* and *sudds* along its length were a consequence of a lower, slower, but more resource-rich waterway effectively making the river narrower and possibly hampering the movement downstream (see Figure 4.3).

So, as flood levels declined and nutrient levels increased over the Old Kingdom, we would expect to see a significant

[191] Zorn & Komac, '*Erosivity*': https://link.springer.com/referenceworkentry/10.1007%2F978-1-4020-4399-4_121 (21/11/2017).
[192] Sukhanovski etal., *Rainfall Erosivity*, 51–57; Diodato & Bellocchi, *Rainfall Erosivity*, 969–970.
[193] Bintliff, Time, *Process & Catastrophism*, 422–423; Alpert etal., *Paradoxical Rainfall*, 135–154; Diodato etal., *Erosive Rainfall Anomalies*; 2078–2093; Diodato & Bellocchi, *Rainfall Erosivity*, 977.
[194] Bintliff, *Time, Process & Catastrophism*, 429–430.
[195] Rzóska, *Nile*, 213; Rzóska, *Upper Nile Swamps*, 2.
[196] Speight & Blackith, *The Animals*, 349; Khan & Ansari, *Eutrophication*, 465–466; Tackholm, *Landscape*, 59–60.

[197] Malek, *Old Kingdom*, 118; Manassa, *El-Moalla to El-Deir*, 5; Coulon, *Famine*, 1.
[198] Author's translation, from Vandier, *Mo'alla*, 163; also Manassa, *El-Moalla to El-Deir*, 8.
[199] Vandier, *Mo'alla*, 220–221; Darnell, *The Message of King Wahankh*, 101–108.

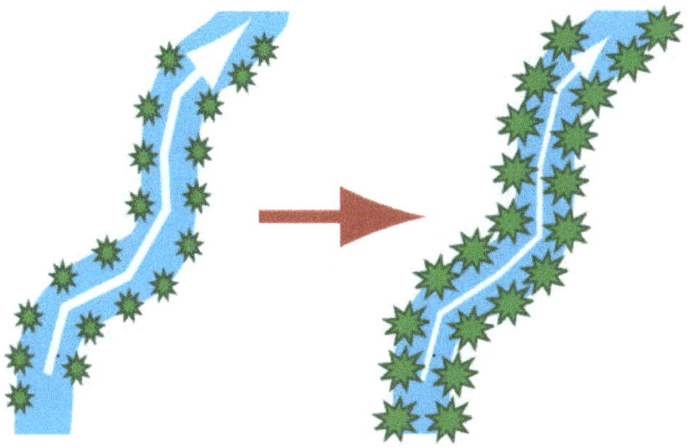

Figure 4.3: Excess growth disrupts flow and hinders transportation.

increase in the amount of marshland developing along the river.[200]

4.10. Summation: Excess Nutrients Cause some Plants to Thrive?

When the river delivers a weak inundation, the nutrients it carries are more likely to remain within and this would affect the plant life of the river. Rainfall events add more nutrients.

The main points determined in this chapter include:

- A weaker lower river would have flowed less powerfully, with a reduced main channel and a greater number of secondary channels.
- These secondary channels should be warmer, saltier, more acidic and contain less dissolved oxygen, changing environmental conditions.
- Nutrients would have remained in the river and rainfall events would have added more nutrients to the river: potentially changing the riverine ecology.
- Some plant species would have responded more favourably to the changing circumstances than others.
- Papyrus and Phragmites display characteristics that suggest that these species would have responded favourably to these conditions.
- Typha (cattails) and Lotus species have characteristics that suggest that they would experience lessening chances of success in these new conditions.
- It was envisaged that the Nilotic channels would have become increasingly clogged with thick papyrus clumps and phragmites mats spreading towards the middle, with the other two significant plant groups, such as the cattails and lotus, diminishing in their relative abundance and distribution.
- Animals, both land, and aquatic would have been attracted to the increased aquatic resources and they would have exploited the burgeoning large papyrus clumps and phragmites mats.

[200] *Project A5*, http://www.uni-koeln.de/sfb389/a/a5/pics/pic_06.gif (21/05/2016).

Part C

The Cultural Response

5

Changes in Tomb Wall Scenes

If the 'normal' biological cycle had become disrupted, then the nutrient glut that would have developed would have had a transformative impact upon the Nilotic ecology. This would have had an impression upon the culture. In this chapter, tomb decoration data will be analysed to test the A.R.I.D. hypothesis. Artistic themes that may shed light on the veracity of the hypothesis will be investigated. The progression of tomb decoration themes will then be examined to investigate the possibility that changing environmental circumstances may have had an impact upon contemporary visual culture.[1]

5.1. Papyrus: Art Suggesting Riverine Change?

From previously being restricted to the royal tombs, fishing and fowling scenes began to appear on the walls of nobles' tombs in the Fifth Dynasty, and then became a regular theme in tomb wall scenes from the mid-Fifth to the late Old Kingdom. Harpur comments on the changing representations of papyrus thickets in tombs in the mid-late Old Kingdom,[2] Kanawati summarises the overall progression[3] and Woods identifies tombs in which scenes depicting papyrus are changing.[4] In the Dynasty Six tomb of Mereruka at Saqqara, for example, the papyrus clumps make up a significant portion of the fishing and fowling scene,[5] and the wall scenes in the tomb of his wife, Waatetkhethor depict papyrus to a much greater extent than other marsh plants.[6] By the second half of the Old Kingdom, marsh scenes depicting fishing and fowling had become one of the focal points, if not the main feature, of tomb scenes, becoming more elaborate and increasingly more detailed.[7] Woods identifies the tendency within the wall scenes to also incorporate more than one activity against the same background with the scenes appearing to flow from one into another.[8] A growing emphasis on riverine scenes may explain why the location of these compositions, initially presented inside the tomb, moved to be within view of the river, with Woods identifying the time of Niuserra onwards as the date for the main change of this positioning.[9]

5.1.1. Marshlands Become a More Crowded Place

In the Fourth Dynasty tombs of Seneb at Saqqara and Sabni at Aswan, the figures appear relatively large compared to the papyrus stands,[10] assigning greater emphasis to the individuals themselves. Conversely, in the tomb of Ty at Saqqara, for example, dating to the reign of Niuserra, the stands of papyrus seem to dwarf the subjects.[11] Similarly, in the tombs of Djau and Ibi at Deir el-Gebrawi,[12] Itisen at Giza, Khunes at Zawiyet el- Maiyetin_and Hetepherakhti at Saqqara, the figures are dwarfed by the papyrus thicket behind,[13] suggesting that that this botanical feature had become more important during the latter half of the Fifth Dynasty and into the Sixth. The marshland depicted in the latter Old Kingdom scenes display a more 'crowded' botanical background than those produced in the Middle Kingdom.[14] At Meir, the Sixth Dynasty tombs of Pepyankh the Middle, Niankhpepy the Black and Pepyankh the Black, for example,[15] represent papyrus in a more congested manner than in the later tombs of Senbi and Wekhhotep.[16]

Earlier representations of papyrus are thought to be stylised renditions.[17] In the tomb of Nebemacht at Giza, for example, the arrangement of the papyrus thicket is very controlled and the ranks of birds atop it are ordered in a regimented manner.[18] This representation of papyrus umbels may be designed to perhaps impose more order on an increasingly disordered setting.[19] However, later, some of these illustrations become less stylised and more irregular. Kantor notes that from the time of Niuserra onwards, the scenes depicting huge papyrus clumps evoke this impression of overgrowth:

> "... the feeling for the massive, overshadowing character of the papyrus swamps, which produced the tremendous background of sedges used in Ti's tomb, appears to be characteristic chiefly of the Old Kingdom. In later

[1] This extends the ecological style 'distribution and abundance' analysis method used to identify changes in tomb decoration emphases: see Burn, *Mehu* and *Pyramid Texts*.
[2] Harpur, *Decoration*, 195–196; Woods, *Marshes*, 40–41; Kanawati, *Papyrus Thickets*, 119.
[3] Kanawati, *Papyrus Thickets*, 119–212.
[4] Harpur, *Decoration*, 189; Woods, *Marshes*, 358.
[5] Kanawati, *Mereruka & Teti*, pl. 107, 110.
[6] Kanawati & Abder-Raziq, *Waatetkhethor*, 21, pl. 65–66.
[7] Harpur, *Decoration*, 197.
[8] Harpur, *Decoration*, 194; Woods, *Marshes*, 377.
[9] Harpur, *Decoration*, 57, 185; Woods, *Marshes*, 38–39. Burn, *Pyramid Texts*, 233–245.
[10] Junker, *Gîza* V, 65–67, fig. 15; de Morgan, *Catalogue*, 146.
[11] Wild, *Ti*, II, pl. 85–86, 119.
[12] For Djau, see Davies, *Deir el-Gebrâwi* II, pl. 5; Kanawati, *Deir el-Gebrawi* III, pl. 57: for Ibi, see Davies, *Deir el-Gebrâwi* I, pl. 4–6; Kanawati, *Deir el-Gebrawi* III, pl. 46, 47.
[13] For Itsen, see Hassan, *Gîza* V, fig. 123, pl. 37; for Khunes, see Lepsius, *Denkmaler*, II, pl. 106a and for Hetepherakhti, see Mohr, *Hetep-her-akhti*, fig. 34, pl. 2.
[14] Harpur, *Decoration*, 257.
[15] Kanawati, *Meir* I, pl. 79–82, updating Blackman, *Meir* IV, pl. 7; Kanawati etal., *Meir* III, pl. 24–28, 69 and Kanawati & Evans, *Meir* II, pl. 34–36, 84 respectively.
[16] Harpur, *Decoration*, 199–201; Kanawati & Evans, *Meir* IV, pl. 78, 80 and 92 respectively, updating Blackman, *Meir* I, pl. 3 and *Meir* VI, pl. 6, 13.
[17] Woods, *Marshes*, 67–68, n. 161, 162.
[18] Hassan, *Gîza* IV, fig. 77.
[19] Woods, *Five Features*, 1901–1902.

times there was a continuous tendency to diminish the size and rigidity of the swamp landscape…." [20]

Woods identifies the tumultuous or irregular arrangement of subjects within marshland scenes as one of her characteristic themes: "#174: *The composition of the water is broken up with vegetation – papyrus buds and lotus flowers shown in the water attached to the baseline.*" Her selection includes tombs dating from the time frame in question and beyond, and mostly from the provinces.[21] This asymmetrical arrangement of the papyrus thickets can be found in the Memphite tombs of Rakhaefankh,[22] Khufukhaef II,[23] Irenkaptah,[24] Kaemnefert II,[25] Niankh-khnum and Khnumhotep,[26] Ty,[27] Iteti,[28] Neferirtenef[29] and Seneb.[30] With the exception of Rakhaefankh, whose date is under contention, all date from after the time of Niuserra.[31] At Sheikh Said, in the tomb of Werirni,[32] the arrangement of the papyrus umbels appears similarly haphazard, giving the impression of a vast and overwhelming mass.[33] Other provincial tombs that display an indiscriminate umbel arrangement include those of Kakhent and Iufi at el-Hammamiya,[34] Sabni at Qubbet el-Hawa,[35] Mery-aa at El-Hagarsa[36] and Hemre/Isi at Deir el-Gebrawi,[37] suggesting a more chaotic, less ordered, 'wilder' marshland. If the Nile's papyrus clumps and reed mats became larger and more extensive as a consequence of a weaker river flow, it would be expected that the plant growth would be as disordered as the scenes appear to show.

Some decorations appear to have been uniquely provincial, suggesting that the artists were depicting truly what was witnessed.[38] The 'open' appearance of some thickets in the latter half of the Old Kingdom, may represent a desire to depict more clearly what lies within. It is difficult to reconcile the view that these scenes were purely symbolic,[39] as some top tomb officials did not depict themselves engaged in marshland-based activities.[40]

5.1.2. Fishing and Fowling in Crowded Marshlands

Of the fourteen tombs lacking a papyrus-backed fishing scene identified by Woods, all are dated to the late Fifth Dynasty and beyond, with the majority in tombs below the Memphite region.[41] In the tomb of Djau/Shemai at Deir el-Gebrawi, the fishing and fowling scene appears to be set beyond the papyrus bank, giving the impression that these activities are occurring outside the congestion of the papyrus zone.[42] Also at Deir el-Gebrawi, in the tomb of Henqu/Ih[…]f, no papyrus thicket is shown, just a few isolated plants,[43] whereas Ibi is depicted among the papyrus while fowling[44] and beyond them while spearfishing.[45] It would be expected that, as marshlands became more crowded, it would be more difficult to approach spearfishing sites from the banks, making it more practical to approach from open water.[46] The representations of the figures of Merefnebef and Mehu, appear to be imposed in front of the papyrus thickets,[47] not within them, suggesting that the activity depicted is occurring beyond the papyrus zone. The act of fowling required some ability to sneak up to the prey. Fowling would necessarily have to take place within the papyrus marshes, using the thickets as a form of 'cover'. This arrangement of fowling near the thicket and spearfishing in open water is similar to that depicted in the tombs of Inumin at Saqqara[48] and Iteti/Shedu at Deshasha.[49] In the tomb of Kakhent and Iufi at el-Hammamiya, Iufi, depicted on her pleasure cruise, is not passing through a thicket where her husband is engaged is his activities, but is depicted beyond them.[50] A similar arrangement is seen in the tomb of Idu/Seneni at Qasr el-Sayad, where the fowler is preparing to hurl the throwstick into a thicket, while the spearfishing image appears to have no papyrus backing.[51] The tomb of Kagemni contains a similarly differentiated image, where the marshland activities of husband and wife appear to be in different locations.[52] While spearfishing and fowling may require some 'cover' for their activities, the act of fishing with nets requires more open spaces to avoid entanglement.[53]

[20] Kantor, *Plant Ornament*, 10; Harpur, *Decoration*, 257.
[21] Woods, *Marshes*, 379.
[22] Lepsius, *Denkmaler*, II. 9 [lower].
[23] Simpson, *Kawab & Khafkhufu*, pl. 37[a], fig. 47.
[24] Moussa & Junge, *Two Craftsmen*, pl. 12–13.
[25] Hassan, *Giza* II, fig. 140.
[26] Moussa & Altenmüller, *Nianchchnum und Chnumhotep*, Abb. 12, Taf. 31.
[27] Wild, *Ti*, II, pl. 87, 110–112.
[28] Badawy, *Iteti*, fig. 17.
[29] van de Walle, *Neferirtenef*, pl. 13, 21 (detail).
[30] Junker, *Gîza* V, Abb. 15, Taf. 6 [a].
[31] Woods, *Marshes*, 362.
[32] Davies, *Sheikh Saïd*, pl. 11.
[33] Moussa & Altenmüller, *Nianchchnum und Chnumhotep*, Abb. 5, 6, Taf. 74, 75.
[34] Mackay etal, *Bahrein & Hemamieh*, pl. 26, updated El-Khouli & Kanawati, *El-Hammamiya*, pls 35–36.
[35] Lhote, *Chefs-d'oeuvre*, fig. 6.
[36] Kanawati, *El-Hagarsa* III, pl. 42, 44–45.
[37] Davies, *Deir el-Gebrâwi* II, pl. 17, updated Kanawati, *Deir el-Gebrawi* I, pl. 59.
[38] Lashien, *El-Qusiya*,150–166.
[39] Vandier, *Manuel* IV, 717.
[40] Kanawati, *Papyrus Thickets*, 119.
[41] Woods, *Marshes*, 358.
[42] Davies, *Deir el-Gebrâwi* II, pl. 3, 5, updated Kanawati, *Deir el-Gebrawi* III, pl. 57, 69.
[43] Davies, *Deir el-Gebrâwi* II, pl. 23, updated Kanawati, *Deir el-Gebrawi* I, pl. 54.
[44] Davies, *Deir el-Gebrâwi* I, pl. 5, updated Kanawati, *Deir el-Gebrawi* II, pl. 47.
[45] Davies, *Deir el-Gebrâwi* I, pl. 3, updated Kanawati, *Deir el-Gebrawi* II, pl. 46.
[46] Kanawati, *Papyrus Thickets*, 122.
[47] Myśliwiec etal., *Merefnebef*, 133, pl. 65; Altenmüller, *Mehu*, pl. 9, 11, 13.
[48] Kanawati, *Inumin*, pl. 46.
[49] Kanawati & McFarlane, *Deshasha*, pl. 44 (spear-fishing) and pl. 48 (fowling).
[50] El-Khouli & Kanawati, *El-Hammamiya*, pl. 50–51; Woods, *Marshes*, 358.
[51] Säve-Söderberg, *Hamra Dom*, pl. 7 (spear-fishing), pl. 8 (fowling).
[52] Harpur & Scremin, *Kagemni*, 353–364, 491[3].
[53] Kanawati & Abder-Raziq, *Teti III*, pl. 69; Davies, *Deir el-Gebrâwi* I, pls 3–6; Kanawati, *Meir* I, pls 79–82; Altenmüller & Moussa, *Nianchchnum & Chnumhotep*, fig. 12.; Kanawati etal; *Mereruka*, pl. 67.

5.1.3. Swampweed and Boats

Most depictions of rivercraft depict the occupant(s) travelling through the marshes, encountering swampweed in the process. The majority of scenes depicting skiffs but not depicting swampweed present are attested to before the Sixth Dynasty.[54] Initially, swampweed was depicted under the prow, perhaps representing the skiff entering the thicket.[55] Over the time frame spanned by this investigation, the depictions of the skiffs and the positions of these weeds under them changed.[56] Swampweed, normally depicted at either prow or stern, increasingly appears under both ends.[57] The images in the Sixth Dynasty tombs of Ibi at Deir el-Gebrawi[58] and Pepyankh the Black at Meir, [59] among others, show swampweed under both ends of the vessel. One instance, the tomb of Iynefert, depicts both ends of the vessel atop water plants, but displays the bow passing across lotus flowers, not swampweed.[60] The dating of this type of marsh scene has been fixed to the time of Teti and continues through to the end of the Old Kingdom.[61] This proliferation of the presence of weed may indicate the development of a more crowded waterway through which to traverse in the final stages of the Old Kingdom. More than two-thirds of all tombs depicting this abundance of waterweed are provincial.[62]

The increasing congestion of the river may also be indicated by changes to the design of the small watercraft used to conduct fishing and fowling activities.[63] Baines and Malek supply a schematic illustration of a basic wooden skiff and its component structures. Tracing the evolution in design of boats as represented as tomb decorations, Woods observes that boats develop a gradually rising and curving stern and prow until:[64]

> "...the final shape is mainly attested in marsh scenes dating to the late Sixth Dynasty in the (Upper Egyptian) provincial cemeteries of Naga ed-Dêr, Aswan and Mo'alla. In these scenes, the papyrus boat has a thick base and the prow and stern project sharply up from the water line in contrast to a gradual curve..." [65]

The thicker base and sharply upturned prow result in a skiff with an apparently shallower draught, as though it was designed to enable it to travel over the water and not through it – another indicator of less-than-smooth travel on the river, perhaps as a result of a more crowded waterway slowing forward progress. In the tomb of Khunes at Qubbet el-Hawa,[66] the skiff depicted has a high prow and stern. In the nearby tombs of Senbi I and Senbi II,[67] the scene also depicts these more rugged craft atop swampweed encroaching both fore and aft. Similar suggestions that these marshland scenes were set in the delta can also be discounted, because some depict sailing boats nearby, something difficult to perform within the delta.[68] Pepy the Black at Meir, mentions the marshlands of Upper Egypt before he mentions the marshlands of the delta.[69]

However, it is important to not overstate the case for total reliance on visual evidence: the tombs of Mery-aa and Wahi at El-Hagarsa,[70] among a few others,[71] depict the tomb owners undertaking marshland activities without any skiff, despite obviously needing one. This has been explained as most likely a result of poor artistic skills rather than by design choice,[72] however, it could also be an indication of the increasing interrelationship or overlap of those activities involved with the acquisition of riverine resources: the men could be standing on the riverbank.[73] With these previous examples, the composition of some tomb scenes appeared to be changing. The possibility exists that these changes may be in response to changing environmental circumstances. Perhaps other compositional changes were occurring.

5.2. Tomb Decoration: Patterns of Change

Since art represents the major preoccupations of a society, it may be that changes to the art produced by that society reveals a pattern of environmental influence. It is important to recognise that any study relying on 'artistic' evidence is only partially reliable because the 'categories' we apply when we study wall scenes have been assigned in hindsight and after the passage of many centuries, without full knowledge of the artists' philosophies. Alongside those considerations, discrepancies within dating criteria mean that that some attestations are not as precisely assigned as one would like. Additionally, not all pieces of evidence are extant. Also, it is difficult to 'allocate' the date of a tomb to one king because most lifespans and tomb constructions schedules did not align with only one king or with service to only one king. For example, very few tombs have been suggested as belonging to the reigns of Shepseskara and Raneferef, respectively, the fourth and fifth kings of Dynasty Five,[74] though this seems to reflect the brevity of these individuals' reigns rather than any significant social

[54] Woods, *Marshes*, Appendix 3, no. 186.
[55] Kanawati, *Papyrus Thickets*, 122.
[56] Harpur, *Decoration*, 201.
[57] Woods, *Marshes*, 145, appendix 3, no. 188; Harpur, *Decoration*, 356–363.
[58] Fowling- Davies, *Deir el-Gebrâwi* I, pl. 5, updated Kanawati, *Deir el-Gebrawi* II, pl. 47: Spearfishing- Davies, *Deir el-Gebrâwi* I, pl. 3, updated Kanawati, *Deir el-Gebrawi* II, pl. 46.
[59] Fowling- see Blackman & Apted, *Meir* V, pl. 28, updated Kanawati & Evans, *Meir* II, pl. 88. Spearfishing- see Blackman & Apted, *Meir* V, pl. 24, updated Kanawati & Evans, *Meir* II, pl. 84.
[60] Schürmann, *Ii-nefret*, pl. 21.
[61] Woods, *Five Features*, 1904–1906.
[62] Harpur, *Decoration*, 201; Woods, *Marshes*, 384.
[63] Harpur, *Decoration*, pl. 24; Woods, *Marshes*, appendix 3, nos. 109–128.
[64] Baines & Malek, *Atlas*, 68–69; Woods, *Marshes*, 61–63.
[65] Woods, *Marshes*, 63.

[66] de Morgan, *Catalogue*, 146.
[67] Harpur, *Decoration*, pl. 24.
[68] Kanawati, *Papyrus Thickets*, 120.
[69] Blackman, *Meir* V, pl. 24, updated Kanawati & Evans, *Meir* II, pl. 84.
[70] Kanawati, *El-Hagarsa* III, pl. 42, 44–45 and pl. 28, respectively.
[71] Woods, *Five Features*, 1902, n.28.
[72] Vandier, *Manuel* IV, 719 suggesting incompetence.
[73] Kanawati, *Papyrus Thickets*, 123–124.
[74] Baines & Malek, *Atlas*, 36; Shaw, *Oxford Encyclopaedia*, 482; Verner, *Dating Old Kingdom*, 31–32.

Table 5.1: Themes 4th → 8th Dynasties: abundance and distribution summary

DATE	IV	IV	IV	IV	IV	IV	V	V	V	V	V	V	V	V	V	VI	VI	VI	VI	VI	OVERALL	
OEE SCENE-DETAILS: THEMES	**1**	**2**	**3**	**4**	**5**	**6**	**1**	**2**	**3**	**4**	**5**	**6**	**7**	**8**	**9**	**1**	**2**	**3**	**4**	**5+**	**ΣT**	**%**
1. Marsh-related activities	11	3	1	1	3	15	4	12	15			123	42	78	130	104	61	14	41	62	720	32%
2. Desert and desert-related activities	5		1									6	2	2	10	4	2	2	6	7	47	2%
3. Agricultural pursuits	4	1				3	1	3	4			27	6	23	24	14	6	1	10	24	151	7%
4. Pastureland and animal husbandry	4					4		3				53	11	34	30	28	21	2	16	38	244	11%
5. Orchard scenes								1	1			7	1	8	3	10			1	1	33	1%
6. Gardening												9		6	4	5	5			0	29	1%
7. Bird procession and poultry yard scenes		1			1	3			1			16	5	6	13	11	11	1	3	1	73	3%
8. Manufacture and storage of wine and oil	1					6			1			21	4	15	12	10	7		4	2	83	4%
9. Food Preparation and Brewing	3					3	1	1	2			36	2	19	20	21	16	8	10	33	175	8%
10. Workshop Activities	2	1			6	10	2	2			1	33	10	20	12	15	10	1	15	13	153	7%
11. Commerce; management of workers & possessions	1			2		5	2	1	2		2	25	8	17	20	25	11	1	8	11	141	6%
12. Dance, music and games		2		1	4	7	6	8	11			34	8	33	41	21	14	10	9	15	224	10%
13. Medical procedures, relaxation etc.						2		1	1			3		1	10	10	5	2	1	5	41	2%
14. Warfare and war-related activities														1	4					0	5	0%
15. Funerary rites and funeral scenes		1		1	6	4	6	3	5		1	28	6	18	23	23	11	4	10	14	164	7%
TOTAL ATTESTATIONS PER REIGN	**31**	**9**	**2**	**5**	**20**	**62**	**22**	**35**	**43**	**0**	**4**	**421**	**105**	**281**	**356**	**301**	**180**	**46**	**134**	**226**	**2283**	

disruption.[75] The life of Ptahshepses, of Abusir, for another example, overlapped at least two kings: his career started as Niuserra's hairdresser, before being appointed Overseer of Upper Egypt under the subsequent king, Menkauhor.[76]

5.2.1. The Sample: Distribution and Abundance of Artistic Themes

A distribution and abundance analysis was performed on the sample data as outlined in Chapter Two, (see Table 5.1; note the full table is in the appendices). The Roman numerals represent the Dynasty, and the Arabic numerals represent the king: King Teti, for example, is identified as VI.1 where 'VI' indicates Dynasty Six and '1' identifies its first king. This data was obtained from artistic scenes contained within the Oxford Expedition to Egypt Scene-details Database, drawing from tombs dating from the Fourth Dynasty to the Eighth, with some, perhaps, falling into the First Intermediate Period. As mentioned in Chapter Two, despite an incomplete record, the 495+ entries should provide an excellent overview of the decoration programme of the time frame under investigation. 'Abundance' was determined by the number of attestations of each decoration theme and 'distribution' was measured by the date as assigned to the decorations. The complete distribution and abundance table is presented in the Appendices.[77]

5.2.2. Old Kingdom Scene Types: An Overview

As is to be expected for a culture that relied so much upon the river and the bounty provided by the annual inundation, almost one third of all the tomb decorations attested depict a scene whose theme can be related to a marshland activity. Despite the consensus 4200 BP date for an increase in dryness, and with increasing aridity seeming to have been a perennial problem throughout the Old Kingdom, it does seem curious that decorations representing this theme persist in such a large proportion. The next most common theme is pastureland and animal husbandry, followed by depictions involving dance, music, and games. Despite the

[75] Verner, *Who was Shepseskara*? 581–602.

[76] Bárta, *Ptahshepses Junior* II, 66.

[77] see ARID_Appendix.

Table 5.2: New scene types: Niuserra onwards

1. Marsh-related activities
1.10. Men gathering reeds or lotuses in a marsh (excluding artificial pools)
1.16. Fishing with a large net, set from boats
1.17. Fishing with a funnel trap
1.18. Fishing with a rounded basket trap, or carrying fish in a basket trap
1.19. Fishing with a hand-held net
1.20. Boatmen spear-fishing in a marsh (excluding the major figure)
1.21. Angling from a boat
1.27. Net-making or repairing a net
1.28. Drying a fishing net after use
4. Pastureland and animal husbandry
4.1. Fighting bulls
4.3. Cattle mating in pasture
4.6. Hobbled and tethered calves
4.9. Force-feeding hyenas
5. Orchard scenes
5.1. Gathering fruit from trees
5.3. Trapping song-birds in a clapnet
5.4. Trapping song-birds in hand-set traps
6. Gardening
6.1. Hoeing, to prepare or maintain a garden
6.2. Watering a garden
6.3. Cultivating and harvesting vegetables in a garden
6.4. Growing papyrus or lotuses in a man-made garden pool
7. Bird procession and poultry yard scenes
7.2. Poultry-yard scene
7.3. Force-feeding poultry
8. Manufacture and storage of wine and oil
8.2. Treading grapes or dates in a vat
9. Food Preparation and Brewing
9.4. Preparing fish for consumption (excluding fish-gutting)
10. Workshop activities
10.2. Linen production
10.7. Pottery making
10.8. Seal engraving
10.9. Staff fashioning
11. Commerce, including management of workers and possessions
11.1. Market scene
11.2. Freight-boat scene
11.6. Tomb owner conveyed on donkeys, probably to inspect possessions
11.10. Officials witnessing a will
12. Dance, music and games
12.4. Lone musician entertaining workers
12.6. Workers beating rhythms with resonant sticks
12.7. Men or women clapping rhythms for movement (excluding dancing)
12.8. A board-game called 'senet', or 'draughts'
13. Medical procedures, relaxation and bodily care
13.4. Manicure scene
13.5. Pedicure scene
13.7. Shaving scene
14. Warfare and war-related activities
14.1. Craftsmen making weapons
14.2. Siege scene
14.3. Soldiers marching or jogging to a battlefield
14.4. Wrestling, perhaps as part of army training
15. Funerary rites and funeral scenes
15.2. Statue or statues conveyed by boat to a tomb
15.5. Jars of oil or wine conveyed by sledge to a tomb

significance of cultivation to the culture of ancient Egypt, fewer than one in twelve scenes attested are related to the agricultural process.

5.2.3. New Iconographical Scenes: Mid-Fifth Dynasty

From the time of Niuserra onwards [V.6+], an evolution (revolution?) in decorations appears to have occurred, with a substantial increase in both the amount and variety of tomb decorations noted, for example, gardening activities are first depicted during this reign; and scenes depicting the Desert Hunt re-enter the repertoire, last appearing early Dynasty Four. As identified in Part A, the 15 broad themes of the Scene-details Database, are broken down into more detailed 'Scene Types'. When the sample was expanded to include those scene types within each category, new scene types that originated from the time of Niuserra emerged (see Table 5.2). The majority of the new scene types have something to do with the acquisition, exploitation, storage, and regulation of food-based resources: of the forty-four new additions, more than two-thirds of the scene types can be classified in this manner. All but one of the nine new additions to Theme 1, *Marsh-related activities*, depict fishing and its associated techniques and technologies. The new scene types added to Theme 13, *Medical procedures, relaxation, and bodily care*, represent data from only four different tombs, perhaps skewing the data. Similarly, the scenes added to Theme 14, *Warfare and war-related activities* come from only a few tombs; these examples may skew the data slightly, but are an important feature to note, nevertheless. Interestingly, some of these scene types newly added to the tomb decoration repertoire in the Fifth Dynasty, such as those in Theme 6, *Gardening*, and those from Theme 14, *Warfare and war-related activities*, do not persist long into the decoration catalogue produced for the next dynasty.[78]

[78] No attestations date to beyond the reign of Pepy I: see ARID_ Appendix.

5.2.4. Niuserra: A Changing Emphasis?

The significance of the reign of Niuserra has been alluded to earlier and is discussed more fully in Part D. Many of the newly arrived scene types that appeared at the time of Niuserra seem to support the A.R.I.D hypothesis, with many themes relating to the exploitation and acquisition of food resources. Food-producing land may have been so important by this time that it was thought vital to record 'proof' of the family's ownership, or more correctly, rights to utilise the resources produced.[79] For example, a decoration representing officials witnessing a will is related to a deed transferring ownership of agricultural land.[80] Other aspects of Theme 11, *Commerce*, that appear as new representations could possibly be linked to dealing with food resources or agricultural land in the owner's possession. There are almost fifty instances where the tomb owners are depicted travelling to visit their possessions. There are three instances of the tomb owners doing so astride donkeys, while the remaining forty-three are transported by palanquins. Interestingly, there is only one attestation of this activity dates to before the time of Niuserra, found in the tomb of Nefermaet at Maidum.[81] All the rest date from Niuserra onwards, suggesting that this activity had become significant enough to be commemorated within many tombs. Similarly, all of the attestations depicting scene type 11.1, 'Market scenes', date from the mid-Fifth Dynasty onwards.

5.2.5. New Iconographical Scenes: Sixth Dynasty Onwards

In contrast from the time of Niuserra to the end of the Fifth Dynasty, very few new scene types were added to the iconography in the Sixth Dynasty (see Table 5.3). Despite the new scene types in Themes 10, 13 and 15 introducing some new activities, the total number of actual decorations depicting these scenes is quite low, and their absolute number when compared to the overall number of representations attested is very small, and the comparative proportion of attestations with these themes (#10, 13 & 15) either did not change or fell markedly. Significantly, none of these themes can be linked to the environment or to resources. The observation made above about depictions relating to Theme 13: *Medical procedures, relaxation, and bodily care,* is equally valid here, with the new scene type coming from a single tomb. The application of a medical connection to the scene depicting circumcision, and not a religious association, will be discussed later. However, as the marshlands enlarged, it would be expected that the incidence of mosquito-borne diseases should increase as well as an increase in medical conditions associated with increasingly poor water quality. The remaining new iconographic representations appeared within those themes identified as increasing in overall proportion of the total sample and can be readily linked in some way to food resources.

Table 5.3: New scene types: Sixth Dynasty onwards

DESCRIPTION
1. Marsh-related activities
1.20. Boatmen spear-fishing in a marsh (excluding the major figure)
1.24. Fishermen registering their catch for distribution by scribes
1.25. Mat-making scene
1.26. Rope-making scene
5. Orchard scenes
5.4. Trapping song-birds in hand-set traps
8. Manufacture and storage of wine and oil
8.5. Decanting oil
10.6. Painting onto a screen
13. Medical procedures, relaxation and bodily care
13.8. Circumcision scene
13.9. Mouth-examination scene
15. Funerary rites and funeral scenes
15.5. Jars of oil or wine conveyed by sledge to a tomb

5.3. A Shifting Artistic Imperative?

Within the gamut of tomb decorations, some themes and scene types seem to have endured over the entire time frame under investigation. Because not every tomb used every type of decoration, it is difficult to claim that have been included for symbolic or religious reasons.[82] It is possible to suggest the apparent change in the decorations depicted upon tomb walls represented a society that had become more concerned with ensuring an adequate and regular supply of food. It may be that the change within the decoration sequence indicates the perception that a regular and reliable food supply was becoming a source of increasing concern. This may have led to a (perhaps unconscious?) increasing pre-eminence applied to decorations that related to the acquisition, exploitation and utilisation of food resources. Particular emphasis remains upon scene types relating to agriculture (even though this declined, somewhat) and activities relating to marshes, the desert, poultry, and the utilisation of cattle.

5.3.1. Tomb Decorations: Proportion as a Variant over Time

Despite a general increase in tomb size *and* a corresponding increase in the total surface area available for scenes to be depicted,[83] some attested themes decreased in their relative comparison: their proportion declined when compared to other themes and at earlier times. While an increased

[79] Thompson, *Tehna* I, 15–20.

[80] See Wepem-Nefert-Wep, in Hassan, *Giza* II, 190–192, fig. 219, pl. 74–76, (dated Dynasty 6–8).

[81] Harpur, *Maidum*, 67–68, 186–187, fig. 77, pl. 8 [a].

[82] Kanawati, *Papyrus Thickets,* 119.

[83] Kanawati, *Administration*, 159; El Khouli & Kanawati, *El-Hammimiya*, 13–14.

Table 5.4: Themes: Pre-Dynasty 6 and Dynasty 6 onwards

DATE	IV	IV	IV	IV	IV	IV	V	V	V	V	V	V	V	V	V	VI	VI	VI	VI	VI	Pre-D6		D6 On	
OEE SCENE-DETAILS THEMES	**1**	**2**	**3**	**4**	**5**	**6**	**1**	**2**	**3**	**4**	**5**	**6**	**7**	**8**	**9**	**1**	**2**	**3**	**4**	**5+**	**ΣT**	**%**	**ΣT**	**%**
1. Marsh-related activities	11	3	1	1	3	15	4	12	15			123	42	78	130	104	61	14	41	62	438	31%	282	32%
2. Desert and desert-related activities	5		1									6	2	2	10	4	2	2	6	7	26	2%	21	2%
3. Agricultural pursuits	4	1				3	1	3	4			27	6	23	24	14	6	1	10	24	96	7%	55	6%
4. Pastureland and animal husbandry	4					4		3				53	11	34	30	28	21	2	16	38	139	10%	105	12%
5. Orchard scenes								1	1			7	1	8	3	10			1	1	21	2%	12	1%
6. Gardening												9		6	4	5	5			0	19	1%	10	1%
7. Bird procession and poultry yard scenes		1			1	3			1			16	5	6	13	11	11	1	3	1	46	3%	27	3%
8. Manufacture and storage of wine and oil	1					6			1			21	4	15	12	10	7		4	2	60	4%	23	3%
9. Food Preparation and Brewing	3					3	1	1	2			36	2	19	20	21	16	8	10	33	87	6%	88	10%
10. Workshop Activities	2	1			6	10	2	2			1	33	10	20	12	15	10	1	15	13	99	7%	54	6%
11. Commerce, including management of workers and possessions	1			2		5	2	1	2		2	25	8	17	20	25	11	1	8	11	85	6%	56	6%
12. Dance, music and games		2		1	4	7	6	8	11			34	8	33	41	21	14	10	9	15	155	11%	69	8%
13. Medical procedures, relaxation etc.						2		1	1			3		1	10	10	5	2	1	5	18	1%	23	3%
14. Warfare and war-related activities														1	4					0	5	0%	0	0%
15. Funerary rites and funeral scenes		1		1	6	4	6	3	5		1	28	6	18	23	23	11	4	10	14	102	7%	62	7%
TOTAL ATTESTATIONS PER REIGN	31	9	2	5	20	62	22	35	43	0	4	421	105	281	356	301	180	46	134	226	1396		887	

number of attestations may result in an increase in total abundance, the *relative proportion* when compared to the entire sample occupied by the thematic categories changes over the time frame in question, with some increasing at the expense of others, sometimes significantly. If no external influence were present, then it is reasonable to expect that the overall proportion of decoration themes should display a consistent distribution throughout the relatively short time period that was the Old Kingdom. Therefore, it was deemed important to ascertain if any change in the thematic proportion occurred over the time frame under investigation.[84]

[84] See ARID_Appendix.

5.3.2. Pre- and Post-Sixth Dynasty: The Impact of 'Drought'

To see if any 'turning point' in the progression of tomb decorations can be identified signalling changing environmental conditions, it was determined that the onset of the Sixth Dynasty be imposed as a 'boundary' within the data. Table 5.4 presents this data.

Table 5.5 displays the overall proportion of each wall scene theme out of the total scene sample over the time frame under investigation and identifies whether it varied significantly over the course of the Sixth Dynasty. A change that is less than significant (< ±10%) is represented (≈). Significant trends are represented in the following manner: +10% as (↑), and –10% as (↓).

Table 5.5: Themes: Pre- versus post Dynasty 6 – proportion and trends

Date	Pre-Dynasty 6		Dynasty 6 On		Trend	
Theme	**T**	**%**	**T**	**%**	**±**	**?**
1. Marsh-related activities	438	31%	282	32%	101%	≈
2. Desert and desert-activities	26	2%	21	2%	127%	↑
3. Agricultural pursuits	96	7%	55	6%	90%	↓
4. Pastureland and animal-husbandry scenes	139	10%	105	12%	119%	↑
5. Orchard scenes	21	2%	12	1%	90%	↓
6. Gardening	19	1%	10	1%	83%	↓
7. Bird-procession and poultry-yard scenes	46	3%	27	3%	92%	≈
8. Manufacture and storage of wine and oil	60	4%	23	3%	60%	↓
9. Food preparation and brewing	87	6%	88	10%	159%	↑
10. Workshop activities	99	7%	54	6%	86%	↓
11. Commerce, including the management of workers and possessions	85	6%	56	6%	104%	≈
12. Dance, music and games	155	11%	69	8%	70%	↓
13. Medical procedures, relaxation and bodily care	18	1%	23	3%	201%	↑
14. Warfare and war-related activities	5	0%	0	0%	0%	↓
15. Funerary rites and funeral scenes	102	7%	62	7%	96%	≈

The change in overall proportion of wall scene types crossing from the Fifth Dynasty into the Sixth can be summarised as follows:

1. Marsh-related decorations continued their importance.
2. Desert-related decorations increased significantly.
3. Agricultural pursuit decorations decreased from the Sixth Dynasty onwards.
4. Pastureland and animal husbandry decorations increased significantly.
5. Orchard-related decorations decreased.
6. Gardening-related decorations decreased.
7. Bird procession and poultry-yard themes did not change significantly.
8. Decorations depicting the manufacture and storage of wine and oil decreased significantly.
9. Food preparation and brewing decorations increased significantly.
10. Workshop activities were represented significantly more often before the Sixth Dynasty.
11. Commerce and management decorations maintained their significance.
12. Dance, music and games were significantly more common before the Sixth Dynasty.
13. Representations of medical procedures increased significantly in the Sixth Dynasty.
14. Decorations depicting warfare and related activities are not attested in tombs dating to the Sixth Dynasty.
15. Decorations depicting various funerary rites and funeral scenes maintained their overall importance.

Six themes showed 'negative' trends where the decrease, in proportional terms, was more than ten percent (10%) over the time frame in question: Theme 5: *Orchard*, Theme 6: *Gardening*, Theme 8: *Manufacture and storage of wine and oil*, Theme 10: *Workshop activities* and Theme 12: *Dance, music and games*.

Depictions relating to Theme 14: *Warfare and war-related activities*, are not attested during this period. For a detailed discussion of the implications of the 'negative' trends, see Part C. Four scene types showed a 'positive' trend where an increase, in proportional terms, was more than ten percent (10%): Theme 2: *Desert-related activities*, Theme 4: *Pastureland and animal husbandry scenes*, Theme 9: *Food Preparation and Brewing* and Theme 13: *Medical procedures, bodily care and relaxation.* Apart from the last category, (see below), these categories can be readily interpreted as an indication of a growing preoccupation with food resources.

5.3.3. Some Result Limitations

Statistically, warfare-related activities recorded a significant decrease, but there are only two tombs in this period decorated with scenes depicting actual warfare.[85] Another tomb, Khunes at Zawyet el-Maiyetin, displays the making of weapons;[86] but these could derive from a hunting context, though the number of spears represented suggests not. Another, that of Ptahhotep II Thefi at Saqqara, depicts wrestling, which may or may not be military in context, since the participants are juvenile in appearance.[87] Since so few tombs contain decorations relating to actual warfare, this small number does not allow for valid use in

[85] Kanawati & McFarlane, *Deshasha*, 24–25, pl. 2, 26, 27; McFarlane, *Saqqara Mastabas*, 33–34, 48, pl. 2 [a], 10, 11 [a-c].
[86] Khunes – Lepsius, *Denkmaler* II, 108.
[87] Davies, *Ptahhetep & Akhethetep* I, 11, pl. 21, 24; Paget & Pirie, *Ptahhetep*, 29, pl. 33.

statistical analysis. Commentary on this decoration theme will be limited to the discussion.

The overall increase of wall scenes relating to Theme 13: *Medical procedures, relaxation, and bodily care*, is difficult to relate to a declining river. As mentioned previously, a few significant tombs with a particular focus in their decorations may skew the statistics. Incorrectly identified as '*the Tomb of the Physician*', the tomb of Ankhmahor[88] in the Teti Cemetery, Saqqara, contains five instances of decorations relating to this individual theme,[89] including the only recorded circumcision scene.[90] This one tomb accounts for more than one third of all the attestations for this theme in the Sixth Dynasty. In an equivalent fashion, the tomb of Khentika, in the same cemetery, also contains many decorations relating to this theme,[91] including a recently identified circumcision scene.[92] Similarly, the tomb of the twins, Niankh-khnum and Khnumhotep, in the Unis Cemetery, Saqqara, personal groomsmen to the king, contain almost one-half of the decorations related to this theme in the Fifth Dynasty.[93] Therefore, three tombs, combined, represent almost two-thirds of the attestations. Because these individual tombs contain such a large share of the data, it seems judicious to partially ignore them when relating the investigation to the hypothesis under examination. Another issue that may skew the data is the increasing tendency for the elite to be buried in the provinces, so that the 'Memphite style' may has travelled with them.[94] This is even more interesting when, as identified later, many of the new themes emphasising resource-based scenes, seem to have become significant initially in these same provincial tombs. Notice that the vast majority of scene types that show an increase in proportion over the Sixth Dynasty exhibit characteristics that suggest a growing reliance on non-agricultural food sources. Other potentially significant positive trends are some selected scenes within those depicting *Orchard* and *Gardening* scenes (Themes 5 and 6), whose combined total are still quite low compared to many of the other types of decorations.

In a similar fashion, the combined total number of scenes depicting *Desert* scenes is still low, when compared to the other categories. However, the significant increase in overall proportion is noteworthy, nonetheless. If the drought were so significant, then it was expected that more tombs should be depicting a desert scene; and this was seen to be the case with *Desert-related activities* (Theme 2) continuing to be depicted right into the end of the Old Kingdom, with many new subtleties and refinements added to the repertoire. Some tomb owners chose to depict only this theme in their tomb,[95] though only one Sixth Dynasty tomb owner, Nehwet-Desher-Meri,[96] chose to do so. An increase in scenes depicting cattle as desert animals, at the same time as a decrease in cultivation of all forms infers that the society may have come increasingly more reliant upon mobile wealth. Scenes depicting the preparation of food and drink increased by almost two-thirds, evidence suggesting a growing concern with an adequate supply of food. Interesting, too, is the fact that, as a proportion of the overall sample, scenes depicting marshland-related activities increased slightly at the same time as did those scenes depicting the desert. If the drought was encouraging a bloom in riverine plants, then this seems reasonable. Also noteworthy is that, in a drought, we appear to have a decrease in the depiction of men utilising large nets set predominantly from boats. At the same time, however, an increase in scenes depicting fishing activity on the river's edge; with people using traps and smaller nets, is noted.

5.4. Resources: A Need to Re-arrange the Data?

While the OEE Scene details Database provided an effective baseline for the initial overview, its value for continued utilisation needed to be further assessed and evaluated. The Database was designed with different thematic imperatives in mind to that performed in this study. Therefore, we are limited by the initial classifications of the designer of the database. This does not suggest that some of the decorative elements may be counted more than once, but entries appear in places where they may be less than helpful in understanding the artistic progression for the current investigation.

For example, there are nine Oxford scene-detail categories relating to birds and poultry, but they have been grouped into four different themes: *Marshland activities*, *Orchard* scenes, *Bird procession and poultry-yard* scenes and *Food preparation* scenes all contains scenes depicting birds. Similarly, some scenes depicting fishing and its associated techniques and technologies have been classified into the *Marshland* theme, whereas those scenes depicting the consequences of fishing have been categorised into the theme relating to *Food preparation*. Similarly, scene details depicting cattle in various stages of their life cycle are categorised into three different themes: *Marshland activities*, *Pastureland* scenes, and *Food preparation*.

5.4.1. A Resource-based Analysis May Be Needed

The categories of themes as it exists is difficult to relate to the A.R.I.D. hypothesis because the flow of the investigation from the ecological evaluation then

[88] Badawy, *Nyhetep-Ptah & Ankhmahor*, fig. 27, pl. 30; Kanawati & Hassan, *Ankhmahor*, pl. 19, 55[b].
[89] These include five scene types: 13.4 Manicure, 13.5 Pedicure, 13.6 Massage, 13.8 Circumcision, and 13.9 Mouth-examination.
[90] Badawy, *Nyhetep-Ptah & Ankhmahor*, 19, fig. 27, pl. 30; Kanawati & Hassan, *Ankhmahor*, 49–50, pl. 19, 55b.
[91] Wreszinski, *Atlas* III; James, *Khentika*, pl. 11. This tomb includes three types: 13.4 Manicure, 13.5 Pedicure and 13.6 Massage.
[92] Ebbel, *p.Ebers*, 115. Megahed & Vymazalová, *Circumcision*, 161–162, suggest the scenes representing circumcision actually represent the ritualised shaving of the pubic areas as a sign of cleanliness, while Spiegelman, *Emergency Surgery?* 91–100, suggests it is a procedure resulting from some injury.
[93] Moussa & Altenmüller, *Nianchchnum & Chnumhotep*, Taf. 21, 27a; Abb. 10.
[94] Swinton, *Dating Tombs*, 171.
[95] See the D4 tombs of Methen and Minkaef.
[96] Kanawati, *Akhmim* VIII, 12, pl. 2a, 8a, fig. 3b.

Table 5.6: The growing, recording, storage and distribution of food

Date	Pre-Dynasty 6		Dynasty 6 On	
THEME	T	%	T	%
3.1. Agricultural sequence, from hoeing to stacking grain	78	5.6%	47	5.3%
3.2. Filling granaries and issuing, measuring or recording grain-seed or grain	18	1.3%	8	0.9%
5.1. Gathering fruit from trees	9	0.6%	5	0.6%
6.1. Hoeing, to prepare or maintain a garden	1	0.1%	0	0.0%
6.2. Watering a garden	7	0.5%	4	0.5%
6.3. Cultivating and harvesting vegetables in a garden	6	0.4%	4	0.5%
6.4. Growing papyrus or lotuses in a man-made garden pool	5	0.4%	2	0.2%
8.1. Gathering grapes or dates	14	1.0%	7	0.8%
8.2. Treading grapes or dates in a vat	14	1.0%	6	0.7%
8.3. Extracting juice in a wine press or date press	14	1.0%	5	0.6%
8.4. Filling, sealing, recording and/or storing jars of wine	11	0.8%	3	0.3%
8.5. Decanting oil	0	0.0%	2	0.2%
8.7. Filling, sealing, recording and/or storing jars of oil	4	0.3%	0	0.0%
8.6. Extracting oil in an oil press, and mixing oil	3	0.2%	0	0.0%
11.3. Estate managers rendering their accounts to scribes	29	2.1%	13	1.5%
SUBTOTAL + PROPORTION	213	15%	106	12%
Variance	**-22%**			

becomes focussed on potential resource changes within the environment. For the purpose of this investigation, it may be of added value to investigate all the scenes that focus upon one basic type of resource together. If the role of cattle were to be investigated by tomb decorations depicting them, it would seem practical to cluster all the cattle-related decorations together. It was therefore resolved to link 'resource-related' scene details together under 'resource group' subject headings. This should enable a more coherent exploration of the importance or otherwise of changing patterns in tomb decorations. Areas to be investigated will be those relating to the acquisition, exploitation and regulation of food-based resources.

The resource groups are as follows:

A. CULTIVATION – The Growing, Recording, Storage and Distribution of Food
B. FISHING – Fishing Techniques, Technologies and Associated Activities
C. AVIAN – Avian and Associated Attestations, including Waterfowl
D. PASTURELAND – Cattle, Pastureland and Animal Husbandry Activities
E. DESERT – Desert Activities

When the data is re-arranged into resource groups, a clearer picture begins to emerge. The narrative of the decoration changes suggest that the consequences of the hypotheses outlined in Chapter Four seems valid. Overall, the relative importance of themes relating to the growing of food decreases over the time frame in question. Scenes representing the agricultural sequence decreases slightly, whereas those decorations relating to orchards and the growing of papyrus decrease significantly (see Table 5.6). Decorations relating to gardening appear to maintain their relative importance. While there was a very slight, insignificant, increase in representations depicting food, depictions of its storage and distribution declined considerably. From the Fifth to the Sixth Dynasty, the relative proportion of food recording, storage and distribution themes decreased by more than one-third when compared to 'pre-drought' times. This activity was thought to be one of the major drivers for the development of administration during the Predynastic era, yet it is declining considerably. Perhaps this can be linked to a decrease in overall production, resulting in less surplus, resulting in a decline in food available for re-distribution? If an increase in attestations suggests an increase in eminence, then the importance of bread to the society is becoming predominant. Representations depicting the stages of bread-making are the only ones in this category that increased both in raw number and proportion of the total.

As identified earlier, a reduced inundation and a weaker river flow would have led to an increased retention of nutrients within the river. This would have led to an increase in some aquatic plant numbers, which would have impacted upon the diversity and variety of habitats that may have developed. This change would affect the fish populations as a consequence. The apparent increasing importance of fishing and its associated activities (see Table 5.7).

A slight decline in overall abundance occurs (Table 5.8). However, if we focus upon the subset relating to only water-dwelling avians, a significant increase in the relative proportion of representations can be noticed (see Table 5.9). Although this large increase in waterfowl/poultry scenes may be eye-catching, it is important to realise

Table 5.7: Fishing techniques, technologies and associated activities

Date	Pre-Dynasty 6		Dynasty 6 On	
THEME	T	%	T	%
1.1. Spear-fishing and/or fowling scene	31	2.2%	59	6.7%
1.15. Dragnet or seine-netting scene	55	3.9%	31	3.5%
1.16. Fishing with a large net, set from boats	3	0.2%	3	0.3%
1.17. Fishing with a funnel trap	3	0.2%	8	0.9%
1.18. Fishing with, or carrying fish, in a rounded basket trap	5	0.4%	4	0.5%
1.19. Fishing with a hand-held net	14	1.0%	10	1.1%
1.20. Boatmen spear-fishing in a marsh (excl. major figure)	0	0.0%	2	0.2%
1.21. Angling from a boat	14	1.0%	19	2.1%
1.22. Fish-gutter cleaning a recent catch of fish	29	2.1%	9	1.0%
1.23. Fish-bearer(s)	25	1.8%	21	2.4%
1.24. Fishermen registering their catch for distribution…	0	0.0%	2	0.2%
1.26. Rope-making scene	11	0.8%	3	0.3%
1.27. Net-making or repairing a net	4	0.3%	7	0.8%
1.28. Drying a fishing net after use	1	0.1%	1	0.1%
9.4. Preparing fish for consumption (excluding fish-gutting)	5	0.4%	1	0.1%
SUBTOTAL + PROPORTION	200	14.3%	180	20.3%
Variance	**+42%**			

Table 5.8: Avian and associated attestations, including waterfowl

Date	Pre-Dynasty 6		Dynasty 6 On	
THEME	T	%	T	%
1.13. Clapnet or bird-trapping scene	16	1.1%	7	0.8%
1.14. Workers placing captured marsh birds in cages	16	1.1%	7	0.8%
5.2. Netting song-birds in an orchard	11	0.8%	4	0.5%
5.3. Trapping song-birds in a clapnet	1	0.1%	0	0.0%
5.4. Trapping song-birds in hand-set traps	0	0.0%	3	0.3%
7.1. Bird-procession scene	40	2.9%	7	0.8%
7.2. Poultry-yard scene	3	0.2%	12	1.4%
7.3. Force-feeding poultry	3	0.2%	8	0.9%
9.3. Preparing fowl for consumption	23	1.6%	22	2.5%
SUBTOTAL + PROPORTION	113	8.1%	70	7.9%
Variance	**-3%**			

Table 5.9: Waterfowl

Date	Pre-Dynasty 6		Dynasty 6 On	
THEME	T	%	T	%
1.14. Workers placing captured marsh birds in cages	16	1.1%	7	0.8%
7.2. Poultry-yard scene	3	0.2%	12	1.4%
7.3. Force-feeding poultry	3	0.2%	8	0.9%
9.3. Preparing fowl for consumption	23	1.6%	22	2.5%
SUBTOTAL + PROPORTION	45	3.2%	49	5.5%
Variance	**+71%**			

that the proportion is an increase over a small sample set; nevertheless, the increase is highly suggestive. An increase in representations of an apparently unimportant and unpretentious animal group such as poultry, as a part of tomb scenes, does warrant some discussion.

In total number, there are almost as many attestations of cattle and pastureland-related depictions in the Sixth Dynasty as there are during the Fourth and Fifth Dynasties combined (see Table 5.10). As an overall proportion, despite the apparent 'decline' at the end of the Sixth

Table 5.10: Cattle, pastureland and animal husbandry activities

Date	Pre-Dynasty 6		Dynasty 6 On	
THEME	T	%	T	%
1.7. Cattle fording a canal or stream	32	2.3%	22	2.5%
4.1. Fighting bulls	1	0.1%	16	1.8%
4.2. Livestock browsing in pasture	6	0.4%	9	1.0%
4.3. Cattle mating in pasture	11	0.8%	7	0.8%
4.4. Cow giving birth	20	1.4%	15	1.7%
4.5. Milking a cow	29	2.1%	19	2.1%
4.6. Hobbled and tethered calves	16	1.1%	3	0.3%
4.7. Goats in or near trees	12	0.9%	15	1.7%
4.8. Feeding domesticated animals by hand	29	2.1%	15	1.7%
4.9. Force-feeding hyenas	1	0.1%	4	0.5%
4.10. Overseer seated near animal-tending activities	14	1.0%	2	0.2%
9.1. Disembowelling and/or skinning an animal	7	0.5%	5	0.6%
9.2. Preparing red meat for consumption	11	0.8%	17	1.9%
SUBTOTAL + PROPORTION	189	13.5%	149	20%
Variance	**+24%**			

Table 5.11: Desert and desert related activities

Date	Pre-Dynasty 6		Dynasty 6 On	
THEME	T	%	T	%
2.1. Desert-hunt or desert-landscape scene	20	1.4%	16	1.8%
2.2. Hunters returning from the desert with game	6	0.4%	5	0.6%
SUBTOTAL + PROPORTION	26	1.9%	21	2.4%
Variance	**+27%**			

Dynasty, this scene type incorporates a greater proportion of tomb wall scenes. With an increase in total volume of papyrus and phragmites plants proposed, it is to be expected that the overall number of grazing animals would have increased in consequence, and therefore, depictions of these should have increased in a corresponding manner. The perceived increased significance of 'mobile sources of food' and the rise of the southern cattle barons will be expanded upon later.

It seems quite consistent for an increase in depictions of the desert hunt to have occurred during the 'drought' (see Table 5.11). As the land began to dry out due to lower-than-average inundations, the ecotone, "*the region where one type of habitat or natural environment meets another*",[97] would have thinned, leading to a merging of the desert and the riverside ecosystems.

5.4.2. Theme Proportion Changes and A River in Drought

During the Sixth Dynasty, the proportion of decorative features related to the actual growing of food decreased slightly. So, too, did those wall scenes relating to the storage, distribution and preparation of food. This may suggest an apparent decline in the relative importance of agriculture. Representations depicting the stages of bread-making are the only ones in this category that increased both in raw number and overall proportion. If an increase in attestations suggests an increase in apparent eminence, then the importance of bread to the society is becoming predominant. The relative proportion of scenes depicting birds and their trapping around orchards, declined. However, representations depicting poultry and waterfowl increased significantly, as did fishing and pastureland activities. Interestingly, the only theme in which *every* decoration sub-scene type proportionately increased was that of *Desert*.

5.5. A Changing Emphasis in Tomb Decoration?

Despite paleo-ecological, paleoclimatic and archaeological evidence of an increasing dryness in the Nile valley, marshland scenes remained the most significant proportion of tomb decorations. If agriculture were declining, it would be expected that the relative proportion of scenes representing activities related to the Desert Hunt might increase, as is the case. The changes in the distribution and abundance of tomb decorations may lead to the development of a potential 'visual narrative' regarding the acquisition, exploitation and distribution of resources over the time frame under investigation. This narrative as depicted by the change in tomb decorations during the

[97] http://dictionary.cambridge.org/dictionary/english/ecotone (31/08/2017).

Table 5.12: Fishing and fowling in the marshes: total numbers

Fishing and Fowling	Date		
	Pre-Niuserra	Niuserra → D6	D6 Onwards
Total numbers of attestations	2	29	59

period of the Old Kingdom and may allow us to infer a changing environmental situation. It appears that, from the data presented, a new preoccupation seems to have arisen amongst those producing decorative features for tombs. A larger than usual proportion of wall scenes depicted scenes about or alongside the river. In times of drought, the normal expectation would be that food becomes scarcer and increasingly precious. As outlined previously, if the river did not flood, leading to a drought upon the land, the resources remaining within the river should encourage a rapid growth of riverine and riverbank plants.

While the potential impact of the drought is thought to have been over-rated, it appears that the effects were being noticed by tomb owners and artists. It is also noteworthy that, though the 4200BP event was a significant factor in driving environmental change, the gradual aridification of the Nile Valley had begun earlier.[98] From the time of Niuserra onwards, administrative reforms were continuously enacted to attempt to improve the state's efficiency,[99] and changes in the art can be spotted from his reign onwards, which may link to an underlying motive for these administrative changes. The relative proportion of decorations relating to the exploitation of resources increased over the time frame in question.[100] It may be relevant to note what was happening in the few generations before the onset of the 'drought'.

5.5.1. Fishing and Fowling in the Marshes in A.R.I.D times

Depictions of the tomb owner fishing and/or fowling increased by more than seven times from the earlier Old Kingdom to the end of the Sixth Dynasty (see Table 5.12). While raw numbers show an absolute increase, the relative proportion of fishing and fowling scenes, compared to other attestations indicates the apparent increasing significance of this feature being added to a tomb's decoration programme (see Figure 5.1), especially considering this is a time when it has been suggested that the river became a more attractive haven for fishes and bird species.

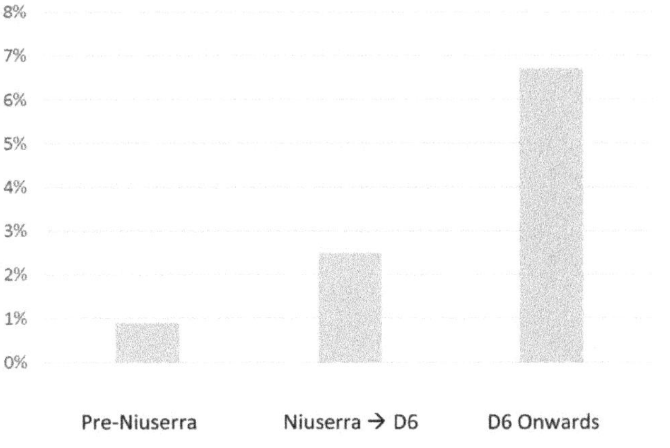

Figure 5.1: Fishing and fowling in the marshes: overall proportion change.

5.6. Summation: An Unfolding Ecological Narrative?

The basic arguments arising from this chapter include:

- A distribution and abundance of tomb scene types in tombs from Dynasty Four to Dynasty Eight showed changes in the progression of the tomb decoration programme.
- Despite average tomb sizes increasing, some themes did not persist throughout the period, while new theme types entered the tomb decoration programme.
- The relative proportion of some scene types, when compared to all attested scenes, changed over the time frame under investigation.
- Changing the analysis to one that investigated the art based on the scenes' subject indicated that scenes depicting the growing, storage and preparation of food declined in overall proportion compared to the total.
- Those scenes depicting activities in the marshlands, such as fishing and catching waterfowl, scenes with the desert as a subject and those scenes relating to the exploitation of cattle increased in apparent importance.

[98] Seidlmayer, *OK Elephantine,* 108–127.
[99] Bárta, *Long Term Short Term,* 183–186.
[100] Bárta, *Punctuated Equilibrium,* 12–13.

6

Cultivation's Failure and A River In 'Drought'

Previously, it was noted that the relative proportion of decorative features relating to the cultivation of food appeared to have decreased significantly over the time frame of this investigation. This led to the inference that the relative prominence of cultivation and the distribution of its produce decreased over the time frame of this investigation. Regular lower-than-usual inundations may have influenced the tomb owners and artists in their choice of subject matter for tomb decorations produced at this time. In this chapter, the chronicle of changing artistic representations of cultivated resources will be outlined in greater depth, and the possible impact of declining resources upon Old Kingdom society will be assessed.

6.1. Cultivation's Declining Importance

Despite agricultural scenes having more than twice as many attestations in the OEE Scene-details Database as those depicting orchards and gardens, more variation in scene types is offered for the latter category. Classification of decorations depicting the agriculture sequence within the database have been broken down into only two scene types, whereas eight scene types have been assigned to those decorations related to orchards and gardening. This does not imply that, to the database designer, the subject was not as important but is a function of the database design itself. The variety of scene types within each category was important. Within the first scene type, 3.1. *Agricultural sequence, from hoeing to stacking grain,*[1] the database offers fifteen scene details, suggesting the possibility of a narrative sequence[2] (see Table 6.1).

Table 6.1: OEE agricultural narrative

3.1.1	Hoeing, to prepare the ground for sowing
3.1.2	Sowing seed on prepared ground
3.1.3	Ploughing, to prepare the ground or to cover sown seed
3.1.4	Workers driving sheep over sown seed
3.1.5	Flax harvest
3.1.6	Tying flax into bundles
3.1.7	Grain harvest
3.1.8	Gleaner, either gleaning or accompanying workers returning home with grain
3.1.9	Tying grain-stems into bundles
3.1.10	Donkeys arriving at a grain field to be loaded with sheaves
3.1.11	Filling or closing sacks of sheaves near a grain field
3.1.12	Loading donkeys with sacks of sheaves, and transporting sheaves on donkeys
3.1.13	Throwing tied or untied sheaves onto a stack ready for threshing
3.1.14	Donkeys, cattle or sheep trampling grain-stems on a threshing-floor
3.1.15	Winnowing, stacking winnowed grain/chaff, and/or decorating stacks of grain

This appears to be a nicely flowing sequence and the progression within the decorations can be traced along the walls of the tombs that contain them. An 'instruction manual' for agricultural production in the afterlife may have been necessary; if so, then tomb decorations would have been placed in an optimal position to recount these guidelines. It would be expected, therefore, that those scenes depicting the process of agriculture should follow a specific sequence.[3]. Within the tomb of Ty, for example, almost the entire progress of activities of the agricultural cycle has been depicted,[4] enabling anyone to follow the sequence and complete the activities themselves – a complete 'narrative', is presented.[5]

With regard to the representations of barley and flax harvesting, most provincial tombs display the harvesting sequence in the correct order, in some form of visual narrative.[6] For the production of linen, flax needed to be harvested 'green' whereas barley was harvested 'dry' in order to facilitate the ease of removal of the crop.[7] If the artists were careful to indicate the correct order of harvesting, then it is reasonable to suspect that their other renditions may have been completed with a similar accuracy.[8]

6.1.1. Changes to the Proportion of Representation of Agriculture

According to the A.R.I.D. hypothesis, a diminishing amount of available water would have led to less irrigation

[1] http://archaeologydataservice.ac.uk/archives/view/oee_ahrc_2006/queryThemes.cfm?section=details&theme=3.1&CFID=31a6a0ea-f5d6-4391-8c19-9e4828e22377&CFTOKEN=0 (12/05/2016).
[2] Kanawati, *Tomb & Beyond*, 88. Also Murray, *Cereals*, 515–526.
[3] http://archaeologydataservice.ac.uk/catalogue/adsdata/ arch-686-1/dissemination/gif/Agriculture/3_1 (21/10/2018).
[4] Wild, *Ti*, II, pl. 78–79 [A, B] (details), 112–113; and III, pl. 136–139 [A, B] (details), 151–154.
[5] Siebels, *Agricultural Scenes*, 55–64.
[6] Davies, *Sheikh Saïd*, 19–23.
[7] Siebels, *Agriculture*, 118–119.
[8] Murray, *Cereals*, 507.

Table 6.2: Agriculture sequence: pre- and post Niuserra

Agricultural Sequence	Date		
	Pre-Niuserra	Niuserra → D6	D6 Onwards
Proportion of overall attestations	6.0%	5.5%	5.3%

water available for cultivation. Attestations for the agricultural sequence become less regular from the time of Niuserra onwards and continue their relative decline into the Sixth Dynasty with the proportion of all scenes depicting agricultural practices declining (see Table 6.2). If the composition of decorative programmes was *not* affected by a changing environment, then it would be expected that no significant change would be noticeable in the data, but during the Sixth Dynasty, the proportion had declined significantly (in statistical terms) from before the time of Niuserra.

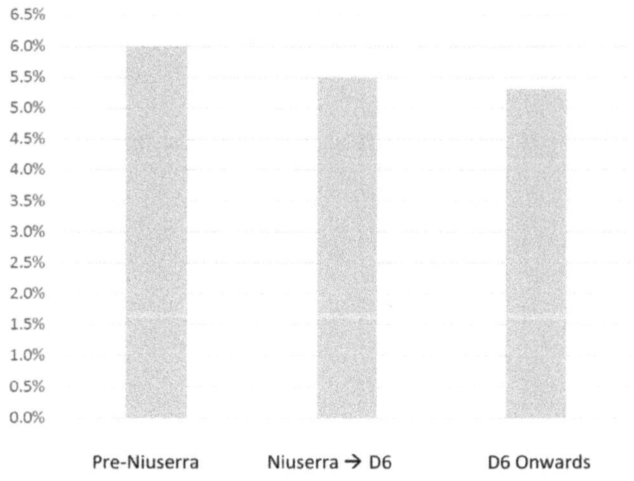

Figure 6.1: Agricultural sequence: proportion change over time.

6.1.2. Cultivation and A River In 'Drought'

The combined data relating to those decorations indicate a small but statistically meaningful decline in the relative importance of this aspect of everyday life into the Sixth Dynasty (see Figure 6.1). Considering the cornerstone role of agriculture in the society, this is a significant factor that needs further investigation. The decline in the apparent importance of cultivation may be explained by the application of the A.R.I.D. hypothesis. Since agriculture relied on the regular inundation to provide a fertile floodplain:[9] if the river did not overflow its banks, reducing its potential benefits to cultivation,[10] the this would have resulted in a lower yield. In times of a less reliable river, it would become, therefore, more important to utilise what water was available in the most effective manner. The impetus to use water more efficiently, for example raising water significantly above low water to aid in the formation of more permanent irrigation channels, was one of the factors driving the adoption of the shadouf in later times.[11]

6.1.3. Barley versus Wheat

Barley (*Hordeum vulgare*) and emmer wheat (*Triticum dicoccum*) have been domesticated for almost 11,000 years.[12] They spread from the Near East to North Africa,[13] with the oldest known sample dated to about 6,500 years ago, but perhaps a little earlier.[14] In Egypt, the archaeological record suggests that the domestication of wheat occurred before that of barley.[15] Wheat is more nutritious than barley, but if barley is not de-husked, its food value increases.[16] While there is a significant negative difference in the nutritional values of the wild versus domesticated version of each of the cereals,[17] it is more important to compare the differences between the two cereals when assessing their use during changing environmental circumstances. As yet, there seems to be no evidence of a genetic change in barley or wheat in response to a changing environment.[18] With increasingly sophisticated technology, the ability to observe more precise changes in the archaeo-botanical record will enable a clearer understanding of any changes to the agricultural cycles that may represent adaptive changes to a changing environment.[19]

Changing environmental circumstances may be the reason that, since textual evidence from the late Old Kingdom suggests that barley had become comparatively more significant than wheat.[20] Despite the cereals being of different species, there does not appear to have been any distinction between barley and wheat artistically.[21] An apparent archaeological abundance of this grain over emmer in certain areas may be an indicator of a larger number of kept animals, rather than human dietary preferences, however, so this factor is needed to be incorporated into future investigations.

6.1.4. Barley More Reliable Than Wheat?

Despite emmer wheat displaying a high pest and disease resistance and with a relatively good tolerance to stress,[22] barley may have been appreciated as a more versatile

[9] Hildebrand & Schilling, *Early Nile Agriculture*, 92–93.
[10] Murray, *Cereals*, 513.
[11] Haldane & Henderson, *Egyptian Shadouf*, 308; Murray, *Cereals*, 513–515.
[12] Zohary & Hopf, *Domestiction of Plants*, 105, 174–176; Zeder, *Origins of Agriculture*, 221–235.
[13] Mascher etal., *Barley Genomics*, 1091–1092, figs 2 & 3.
[14] At least 7,000 years: Madella etal., *African Domestic Cereals*, 6, updating Wendrich etal., *Fayum Neolithic*, 999–1002.
[15] Wetterstrom, *Foraging & Farming*, 203–213.
[16] Eitam etal., *Barley Flour Production*, 5.
[17] Hebelstrup, *Wild versus Domesticated Grains*, 1–4.
[18] Mascher etal., *Barley Genomics*, 1092.
[19] Day, *Botany meets Archaeology*, 5807.
[20] Dr Joyce Swinton, Macquarie (work in preparation).
[21] Murray, *Cereals*, 512.
[22] Bilalis etal., *Mediterranean Innovation*, 328–329.

crop, with husked barley making good livestock feed.[23] Barley is a more flexible crop, usable for fodder, food and brewing;[24] perhaps this versatility led to its increasing use,[25] with barley seen as more valuable for its brewing potentialities than its nutritional value.[26] There does not appear, however, any direct correlation between the availability of cereal and its 'price';[27] one reason perhaps why attempting to guarantee supply became more problematic for administrations at the time.[28] While wheat is a better competitor in ideal conditions, with a greater yield and a more easily harvestable seed,[29] barley displays particular properties that make it more attractive when the environmental circumstances are less than ideal.

As cultivation became more difficult due to a decrease in surplus water, wheat production declined[30] but barley production seems to have increased. Barley tolerates greater ecological extremes than wheat and can thrive in poor soils with low nutrition levels.[31] Compared to wheat, barley is more resilient in disturbed environments,[32] allowing it to more rapidly recover in soils turned over by foraging cattle, for example. Wild barley was found in a broader geographical range than wild varieties of wheat, and cultivated barley can endure a wider range of environmental conditions than wheat,[33] a situation to be expected if the Nile did not flood as regularly. With an irregular supply of water and an increasingly arid habitat, perhaps the environmental situation now favoured barley cultivation over wheat.[34]

6.1.5. Flax Failures?

Despite the origin of flax being unclear,[35] there exists almost ten thousand years of archaeological history of this resource.[36] Not a native to Egypt, with most evidence suggesting that it arrived in Egypt from the Levant in Pre-dynastic times,[37] flax was established in a time when excess water was more readily available, and was well-known before Pharaonic times.[38] As a taxable commodity, flax made its presence felt in the archaeological record.[39] While the new technique of spinning thread for cloth would be depicted early in the Middle Kingdom,[40] this was not practiced earlier; and a decline in quality of weaving and a decrease in quantity of flax production during the First Intermediate Period has been noted.[41] The 'retting' of linen, where the outer bark is removed from the material, required large amounts of running water.[42]

Poor soil nutrition, for example, results in a coarser linen,[43] making it a less desirable commodity. The demand for such products should not have diminished,[44] so other factors must be suggested to explain this drop-off. Flax's need for high soil moisture suggests that its growing circumstances may have been hindered during the time frame of our investigation.[45] Flax requires minimum disturbance to maximise its yield.[46] A low-disturbance environment would become less likely as the river became weaker and more human and animal activity disturbed the ecotone. Flax experiences 'drought-stress', leading to a decline in productive output during periods of low water supply,[47] and suffers when the water supply is slightly saline.[48] Flax's relatively small genome size[49] suggests that it does not have the intrinsic variability of other crops more readily able to adapt to changing environmental conditions. Saline environments cause flax to experience delay in its germination and growth phases.[50] All these requirements may have led to an interruption in the flax-making process as a riverine 'drought' took hold and may explain why its depictions decreased over the time frame under investigation.[51]

6.2. Gardens, Orchards and A River In 'Drought'

An appreciation of the type of horticulture undertaken or represented in tomb scenes has implications for our understanding of the culture and the ongoing expression of continuity and change.[52] Initially, the role of an ancient Egyptian garden was to simply provide fruit and vegetables, irrigating it with water from the Nile. The earliest gardens were composed of planting beds divided into grids by earthen walls, so the water could soak into the soil rather than run off.[53] Gradually as the country became richer, gardens evolved into pleasure gardens, with flowers, ponds and alleys of fruit and shade trees.[54] By the time of the Old Kingdom, temples, palaces and private residences had their own gardens, and models of

23 Kumar, *Health at Hierakonpolis*, 41.
24 McCorriston, *Barley*, 82, 86–88.
25 Willems etal., *Wadi Zabayda*, 315.
26 Katz & Voight, *Bread & Beer*, 23–34.
27 Murray, *Cereals*, 528.
28 Bard, *Archaeology of Ancient Egypt*, 138–139, 176–177.
29 Ladizinsky, *Wild Cereals Collection*, 264–267.
30 Murray, *Cereals*, 506.
31 McCorriston, *Barley*, 86–87; Hillman & Davies, *Domestication Rates*, 157–159.
32 Zohary & Hopf, *Domestiction of Plants*, 27–31.
33 Zohary & Hopf, *Domestiction of Plants*, 39–47, 51–58.
34 Sherratt, *Water, Soil & Seasonality*, 319.
35 Nag etal., *Flax Overview*, 807.
36 Zohary & Hopf, *Domestiction of Plants*, 105–106, map 1.
37 Vogelsang-Eastwood, *Textiles*, 269.
38 Germer, *Flora Pharaonischen*, 101; Lucas & Harris, *Materials & Industries*, 143.
39 Currie etal., *Agricultural Productivity*, 44; Bard, *Archaeology of Ancient Egypt*, 138–139, 176–177.
40 Vogelsang-Eastwood, *Textiles*, 269–270; Elsharnouby, *Linen*, 15.

41 Tata, *Egyptian Textiles*, 78.
42 Elsharnouby, *Linen in Ancient Egypt*, 5–6.
43 Vogelsang-Eastwood, *Textiles*, 270.
44 Tata, *Egyptian Textiles*, 58.
45 Nag etal., *Flax Overview*, 807–808; Moghaddam etal., *Salinity & Flax*, 564; Heller & Byczyńska, *Environmental Factors & Flax*, 26, 30–31, 37.
46 Bilalis etal., *Mediterranean Innovation*, 327–328.
47 Heller & Byczyńska, *Traits of Flax Fiber*, 27, 31–34.
48 Moghaddam etal., *Salinity & Flax*, 571.
49 Brown etal., *DNA & Archaeobotany*, 212, table 1.
50 Moghaddam etal., *Salinity & Flax*, 572–573.
51 Swinton, *Resources*, 156–158.
52 Jones, *Garden Cultivation*, 171–172.
53 Wilkinson, *Symbolism & Design*, 10–13; Verner, *Ptahshepses*, 67, photos 55–6, pl. 34.
54 To trace this development, see Zohary & Hopf, *Old World Plant Domestication*.

gardens were sometimes placed in tombs so their owners could enjoy them in their afterlife.[55]

Garden design theory has been theorised as a combination of three basic aspects: meaning, function and form.[56] Secular gardens were located near the river or canals and were used mainly for growing vegetables. The function of temple gardens was to provide floral, vegetable, and animal offerings and provisions.[57] It is thought that the primary motivator in garden design in ancient Egypt was the representation of its symbolic attributes,[58] with some becoming quite large, and many developing into expressions of individual power. The symbolic function, or its meaning, of tomb gardens was to be available to the spirit of the deceased as a place of shade and refreshment, and to that end they were sometimes illustrated on the walls inside the tomb.[59] The gardens of each king were different, suggesting these were designed to emphasise different attributes of the king,[60] where design foundation had graduated to beyond the simply functional.

6.2.1. Gardens and Orchards at the End of the Old Kingdom

Much that has been written about ancient Egyptian gardens relates to a time later than the Old Kingdom; in describing the design, form and function of Egyptian gardens, Wilkinson uses those of Hatshepsut as his model, with fewer than six lines devoted to Old Kingdom gardens.[61] These gardens that he comments upon represent a sophisticated later stage in the evolution of garden design and do not indicate the progression of this evolution.[62] However, they may be helpful in developing an understanding of the impetus behind the original roles and designs of earlier gardens.[63] While walls curving around individual fruit trees may seem decorative, this feature has been shown to have had the function of allowing a beneficial micro-habitat to develop for each tree:[64] distinct microhabitats for individual organisms increases the odds of them reaching productive maturity.[65] Gardening was hard work, and much effort was also needed to keep birds from eating the crops, with many ingenious traps devised to catch the feathered invaders.[66] Perhaps other means of growing food became more dependable and this may explain the relatively sudden increase in depictions of orchards and gardens.[67] Possibly it was a more efficient use of water to direct it in a more precise manner: adding water to an orchard or garden guaranteed more control over the water supply.

Table 6.3: Gardens and orchards: pre- and post Niuserra

Gardens, orchards etc.	Date		
	Pre-Niuserra	Niuserra → D6	D6 Onwards
Proportion of overall attestations	0.4%	3.5%	2.5%

Decorations depicting orchard and gardening scenes appear to be only recent additions to the tomb decoration repertoire by the late Old Kingdom. Very rapidly, through the latter parts of the Fifth Dynasty, these scenes became relatively more popular. By the start of the Sixth Dynasty, however, garden scenes' relative importance began to decline, by almost one-third compared to what was depicted at the end of the Fifth Dynasty (see Table 6.3).

While garden and orchard scenes make up a minimal proportion of overall scene types, the fact that they were introduced at all should be seen as meaningful. Since these tomb scenes were added to the tomb decorative repertoire at a time when ecological science has suggested that drought and famine should have been impacting upon the society, they must have had some significance (see Figure 6.2). The fact that their proportion increased so significantly during the Fifth Dynasty could be interpreted as indications of significant importance being attributed to this activity. The decline afterwards may be attributable to the changing environmental situation making these activities increasingly difficult to maintain.

An underlying structural element in all garden design was the relationship to and the use of water.[68] Without water, the structures designed to make use of it fail. If the river did not provide an inundation as 'usual', then the likelihood of a garden or orchard failing increases. All of the decorations classified in the OEE Database as depicting the watering of a garden[69] date from the time of Niuserra and onwards, as do those that depict irrigation channels,[70] indicating the perceived importance of a ready supply of water. However, if knowledge of earlier irrigation activities had persisted in the culture, for example as depicted in the representations that van Lepp suggest are depicting irrigation channels,[71] then these activities should have enabled gardens and orchards to be maintained. Perhaps there was not enough water available for rudimentary irrigation.

[55] Baridon, *Les Jardins*, 103; Wilkinson, *Symbolism & Design*, 4–6.
[56] Wilkinson, *Symbolism & Design*, 1.
[57] Wilkinson, *Symbolism & Design*, 2.
[58] Wilkinson, *Symbolism & Design*, 1, 14–15, #3–#6, trace garden evolution.
[59] Wilkinson, *Symbolism & Design*, 4.
[60] Wilkinson, *Symbolism & Design*, 2. Wilkinson, *Symbolism & Design*, 5, suggests Niuserra raised song birds in his Sun Temple; contra view: Dr. L. Evans, Macquarie (13/08/2019).
[61] Wilkinson, *Symbolism & Design*, 6–10, the shadouf, for example, is not an Old Kingdom tool.
[62] Wilkinson, *Symbolism & Design*, 14.
[63] Wilkinson, *Symbolism & Design*, 9.
[64] '*Microhabitats*', https://www.anbg.gov.au/lichen/ecology-habitats-micro.html (02/09/2016).
[65] Callaway, *Positive Interactions*, 306–349, Holmgren etal., *Plant Communities*, 1966–1975; Norfolk etal., *Arid South Sinai*, 659–669.
[66] Baridon, *Les Jardins*, 97.
[67] Murray, *Fruits*, 615–616.
[68] Wilkinson, *Symbolism & Design*, 14.
[69] Scene type 6.2 Watering a garden.
[70] Scene type 6.3 Cultivating and harvesting vegetables in a garden; Scene detail: 6.3.8. Irrigation trenches in a garden, depicted frontally; Bárta, *Abusir* V, 79, 93–94, figs. 3.10, 3.16; Duell, *Mereruka* I, pl. 20–21; Kaiser, *Ägyptisches Museum Berlin*, 32 [295], Abb. 295.
[71] Van Lepp, *Artificial Irrigation*? 197–209.

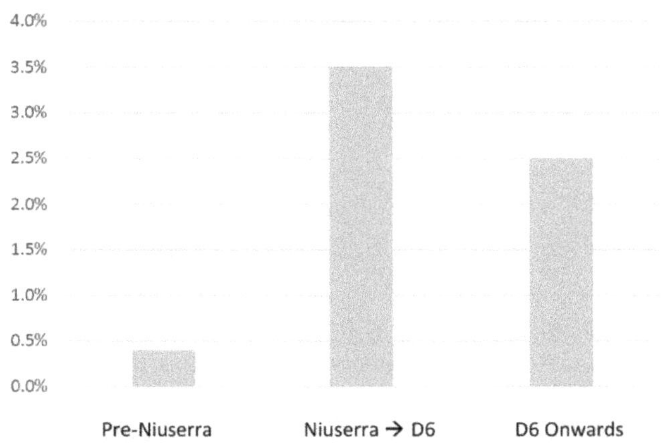

Figure 6.2: Gardens, orchards and the growing of food: proportion change over time.

The scene types may in fact display a narrative tracing the steps needed to result in a bountiful garden,[72] with the tomb of Niankhkhnum and Khnumhotep, for example, depicting the sequence from Hoeing to Watering to Harvest.[73] In trying to relate the decoration themes to the hypothesis under investigation, it is important to note that only one of the gardening scene types in the OEE Database (6.4. *Growing papyrus or lotuses in a man-made garden pool*) could be related to anything other than the simple process of maintaining a successful food garden. Perhaps the gardening sequence represents another example of a narrative-type 'instruction manual' for a time when there is enough water to start gardening again.

6.2.2. The Growing of Lotuses

If the A.R.I.D. hypothesis is correct, then the growing conditions for lotuses would have deteriorated. The first depiction of the gathering of lotus dates to the late Fifth Dynasty.[74] Interestingly, the growing of lotuses in a 'garden' enters the iconography during the time frame under investigation and is first recorded in the tomb of Niankhkhnum and Khnumhotep, dating to the time of Niuserra. Does the inclusion of this scene type indicate the lotus's increasing scarcity as consequence of an environmental circumstance changing to a situation less conducive to its growth? This may have forced the adoption of cultivated lotus for religious activities.

6.3. The Keeping of Records

With a less successful agricultural output, the surplus upon which good government relied, declined. As a consequence, administrative changes were needed. In the literature, these changes have been discussed widely, and primarily focus on reforms instituted from the reign of Niuserra onwards.[75] Despite the best intentions, the reforms do not seem to have made the government more effective. Declining agricultural output may also explain the increasing relative presence of representations of recording and punishing that began to be associated with agricultural and growing scenes.[76] From the time of Niuserra onwards, depictions of administrative activities, such as measuring, recording, and sealing of produce, decline in abundance. (see Table 6.4). If improved efficiency was important, then from the time of Niuserra onwards, it would be expected that depictions of administrative activities, such as measuring, recording and sealing of produce should have increased in their relative importance. It would be expected that, in times of finite resources more oversight would have been needed; yet this does not seem to be the case (Figure 6.3). This proportion declined significantly over the time frame of this investigation.

Table 6.4: Record-keeping: pre- and post Niuserra

Record-Keeping	Date		
	Pre-Niuserra	Niuserra → D6	D6 Onwards
Proportion of overall attestations	6.4%	5.4%	3.4%

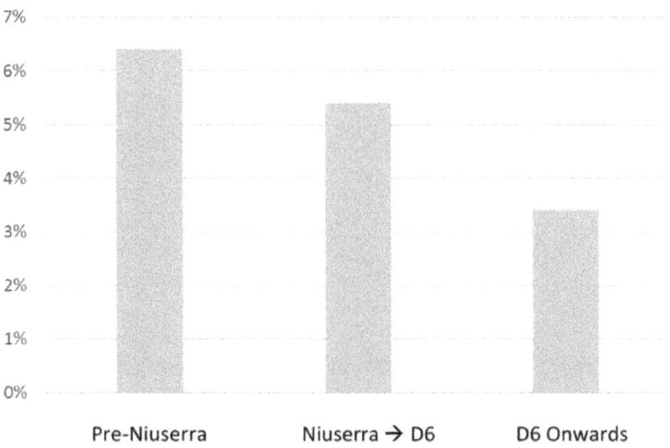

Figure 6.3: Record-keeping: proportion change over time.

6.3.1. The Storage and Distribution of Food Types

When food storage and distribution scenes are examined by produce type, a clearer picture emerges. In comparative terms, the relative proportion of scenes related to wine and oil declines, whereas those depictions representing the administration of grain increase twelve-fold (see Table 6.5).

By the time of the Sixth Dynasty, of those decorations depicting the recording of major produce, those relating to grain became even more relatively significant. Before the time of Niuserra, two-fifths of the depicted of all stored foodstuffs portrayed grain. By the end of the Old Kingdom, almost three-quarters of attestations of recorded foodstuffs depicted grain(see Figure 6.4). This clearly

[72] The database has conveniently ordered them in that sequence: 6.1. Hoeing, to prepare or maintain a garden, 6.2. Watering a garden and 6.3. Cultivating and harvesting vegetables in a garden.
[73] Moussa & Altenmüller, *Nianchchnum und Chnumhotep*, 76–77, Taf. 20-1, Abb. 8.
[74] Moussa & Junge, *Two Craftsmen*, 40, pl. 11.
[75] Krejčí, *Niuserra Revisited*, 513; Bárta, *Niuserra Innovations*, 105.
[76] Müller-Wollermann, *Crime & Punishment*, 228–235.

Table 6.5: Recorded produce: Relative proportion pre- and post Niuserra

Relative proportion of recorded foodstuffs	Date		
	Pre-Niuserra	Niuserra → D6	D6 Onwards
Grain	40%	58%	72%
Wine	20%	35%	28%
Oil	40%	8%	0%

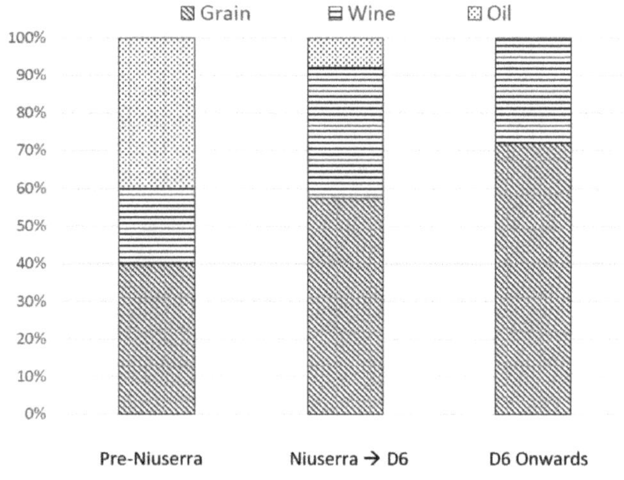

Figure 6.4: Recorded foodstuffs: proportion variance change over time.

Table 6.6: Punishment as a relative proportion of record-keeping scenes

Punishment	Date		
	Pre-Niuserra	Niuserra → D6	D6 Onwards
Percentage of all record-keeping scenes	12%	22%	12%

suggests an increasing preoccupation with the amount of grain available, with oil becoming less regularly depicted, declining to a lack of attestations at all by the end of the period under investigation. This allows for the inferences to be made that accurate records of agricultural produce (and more importantly, seed grain for the next season) had become a more significant consideration to the elite.

6.3.2. Indicators of Food Stress: Punishment Scenes

Very few tombs depict the actual beating of managers and administrators, so it is difficult to assign any significance to the numbers, however, what it symbolises is important. The depictions of estate managers being beaten is first recorded in the tomb of Tepemankh,[77] dated by Harpur to Dynasty Five, to either Neferefra or Niuserra.[78] A slight preponderance of reward and punishment scenes may be linked to the increasing interference of the latter Fifth Dynasty kings in their efforts to improve the efficiency of the bureaucracy and to increase the harvest. Most depictions occur in the Memphite region, and mostly in the tombs of the viziers.[79] Table 6.6 shows the relative proportion of record-keeping scenes where the main focus is upon punishment.

When compared to the mild remonstrance displayed in some late Fifth Dynasty tombs, such as that of Kahai,[80] the scenes depicted on the walls of the Sixth Dynasty tomb of Mereruka display a harsh brutality whose appearance contrasts jarringly to some of the beautiful compositions in the same tomb.[81] The overall proportion of scenes displaying the punitive side of rendering of accounts peaked from the time of Niuserra until the start of the Sixth Dynasty, however they are still a low value. We could interpret these representations as representing the punishment of individuals who, due to incompetence, corruption or disloyalty,[82] have failed to ensure adequate management of the agricultural system. The importance attributed to the administrative minutiae related to the issue and recording of seeds may be an indicator of an increasingly struggling agricultural system and may explain the increasing violence inflicted upon those officials who roused Mereruka's ire.[83]

The scenes on the tomb walls of Khentika also display this unforgiving reality.[84] The punishment theme appears in the later provincial tombs of Pepyankh the Middle and Pepyankh the Black at Meir,[85] where the figures to be punished appear quite fearful and are represented prostrate to receive their beatings.[86] Punishment and humiliation of errant managers may indicate how serious this breach must have been, in the tomb of Khunes at Qubbet el-Hawa, the estate manager is further degraded by apparently having his head pressed down into the dirt.[87]

Earlier southern provincial tombs, at the site of El-Hammamiya, for example, do not depict decorations with this theme, despite these locations having the potential of less successful agricultural prospects. As food surpluses became less reliable, the increasing severity of these representations may be depicting the hardening resolve of

[77] Smith, *Origins Unidentified Reliefs*, 516, fig. 5; MMA, *Egyptian Art*, 404–405, fig. 126.
[78] Harpur, *Decoration*, 169, figs. 160–166.
[79] Harpur, *Decoration*, 220.

[80] Lashien, *Kahai*, pl. 12a.
[81] Duell, *Mereruka*, pl. 37–38, updated Kanawati, *Mereruka & Teti*, pl. 112a–c.
[82] Shafik, *Disloyalty & Punishment*, 181–191.
[83] Kanawati, *Mereruka*, 59.
[84] James, *Khentika*, pl. 9; updated Kanawati, *Mereruka & Teti*, pl. 115.
[85] Blackman & Apted, *Meir* V, pl. xxii [1] and Blackman, *Meir* IV, pl. xv, respectively; Kanawati & Evans, *Meir* II, pl. 83a and Kanawati, *Meir* I, pl. 83.
[86] Kanawati, *El-Hawawish* IV, 21, figs. 12–13 and Kanawati, *El-Hawawish* I, fig. 9, respectively. See also Thompson, *El-Hawawish*, 232; Davies, *Deir el Gebrâwi* I, pl. 8, updated Kanawati *Deir el-Gebrawi II*, pl. 50; Davies, *Deir el Gebrâwi* II, pl. 9, updated Kanawati, in *Deir el-Gebrawi* I, pl. 55 and *Mereruka & Teti*, pl. 113; Davies, *Sheikh Saïd*, pl. xvii.
[87] de Morgan, *Catalogue* I, p. 160 [left]; Harpur, *Decoration*, 220; Kanawati, *Deir el-Gebrawi* II, 36–37.

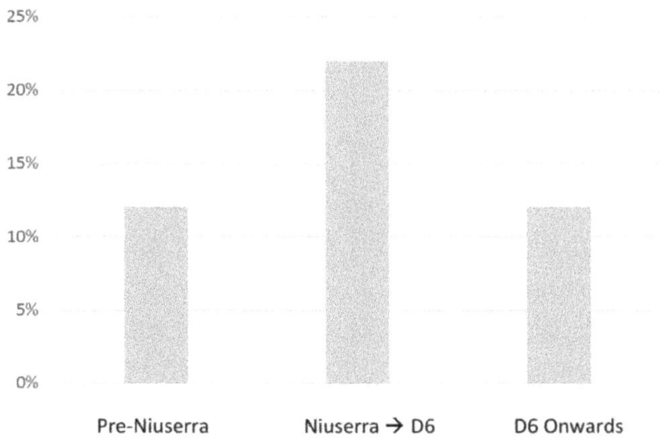

Figure 6.5: Punishment as a relative proportion of record-keeping change over time.

administrators to husband this important resource. These matters were serious enough for the king to re-assign a designated tomb, while leaving the previous owner's name on the outside as a reminder of his disgrace. The unfortunate Ishfu, at Saqqara, for example,[88] was 'punished' in this manner for a misdemeanour of his brother's. However, it is interesting to compare the changing proportion of punishment scenes to that displayed for the decline in representations of agriculture (see Figure 6.5). As noted previously, representations of the storage of agricultural produce decline in total abundance over the time frame in question, following the same trend as that in the aforementioned figure.

6.3.3. Indicators of Food Stress: Granary Shape Changes

While most discussion on granaries is dated to timeframes later than those under investigation,[89] some depictions do exist dating to the Old Kingdom.[90] Dome-shaped granaries are attested in Old Kingdom reliefs from about the time of Niuserra: these are higher and appear more constrained than those depicted earlier, which display a low base, easily reached by the workers. This change of granary shape may indicate changes in the social order, with the higher dome-shaped granaries appearing more difficult to access (and steal from?).[91] This may be another sign of more care taken in the husbandry of this resource. The granaries depicted in the tombs of Ankhtify and Sobekhotep at el-Mo'alla illustrate men climbing ladders to deposit the grain. While at the end of the time frame of the current investigation, this may indicate an increased care in response to famine or the stockpiling of foods as an indicator of preparation for conflict.

6.4. Summation: Changing Climate = Changing Cultivation?

The basic points arising from this chapter include:

- As the river became weaker, less alluvial soil was delivered to the land making large scale cultivation more problematic.
- Less water availability caused irrigation to become less efficacious.
- The rapid decrease in the number of tomb scenes depicting garden and orchards may be an indication of a practical response to a deficiency of available excess water.
- Gardens and orchards, while providing some substantial produce, would not have provided food of the necessary volume to replace that decrease in grain production that developed as a result of water shortage.
- Despite the increased vigilance and agricultural stewardship, suggested by the increased punishment scenes, agricultural success appears to have weakened.

The changes in distribution and abundance of tomb decoration themes may be used to trace the changing agricultural chronicle of the society over the time frame under investigation.[92] A pictorial narrative can be observed depicting how the society responded to changing ecological circumstances leading to a different biological imperative developing within the river. With declining water reliability, the ability to produce excess agricultural products would have declined, [93] so alternative food resources would have been needed.

[88] Shafik, *Disloyalty & Punishment*, 182, 185–186.
[89] Murray, *Cereals*, 526–528.
[90] Siebels, R., 2001. "Representations of granaries in Old Kingdom tombs," BACE 12: 85–99.
[91] Harpur, *Decoration*, 220.

[92] Janick, *Origins of Horticulture*, 28.
[93] Morenó García, *'Irrigation/Irrigations in Pharaonic Egypt: The Interplay between Institutions and Particulars'*, lecture presented at the Twelfth Annual University of Chicago Oriental Institute Seminar Series: 'Irrigation in Early States: New Directions' (03/03/2016).

7

Fishing and A River In 'Drought'

As noted earlier, tomb decoration themes associated with fishing increased significantly in both abundance and overall proportion of total scenes from the latter half of the Old Kingdom. While this assertion may seem counter-intuitive given that the investigation is referring to some form of 'drought', it has been suggested that this was a time of nutrient excess in the river. In this section, the relationship between the river's proposed nutrient retention and an increase in its fishing stocks will be investigated.

7.1. Fishing: Techniques and Related Technologies

Notwithstanding the fish's propensity to generate unpleasant smells,[1] there appears to have been no taboo on its use; the proscription reported by Herodotus, and often quoted… "*…it is not permitted to their priests to taste of fish*"[2] dates from Hellenistic times. Despite Bates claiming "*…it is not a noble food…*"[3] fish appears to have been widely consumed.[4] There may, however, have been some regulations pertaining to who ate it.[5] Fishes may even have been worshipped at Elephantine, as the "*architect or bringer of the flood*",[6] though it seems that no fish were mummified until much later.[7]

In Chapter Five, the rapid increase in proportion of scenes depicting tomb owners fishing and fowling in the marshes was identified, with its apparent importance inferred. The significant change in proportion of the number of scenes depicting the tomb owner spearfishing and fowling may have been influenced by the adoption of royal iconography by the elite, a process started during the time in question. One fish, the Nile perch (*Lates niloticus*), lives in the deeper waters of the river and the other, the Nile tilapia (commonly identified as *Tilapia niloticus*, but more precisely *Oreochromis niloticus,*), is found in shallower, more brackish waters,[8] nor do they inhabit the same locales,[9]

Because most of the figures depicted in tombs are of the elite,[10] and many of the fish depicted on the end of the spear belong to species that do not inhabit areas conducive to spearfishing, these scenes have been interpreted in a symbolic context; with Brewer and Friedman suggesting that the fishing and fowling stance mirrors that of smiting scenes.[11] Since most of the studies about the fishing techniques that were used to catch the ichthyologic species have been based mainly on the funerary iconography, a religious symbolism has been inferred.[12] While there are many fishing scenes on Old Kingdom tomb walls, very few funerary offering scenes depict fish,[13] or offering lists mention fish,[14] so fish may have been reserved for workers and lower-ranking courtiers and not for consumption by the highest officials.[15]

7.1.1. Spearfishing

Despite its limited efficiency, spearfishing is attested throughout the archaeological record of ancient Egypt and dates from the earliest times.[16] Perhaps, then, those representations of spearfishing could be more associated with leisure than religion. Whether this imagery depicts actual spearfishing or something more symbolic is a matter for further development by future research. Whatever the reason, however, spearfishing is a representation of the active collection of food. Despite some localised differences, many artists deviated from the traditional composition.[17] There are numerous spearfishing scenes found in the provincial cemeteries dating to the late Fifth Dynasty at Akhmim, and the early to mid-Sixth Dynasty at Deshasha, Deir el-Gebrawi, Zawiyet el-Maiyetin, Quseir el-Amarna, Meir, Dendera and Aswan,[18] as well as at the late Sixth and early Eighth Dynasty provincial cemeteries of Deir el-Gebrawi, Meir, Akhmim, Dendera, Thebes, Qasr el-Sayad, Kom el-Ahmar, El-Gozeriya, Mo'alla, Gohaina, El-Hagarsa and El-Mahasna.[19]

7.1.2. Techniques and Technologies

Many scenes depicted in the latter half of the Old Kingdom have elicited responses from commentators who have recognised similar fishing behaviours among the

[1] Brewer & Friedman, *Fishing*, 15.
[2] Herodotus, *Histories* 7.
[3] Bates, *Ancient Egyptian Fishing*, 211; also Linseele & Zerboni, *Done with Fish*, 237.
[4] Ikram, *Choice Cuts*, 35–36.
[5] Blackman, *Meir* V, 32; Brewer & Friedman, *Fishing*, 15. Willems etal., *Al-Shaykh Sa'id*, 321.
[6] Edel, *Kult im Elefantengau*, 35–42.
[7] Vogelsang-Eastwood, *Textiles*, 296.
[8] Froese, etal. *Lates Niloticus and Oreochromis* niloticus, in http://www.fishbase.org/home.htm (16/11/2015).
[9] Woods, *Marshes*, 5; Brewer & Friedman, *Fishing*, 11.
[10] Non-elite spearfishing is only depicted twice in Old Kingdom tombs, compared to 93 elite depictions.
[11] Brewer & Friedman, *Fishing*, 21; Gamer-Wallert, *Fischkult*, 128–130.
[12] Robins, *Interpreting Egyptian Art*, 50–52; Binder, *Fishing & Fowling*, 116; Soria-Trastoy, *Fishing Tackle Analysis*, 13.
[13] Brewer & Friedman, *Fishing*, 12.
[14] Bates, *Ancient Egyptian Fishing*, 210.
[15] Ikram, *Meat Processing*, 656; Kanawati, *Tomb & Beyond*, 93.
[16] von den Driesch, *Fish Archaeozoology*, 87; Brewer & Friedman, *Fishing*, 21–22.
[17] Vischak, *Locality & Community*, 238. Woods, *Marshes*, 214–216.
[18] Woods, *Marshes*, 192–193.
[19] Woods, *Marshes*, 191–192.

Table 7.1: Fishing & associated technologies: pre- and post Niuserra

Fishing etc.	Date		
	Pre-Niuserra	Niuserra → D6	D6 Onwards
Proportion of overall attestations	10.7%	14.1%	20.0%

more modern consumers of Nile resources,[20] indicating a historical accuracy with regard to the techniques and technologies depicted, with Paterson and Chapman noting the "*potent ecological force*" displayed by fishing in its potential to produce cultural and environmental impacts.[21] Despite some difficulty in identifying the exact species in some of the depictions, it is still possible to trace the changes in depictions of fish and fishing as a whole.[22] Tracing the adoption of new techniques and technologies, the role of fish and fishing within the resource base of the Old Kingdom has been summarised and updated by Swinton[23] and Soria-Trastoy.[24]

The earliest attestation of an individual presenting a fish for examination by another individual is dated to the early Fourth Dynasty, in the tomb of Rahotep.[25] However, representations of fishing and its associated techniques and technologies are relatively rare until the time of Niuserra in the Fifth Dynasty, after which they become increasingly common until the end of the Old Kingdom. By that time, the proportion of depictions representing fishing had almost doubled from that of the first six reigns of the Fifth Dynasty.

The number of scenes representing all forms of fishing and related technologies increased from less than one-seventh (~14%) of all scene types attested at the end of the Fifth Dynasty to one-fifth (~20%) by mid-Dynasty 6 (see Table 7.1). Importantly, depictions of "*Fishermen registering their catch for distribution by scribes*" (OEE Scene Type database category 1.24) enter the repertoire at this time. This apparent importance is noted in the Sixth Dynasty, because it is at that time that the only two recorded attestations of fisherman accounting for their catch to scribes occur.[26] If the environmental situation were such that a 'drought' resulted in the rapid growth of biomass in the river, as proposed, then the proportion of marshland should increase even more than expected.

While it may seem counter-intuitive to suggest that an aquatic activity like fishing increased in perceived

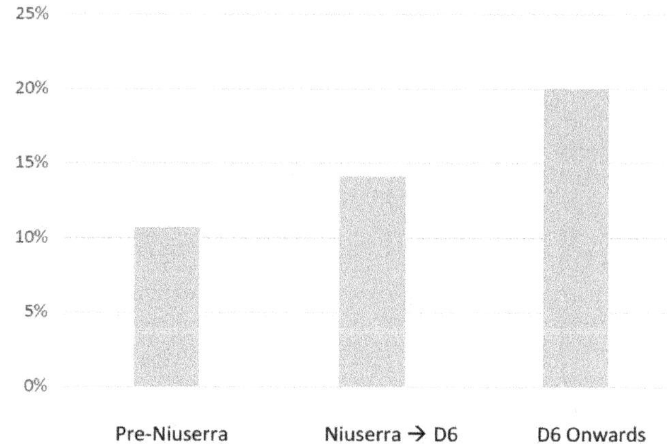

Figure 7.1: Fishing and associated technologies: proportion change over time.

importance during a 'drought', (see Figure 7.1), then the change can be explained as part of the new societal emphasis upon taking advantage of the river's changing environmental circumstances.

Bunbury and Jeffreys suggest that the Egypt of the Pre and Early Dynastic period would have been considerably more 'marshy' than it is today.[27] This suggest that conditions for fish populations would have been more favourable in these times. As the water's edges become more densely populated with aquatic plants such as reeds and papyrus, sources of food for baitfish increases as does potential refuge areas for them increase.[28] This then impacts upon the predator/prey behaviour within fish populations,[29] which could in turn lead to changes in the fishing culture of the local human populations.[30]

As larger fish become increasingly attracted to growing numbers of baitfish, the benefit of an increase in fishing activity increases. Linseele and Zerboni, for example, in their investigation of fish consumption in upper Sudan, identify changes in fishing strategies and exploitation that occurred as a result of changing environmental conditions such as a weaker river and more extensive plant growth along the edges.[31] As environmental stress[32] or rates of exploitation increase,[33] the expected size of fish should decrease due to the increased rates of fishing that apparently occurred. This contrasts significantly to data collected from an earlier time when the river was flowing more swiftly, where the bones were fewer in number but larger in size, suggesting they were from open-water fishes.[34]

20 Bates, *Ancient Egyptian Fishing*, 220–221, 241, figs 85–86, 251; Rzóska, *Nile*, 336–346.
21 Paterson & Chapman, *Nile Perch*, 308. See also Bates, *Ancient Egyptian Fishing*, 203.
22 van Elsbergen, *Fischerei*, 133–137.
23 Swinton, *Resources*, 181–182.
24 Soria-Trastoy, *Fishing Tackle Analysis*.
25 Harpur, *Maidum*, 104, 114, 209–210, fig. 95, pl. 54–55.
26 von Bissing, *Gem-ni-kai*, Taf. 18–19; Altenmüller, *Mehu*, Taf. 34, 36–37.

27 Bunbury & Jeffreys, *Ancient Landscapes*, 66; Stanley & Warne, *Nile Destruction Phase*, 628–634.
28 Paterson & Chapman, *Nile Perch*, 316.
29 Paterson & Chapman, *Nile Perch*, 315.
30 Paterson & Chapman, *Nile Perch*, 308.
31 Linseele & Zerboni, *Done with Fish*, 237–238.
32 Luff & Bailey, *Catfish Growth Structures*, 821–822.
33 Luff & Bailey, *Catfish Growth Structures*, 831–832.
34 Linseele & Zerboni, *Done with Fish*, 235–238.

If the rise in fishing-related scenes in the Old Kingdom reflected environmental changes at the time, then the A.R.I.D. hypothesis could explain changes not only in the artistic record but also in the archaeological record. While identifying the actual species of some fishes in the archaeological evidence poses difficulties,[35] the proportion of fish bones at a site gives an indication of the importance of fish as a source of protein.[36] Over the time frame in question, an increase in the abundance of fish bones in refuse dumps has been anecdotally noted in a number of archaeological sites, for example Edfu and Deir el Barsha,[37] suggesting the increasing prominence of fish in the diet of the local communities:[38] unfortunately, this data has yet to be published. It corresponds to artistic evidence indicating an increased emphasis upon the preservation of fish.[39] Some studies provide a wide-ranging overview of ancient Egyptian dietary patterns; however, the data is taken from widely disparate eras, with some studies not displaying data from the Old Kingdom,[40] or presenting data from only one location or from only one tomb.[41] This is unfortunate in that it only allows for a generalist conclusion such as "*no discernible change over the entire history*"[42] to be made, not allowing for potential changes within smaller time frames.

7.1.3. Adaptations to Fishing Performance and Practice

Of all the fishing and related scenes depicted before the time of Niuserra, the only non-elite representations were scenes depicting seine/dragnet activities. From the time of Niuserra till the end of the Fifth Dynasty, this proportion changed, while during the Sixth Dynasty, while the majority of representations are scenes depicting small groups or individuals in the act of fishing using various techniques (see Table 7.2).

As a consequence of changing environmental circumstances, the increasing variety of habitats that would have developed, as suggested by the A.R.I.D. hypothesis, may explain the increasing assortment of techniques and technologies adopted and adapted at this time, including representations of non-elite individuals spearfishing.[43] At a time when other indicators suggest that ancient Egypt was (supposedly) in the grip of a severe and long-term drought, the summary artistic data presented so far in this study suggests that there was an increase in both the amount and variety of marshland-themed decorations, as well as those scenes depicting fishing and fishing techniques and technologies.

Table 7.2: Fishing techniques: change in overall proportion

Technique Groups	Date		
	Pre-Niuserra	Niuserra → D6	D6 Onwards
Seine/ Dragnets	4.7%	3.8%	3.5%
Traps & other nets	0%	2.1%	2.8%
Angling	0%	1.2%	2.1%
Spearfishing (non-elite)	0%	0%	2.0%

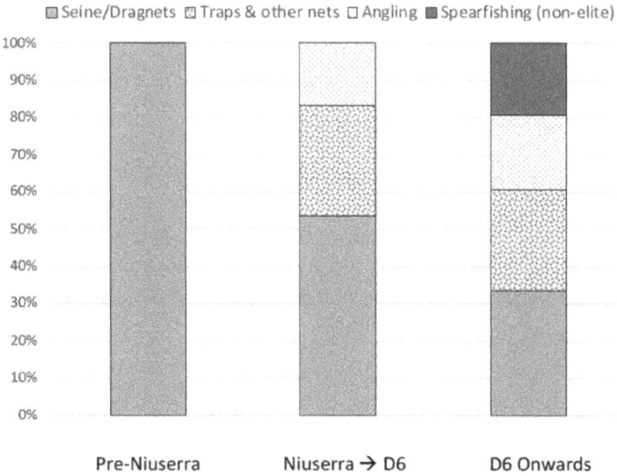

Figure 7.2: Fishing technique groups: relative proportion change over time.

7.2. The Narrative of the Nets

From the time of Niuserra onwards, other techniques are added to the artistic lexicon. Large nets set from boats enter the repertoire at the time of Niuserra, first attested in the tomb of Ptahshepses, at Abusir.[44] Large nets appear to be always set from boats, for example those found in the tombs of Ptahshepses and Mehu (late Pepy I) at Saqqara,[45] have up to eight individuals depicted involved in the activity. Small nets and traps as well as representations of individuals fishing without nets are added to the scene type collection at this time. Small nets and traps require fewer people to operate, whereas large, seine, and dragnets described require many more men. Of all decorations depicting nets, representations of smaller nets increased proportionately across the time frame under investigation (see Figure 7.2).

The significant increase in the number and proportion of images representing certain fishing techniques and technologies can perhaps be explained by the same forces that saw representations of seine and drag nets decline

[35] Greenwood, *Nile Fish Fauna*, 127–133; Linseele & Zerboni, *Done with Fish*, 236.
[36] Nash, *Aquatic Animals*, 456–457; Sahrhage, *Fishing in Ancient Egypt*, 922; Thompson etal., *Egyptian Palaeodiet*, 453.
[37] Gregory Marouard (pers. comm.) and Marleen de Meyer (pers. comm.), respectively.
[38] Luff & Bailey, *Catfish Growth Structures*, 831–832.
[39] Linseele & Zerboni, *Done with Fish*, 235.
[40] Thompson etal., *Egyptian Palaeodiet*, 454, table 1.
[41] Thompson etal., *Egyptian Palaeodiet*, 455, table 2. Stewart, *Fossil Fish*, 678; Osman & Kloas, *Catfish Water Quality*, 390–391.
[42] Thompson etal., *Egyptian Palaeodiet*, 461; Touzeau etal., *Ancient Egyptian Diet*, did not include samples from the First Intermediate Period.
[43] Duell, *Mereruka* II, pl. 127–128.
[44] Vachala, *Ptahshepses*, 138–139 [J1112 + 1113 (2613)].
[45] Altenmüller, *Mehu*, 136, Taf. 35 [a].

at the same time. Despite the use of smaller cast nets having been recorded in ethnographic studies along more southern parts of the modern Nile,[46] this activity has not been recorded on surviving tomb walls.[47] The upkeep of the apparatus of fishing must have been a continuous endeavour.[48] Evidence of the construction, repair and maintenance of nets is found in the archaeological record before Dynastic times,[49] yet representations of these activities were only introduced into the iconographic record at about the time of Niuserra.[50] Smaller and more diverse nets enter the decorative programme from the time of Niuserra onwards, with the tomb of Niankhkhnum and Khnumhotep, for example, containing depictions of all the types of fishing techniques and technologies in the OEE database.[51]

7.2.1. Dragnetting

As identified earlier, and worthwhile mentioning here, the only fishing techniques using nets depicted before the time of Niuserra were those portraying large seine or drag nets. Drag /seine net scenes depict many men working in unison to haul up the harvest: both Ptahshepses and Mehu, among others, also depict this activity.[52] Depictions of seine or dragnetting significantly decrease in proportion over the time frame in question, (see Table 7.3), whereas a variety of new fishing techniques and technologies enter the decorative programme. Of all the scenes associated with fishing technologies, dragnets and seine nets decline from accounting for all of the net-based representations depicted before the time of Niuserra, to about two-thirds of the sample by the time of Niuserra to the end of Dynasty Five. From Dynasty Six onwards, they comprise just over one-half of the representations.

A decline in representations of these activities (see Figure 7.3) can be explained by the hypothesised ecological processes of declining open-water spaces and increasing marshland too shallow or plant-clogged for net utilisation. Some of the important requirements for successful dragnet or seine net operation include relatively open spaces of water in which to lay out the nets, and a waterway edge clear enough to drag the net onto shore.

It seems that seine netting occurred when the water was at its lowest,[53] but with no inundation and a comparative increase in vegetation, perhaps a lack of wide, open-water spaces made this activity more difficult. With an increase in vegetation along the riverbanks, it would also seem

Table 7.3: Dragnet and seine net: proportion change over time

Drag/seine nets	Date		
	Pre-Niuserra	Niuserra → D6	D6 Onwards
Relative proportion of all net scenes	100%	64%	56%

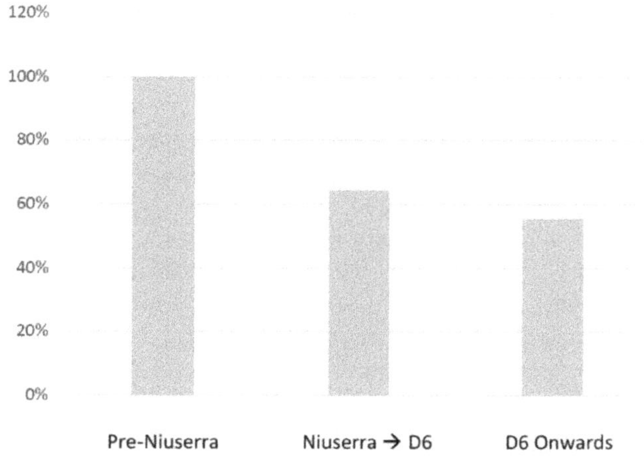

Figure 7.3: Drag and seine net: Relative proportion change over time.

problematic to land a large seine net without entanglement. While an increasing emphasis on fishing was displayed on tomb walls,[54] dragnet scenes as a proportion of total fishing scenes declined, and slight changes in the composition can be recognised.[55] Despite the decline in overall abundance of dragnet scenes, however, the portrayed size of the actual dragnet itself increased, taking up more room within the tomb,[56] and the numbers of haulers also tended to increase.[57] In the tomb of Ty, (from the time of Niuserra), the dragnet haulers number twelve, compared to nineteen in one scene and twenty-three in another in the tomb of Mereruka, (from the time of King Teti) for example.[58] An increased nutrient load in the river should increase the overall fish biomass. Therefore, improved catches would be expected, so the nets should be fuller, leading to the haulers requiring extra assistance.

With the decline in the relative number of scenes depicting dragnetting, another feature enters the iconography: from the time of Niuserra onwards, some dragnet figures are depicted with a strap across their shoulders,[59] presumably to aid them in their hauling. This addition to the decorative repertoire may indicate that these larger dragnets were being increasingly set away from the riverbank, away from the entanglement of the encroaching marshes. On land, friction between feet and the ground provide enough

46 Elster & Jensen, *Fishery Investigation*, 91.
47 Brewer & Friedman, *Fishing*, 41.
48 Greenwood, *Kenyan Fishes*, 56.
49 van Neer, *Prehistoric Fishing*, 251–269. 49–53; Barich, People, *Water & Grain*, 47–48.
50 Soria-Trastoy, *Fishing Tackle Analysis*, 13–18.
51 Moussa & Altenmüller, *Nianchchnum & Chnumhotep*, 96–101, Taf. 30–32, Abb. 12–13.
52 Vachala, *Ptahshepses*, 130–131 [A559 + J466], 132-3 [C814], 134-5 [E820-2, E1438], 136-7 [I138, J635, J563, J482]; Altenmüller, *Mehu*, 136–137, Taf. 32 (part), 31 [a] (part), 35 [a] (part), 35 [b].
53 van Neer, Prehistoric Fishing, 54.
54 Kanawati & Abder-Raziq, *Waatetkhetor*, 21.
55 Harpur, *Decoration*, 143–145.
56 Harpur, *Decorations*, 189–196.
57 Harpur, *Decorations*, 179, 259.
58 Wild, *Ti*, II, pl. 40–41, Duell, *Mereruka* II, pl. 43, 55; Kanawati *Mereruka & Teti*, pl. 107, 110.
59 Harpur, *Decorations*, 189.

leverage to aid the fishermen in hauling in their nets. From personal observations of modern fishermen in Kenya and Uganda,[60] these straps are commonly worn by fishermen hauling large nets onto boats, to compensate for the lack of purchase that solid land may give.

The use of the strap may suggest that the catch was bigger, and the strap added stability and provided extra leverage which helped the haulers land their catch. Dragnet haulers wearing supportive straps are also depicted in the Giza tombs of Iynefert,[61] Khufukhaef II,[62] Senedjemib/Mehi,[63] and Sekhemka;[64] and the Saqqara tombs of Akhethotep,[65] Irenkaptah,[66] Ty,[67] Ptahhotep/Iyniankh,[68] Mereruka,[69] Waatetkhetor,[70] Niankhkhnum and Khnumhotep,[71] Nebkauhor/Idu,[72] Hetepherakhti,[73] Kaemrehu.[74] Tombs further south displaying the use of straps by dragnet haulers include In-snefru-ishtef from Dahshur,[75] Werirni from Sheikh Said,[76] Pepyankh the Black,[77] Pepyankh the Middle[78] and Nyankhpepy[79] from Meir, and Ibi[80] and Djau[81] from Deir el-Gebrawi.

7.2.2. Large Nets Set from Boats

Depictions showing the setting of large nets from boats enter the artistic lexicon from the time of Niuserra.[82] Perhaps the adoption of large boat-set nets may have been a transitional strategy to maintain the 'industrialised' process of large-scale fishing. These include the tombs of Ptahshepses,[83] Ty,[84] Kagemni[85] and Mehu.[86] The combined tomb of Niankhkhnum and Khnumhotep also displays this scene.[87] Further south, the tombs of Ibi[88] and Djau/Shemai[89] at Deir el-Gebrawi also portray this scene type. In the tomb of Waatetkhetor at Saqqara, the haulers are shown standing, not on a boat, but depicted on a structured surface[90] that does not appear to be the riverbank, with the surface apparently supported by another structure; a similar depiction occurs in the tomb of Werirni at Sheikh Said.[91] The right-most figure hauling in the net is on his knees and lifting, rather than pulling, the net: this suggests that the hauler is on a boat, not the land. Similar representations are shown in the later tombs at Meir of Pepyankh the Middle,[92] where the haulers appear in front of, not within, the papyrus band. The scene in the tomb of Niankhpepy the Black,[93] where there is no representation of plants at all, perhaps indicates that the fishing is done beyond the plant zone; and Pepyankh the Black,[94] who also does not depict plants.[95] At Deir el-Gebrawi, the tombs of Ibi[96] and Djau/Shemai[97] also display similar indicators of fishing from a boat, not the land. The addition of large nets to the decoration repertoire may suggest that fishing was increasingly carried out away from a riverbank that had become increasingly clogged and impenetrable, and this increasing reliance on large nets may have necessitated a change in the techniques of using them.

7.2.3. The Need to Re-classify all Large Nets used for Fishing?

To avoid obstruction from blockages below the surface, floats are important.[98] Similarly, perhaps in response to the increased obstruction that would be present below the surface in a plant-clogged river, some nets are now not depicted with sinkers.[99] If no sinkers are needed, neither are floats. Kanawati identifies that this phenomenon is depicted in a number of tombs dating from mid- to late Dynasty 6,[100] which is interesting because the earlier tombs of Gehesa/Nebi and Theti/Ikr do depict these accoutrements.[101]

The representations changed at around the same time as suggested by the A.R.I.D. hypothesis. While this may be

[60] Persona lobservations. Also Bates, *Ancient Egyptian Fishing*, 208–209, 258–262; Kanawati, *Deir el-Gebrawi* III, 30.
[61] Schürmann, *Ii-nefret*, Abb. 9 [a, b], 21.
[62] Simpson, *Kawab*, pl. 36[a], 37[a], figs. 43, 47, 49.
[63] Brovarski, *Senedjemib* I, pl. 115–116, figs. 114–116.
[64] Simpson, *Western Cemetery* I, pl. v [c-e], fig. 4.
[65] Ziegler, *d'Akhethetep*, 82–84, 142–143.
[66] McFarlane, *Irukaptah*, pl. 16, 41, 46.
[67] Wild, *Ti*, II, pl. 90 [A, B] (details), 91 [A] (detail), 123. See also Woods, *Marshes*, 87, n. 67; 114, n. 109; 117, n. 143.
[68] Hassan, *Saqqara* II, figs. 36–7, pl. 77 [A].
[69] Duell, *Mereruka* II, pl. 55 = Scene Two.
[70] Kanawati & Abder-Raziq, *Waatetkhetor*, pl. 56, updating Wreszinski, *Atlas III*, Taf. 95.
[71] Moussa & Altenmüller, *Nianchchnum & Chnumhotep*, Taf. 31, 36 [b], Abb. 12.
[72] Hassan, *Saqqara* I, pl. 5 [A], 24 [A].
[73] Mohr, *Hetep-her-akhti*, fig. 29, pl. 1.
[74] Mogensen, *Le Mastaba*, fig. 3, pl. 1.
[75] de Morgan, *Dahchour* II, pl. 23.
[76] Davies, *Sheikh Saïd*, pl. 12.
[77] Blackman & Apted, *Meir* V, pl. 30, 60 [2].
[78] Blackman, *Meir* IV, pl. 8.
[79] Blackman & Apted, *Meir* V, pl. 13.
[80] Davies, *Deir el-Gebrâwi* I, pl. 4.
[81] Davies, *Deir el-Gebrâwi* II, pl. 5.
[82] Vachala, *Ptahshepses*, 130–137, figs. A559, C814, E820, E822, E1438, I138, J635, J563, J466, J482; Moussa & Altenmüller, *Nianchchnum und Chnumhotep*, 96–97, Taf. 31, 36 [b], Abb. 12; Wild, *Ti*, II [1], pl. 90, 91, 123; von Bissing, *Gem-ni-kai*, I, 14, 20, Taf. 19, 21; Altenmüller, *Mehu*, 136–137, Taf. 31, 32, 35a, 35b; Davies, *Deir el-Gebrâwi* I, 12–13, pl. 4, updated Kanawati, *Deir el-Gebrawi* II, 29, pl. 45.
[83] Vachala, *Ptahshepses*, 138–139, [J1112 + 1113 (2613)].
[84] Wild, *Ti*, II, pl. 77 [A] (part), 61.
[85] von Bissing, *Gem-ni-kai* I, 14, 20 [4749], Taf. 18–19.
[86] Altenmüller, *Mehu*, 136, Taf. 35 [a].

[87] Moussa & Altenmüller, *Nianchchnum & Chnumhotep*, Taf. 31, Abb. 12.
[88] Davies, *Deir el-Gebrâwi* I, 14, pl. 6; Kanawati *Deir el-Gebrawi* II, 29, pl. 47 confirms this suggestion.
[89] Kanawati, *Deir el-Gebrawi* III, 30.
[90] Duell, *Mereruka* II, pl. 43, 55; Kanawati & Abder-Raziq, *Waatetkhetor*, pl. 56.
[91] Davies, *Sheikh Saïd*, pls 5, 12.
[92] Blackman, *Meir* IV, pl. 8; updated Kanawati, *Meir* I, 37, pl. 79.
[93] Blackman & Apted, *Meir* V, pl. 13; updated Kanawati etal., *Meir* III, 45, pl. 69.
[94] Blackman & Apted, *Meir* V, pl. 30, 60 [2]; updated Kanawati & Evans, *Meir* II, 50, pl. 90.
[95] Note, Kanawati, does not suggest these interpretations.
[96] Davies, *Deir el-Gebrâwi* I, pl. 4.
[97] Davies, *Deir el-Gebrâwi* II, pl. 5.
[98] Dulíková, etal., *Ankhires*, 10.
[99] Kanawati, *El-Hawawish* I, fig.12; Kanawati, *El-Hawawish* II, figs. 21–22.
[100] Kanawati, *El-Hawawish* II, 45, no. 169.
[101] Kanawati, *El-Hawawish* VII, fig. 30 and Kanawati, *El-Hawawish* I, fig. 12, respectively.

just carelessness on behalf of the artist, it may also be an indication of accuracy on his part. Perhaps this explains why the numbers of haulers decreases as the nets became smaller in order to fit into the diminishing areas of open water. A more careful analysis of all dragnet scenes may identify if the nets themselves are drawn from land or boat-set, as some of the scenes classified as those identified as one particular type of large net, either seine or drag may need to be re-classified.[102]

7.3. Angling, Trapping and Small Nets

As a consequence of a changing environment, it is possible that fishing transformed from a pleasurable activity[103] to a necessary one. In the Fifth Dynasty, a new type of marsh composition was introduced into royal and elite decoration, known as a 'composite fishing and fowling scene'.[104] Woods identifies a royal relief dating to the time of Userkaf but points out its adoption by the elite was not until the mid-Fifth Dynasty.[105] Perhaps the increasing instances of the removal of wives from the composition of spearfishing scenes from the mid-Fifth Dynasty onwards[106] is an indicator of the less-than-gratifying aspects of the activity and an increasing focus on the practical necessities of the task.

7.3.1. New Scene Types: New Techniques and New Technologies

As a consequence of changing environmental circumstances leading to the development of more varied habitats, the adoption of smaller nets and nets more suitable to narrower channels can be explained by the application of the A.R.I.D. hypothesis outlined in Part B. A greater variety of techniques and technologies was adopted to utilise this resource more effectively. A number of new fishing techniques entered the decoration lexicon in the reign of Niuserra – angling, new net forms and new types of traps (funnel and basket). Fishing equipment appears to be less complex and less sophisticated from the time of Niuserra onwards, suggesting minimal maintenance (see Table 7.4). The types of traps depicted are smaller and need fewer people to operate them effectively. The contention is not that it was impossible to use dragnets, seine nets or large nets, rather that in many places along the river, it was more convenient to use smaller, more manoeuvrable ones. The representation of hand nets, which are quite effective and do not require many operators,[107] changed little from the Old Kingdom to the Middle Kingdom.[108] Depictions of funnel traps in the tomb of Ptahshepses,[109] basket traps in the tomb of Ty,[110] hand-held nets in the tomb of Kagemni[111] and large nets in the tomb of Mehu,[112] for example, all depict activities that could be undertaken by a smaller group, rather than the larger body of men who needed to be organised for dragnetting or seine netting,

Interestingly, the combined tomb of Nianchchnum and Khnumhotep depicts all of the new fishing technologies as well as those more traditional. Even the tomb of Ty, from the same era, only depicts four of the new activities. It is not until the Sixth Dynasty that tombs are prepared depicting more than two of these new scene types. The large tombs of Kagemni and Mereruka depict most of the new fishing technologies as well as the more traditional dragnetting scene. Similarly, at Saqqara, the tomb of Hesi, despite its small size, contains most of these themes, as does the tomb of Mehu. At Dahshur, the tomb of Snefru-in-ishtef displays indications of a more individualised fish acquisition programme, including the use of large and small nets, handheld nets, as well as scenes depicting a (non-elite) individual angling.[113] Further south, the tomb of Ibi, at Deir el-Gebrawi, also portrays these innovations, and includes illustrations of the use of funnel nets.[114]

7.3.2. Smaller Traps Take Advantage of New Channels

If a rapid growth of papyrus and reeds at the edges of the river took place, as Chapter Four proposes, large dragnets would have been difficult to land on the increasingly crowded riverbanks. Smaller dragnets that could be pulled onboard rivercraft may have been adopted, necessitating the fishermen to venture further towards the middle of the river. Individuals unwilling to venture so far from safety may have been forced to adapt or modify the apparatus that they were currently using to take advantage of the new situation.

It is possible that the adoption of the funnel trap was a pragmatic response to the increasing number of smaller canals and secondary channels developed as a result of a weakening river. Smaller, more manageable traps and nets may have been a more pragmatic option for people relying on fish meat as one of their major sources of protein. Intriguingly, of those individuals identified above, all contain a funnel trap scene in addition to scenes depicting the setting of large nets from boats.[115]

[102] Woods, *Marshes*, 388.
[103] Davies, *Deir el-Gebrâwi* I, 12; Harpur, *Decoration*, 181, n. 131; Decker, *Sport & Games*, 158–167; Kanawati, *Tomb & Beyond*, 92–94.
[104] Harpur, *Decoration*, 140.
[105] Woods, *Marshes*, 13, Appendix 3: #4.
[106] Woods, *Marshes*, 54. See Roth, *Absent Spouse*.
[107] Brewer & Friedman, *Fishing*, 40.
[108] Brewer & Friedman, *Fishing*, 38–39.
[109] Vachala, *Ptahshepses*, 138–139 [J1092].
[110] Épron & Daumas, *Ti*, I, pl. 9.
[111] von Bissing, *Gem-ni-kai I*, Taf. 4 [1, 2].
[112] Altenmüller, *Mehu*, 136, Taf. 35 [a].
[113] Borchardt, *Cat. Caire II*, Bl. 103 [CG 1771]; de Morgan, *Dahchour*, II, pl. 24.
[114] Davies, *Deir el-Gebrâwi* I, 12–14, pl. 4, 6, 11; Kanawati *Deir el-Gebrawi* II, 28–29; pl. 7–10, 45–46.
[115] Vachala, *Ptahshepses*, 138–139 [J1092]; Moussa & Altenmüller, *Nianchchnum & Chnumhotep*, 100, Taf. 31, Abb. 12; Wild, *Ti*, II, pl. 77 [B], 61; von Bissing, *Gem-ni-kai* I, 13–14, 20 [43–6], Taf. 17–18; Altenmüller, *Mehu*, 140–141, Taf. 35 [a], 38; Davies, *Deir el-Gebrâwi* I, 14, pl. 6.

Table 7.4: Angling, traps and nets: Niuserra onwards

Name	Date	LARGE NET	ANGLING	SMALL NET	FUNNEL TRAP	BASKET TRAP
Ramaka	V.6					X
Ptahshepses	V.6L-7	X			X	
Niankhkhnum/Khnumhotep	V.6L-7	X	X	X	X	X
Hetepherakhti	V.6-8E		X			X
Ty	V.7-8E	X	X		X	X
Kaemnefert	V.7-9E		X	X		
Senezemib : Inti	V.8M-L		X	X		
Akhethotep	V.8L-9E		X	X		
Hesimin (M 22)	V.8-VI.1		X			
Khenut	V.9		X			
Iynefert : Shanef	V.9		X			
Senezemib : Mehi	V.9		X	X		
Seshseshet : Idut	V.9		X	X		
Ptahhotep : Thefu	V.9L		X	X		
Seshemnefer [IV]	V.9-VI.1M		X			
Nefer [I]	V.9-VI.1			X		
Nimaetre	V.9-VI.1		X	X		
Kagemni : Memi	VI.1E-M	X	X	X	X	
Mereruka : Meri	VI.1M		X	X	X	X
Snefruinishtef	VI.1?		X	X		X
Hesi	VI.1L-2E		X	X	X	
Uzahat..Nefer..Sheshi	VI.1L-2E		X			
Nikauisesi	VI.1L-2E		X			
Kairer	VI.2			X		X
Mereri	VI.2E		X			
Meryre-meryptahankh : Nekhebu	VI.2		X			
Mehu	VI.2M-3?	X	X		X	X
Tomb G 97	VI.2L-3		X			
Bawi (?)	VI.2-4E		X			
Seshemnefer : Ifi	VI.2-4?			X		
Ibi	VI.4E	X	X	X	X	
Djau:Shemai	VI.4L		X	X		
Pepyankh : Heny-kem	VI.4-5		X	X		
Kahep : Theti-iker	VI.5-6		X	X		
Shepsipumin : Kheni	VI.6		X	X		
Unisankh	VI.6		X			

Mereruka's tomb contains one of these scenes, as does the tomb of Hesi.[116]

An area for further investigation may be the increased likelihood of the adoption of an attendant (incidental?) form of aquaculture during this time of riverine nutrient enrichment. This will be discussed further in Chapter Nine, where the role of cattle in the development of 'incidental aquaculture' will be discussed.

[116] Duell, *Mereruka* I, pl. 42–43, 45; Kanawati & Abder-Raziq, *Hesi*, pl. 21–22, 24–25.

7.4. Fighting Boatmen

The image of the fighting boatman is one of the typical marshland scenes depicted in ancient Egyptian tomb paintings,[117] (see Table 7.5). While it may represent a sport or physical activity,[118] others suggest a deeper symbolic

[117] Harpur, *Decoration*, 141–155 and 339–350, tables 6.19–6.24; Bolshakov, *Boatmen Jousting*, 30–37. Herb, *Der Wettkampf in der Marschen*, 28–51.

[118] Montet, *Les scenes*, 81; Vandier, *Manuel* V, 510–531; Decker, *Bildatlas zum Sport*, 533–545; Harpur, *Decoration*, 153.

Table 7.5: Boatmen depictions: pre- and post Niuserra

Boatmen depictions	Date		
	Pre-Niuserra	Niuserra → D6	D6 Onwards
Proportion of overall attestations	5.2%	3.7%	2.4%

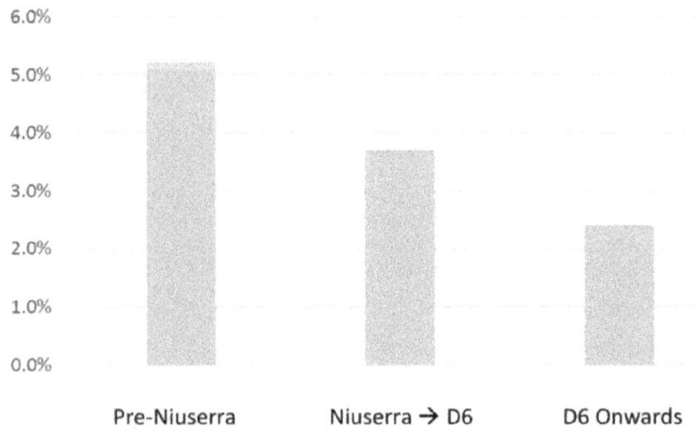

Figure 7.4: Boatmen depictions: proportion change over time.

meaning, that the scene is thought to represent.[119] There are two basic scene types: those depicting boatmen returning peacefully and those depicting fighting or jousting.

Interestingly, in the light of the A.R.I.D. hypothesis, the overall proportion of scenes depicting boatmen decreases over the time frame in question with the overall proportion of decorations that depict boatmen declining to less than half of the proportion before Niuserra (see Figure 7.4). This rapid decline in representations may suggest a decreasing relative importance to the tomb owner This may also indicate that this activity may have diminished as a consequence of a less open, more crowded river. However, it would be expected that a more crowded river should have led to increased competition for open space and precipitated more aggressive responses in the boatmen.[120] Before the time of Niuserra, the majority of scenes representing boatmen depict them returning peacefully. From the time of Niuserra onwards, there is a significant decrease in the proportion of scenes depicting boatmen engaged in peaceful interactions (see Table 7.6).

7.4.1. Boatmen and A River In 'Drought'

As the overgrowth of papyrus clumps and reed mats entered the river proper, they would have restricted clear passage for any large watercraft using them, just as the proposed space restrictions would have limited the use of large nets. Using the term 'fighter' to describe boatmen fighting or jousting may seem inappropriate, as it suggests

Table 7.6: Boatmen activities: pre- and post Niuserra

Boatmen Activities	Date		
	Pre-Niuserra	Niuserra → D6	D6 Onwards
Proportion overall	5.2%	3.7%	2.4%
Proportion fighting	1.3%	2.0%	1.2%
Proportion peacefully returning	3.9%	1.7%	1.2%

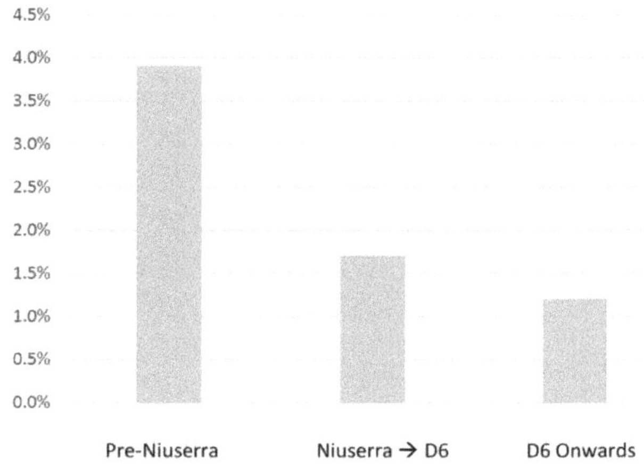

Figure 7.5: Boatmen fighting: proportion of boatmen returning peacefully over time.

something beyond friendly rivalry and competition (see Figure 7.5).

However, with less open space, perhaps there developed increased competition for clear water in which to fish or lay the larger nets which were the most common form of nets depicted before the time of Niuserra. It may be that good-humoured rivalry developed into something more antagonistic as fishing intensified as a major resource activity. Whether the representations were metaphoric or more visceral, the composition changed over the time frame in question.[121] Perhaps competition for resources and space occurred alongside the competition to be the first crew to deliver the gifts to the tomb owner. Perhaps the testicular manipulation depicted in some tombs[122] is an indication of the increasing fierceness of the competition for resources.

7.5. Catfish, Drought and Art

The only fish represented from a (mostly) top-down, as opposed to a side-on, view, is the 'walking' catfish (*Clarias gariepinus*); with both eyes of this catfish usually represented in Egyptian art.[123] The other two main catfish species depicted in Egyptian art are the 'African'

[119] Bolshakov, *Boatmen Jousting*, 37–39.

[120] Duell, *Mereruka* II, pl. 13; Kanawati & Abder-Raziq, *Unis*, II, pl. 62; Moussa & Altenmüller, *Nianchchnum & Khnumhotep*, pl. 74–75; Simpson, *Kaiemnofret*, pl. G; Moussa & Altenmüller, *Nefer & Kahay*, 26, pl. 10–11; Brovarski, *Senedjemib*, I, 44, figs. 38–39. Gardiner, *Ancient Athletics*, 8, calls it a sport but describes the pole as a 'weapon'.

[121] Harpur, *Decorations*, 154–155; Woods, *Marshes*, 43–45.

[122] Kanawati, etal., *Inumin*, pl. 17, 48; El-Khouli & Kanawati, *Saqqara* II, pl. 40; Kanawati & Woods, *Artists*, 56, pl. 99; Kanawati, *Mereruka & Teti*, 40–41, pl. 70 [a–b].

[123] Duell, *Mereruka*, pl. 43 and Kanawati & Abder-Raziq, *Hesi*, pl. 53.

catfish, *Synodontis schall*, and the 'upside-down' catfish, *Brachysynodontis batensoda,* more commonly identified as *Synodontis batensoda* in Egyptological literature; and these are always represented side-on.[124] Catfish have an organ that allows them to live for short periods of time on air.

7.5.1. Upside-down and Top-down/Both-eyes Catfish Scenes

The upside-down catfish is drawn in an eccentric manner, allowing it to be distinguished from other catfishes in wall scenes.[125] Depiction of upside-down catfish appears to have begun during the reign of Niuserra.[126] The image displayed on the walls of the tomb of Waatetkhetor shows the animal upside down at the surface of the water.[127] While some of this creature's main food sources are on the surface of the water, it is important to note that the surface of the water, the boundary between the air and liquid environments, also contains a higher concentration of oxygen.[128]

It would be expected that the greater number of smaller secondary channels would have experienced a faster-than-normal diminution in the levels of oxygen, leading to more instances of catfish gathering at the surface. Contamination due to excess nutrients could have also become a source of distress.[129] Increasingly crowded waters, due to the papyrus/reed bloom that developed in a weakening river, could have led to a decline in oxygen content, forcing the catfish to migrate between now unconnected secondary channels.[130] Despite the potential formation of environmental oxygen stress due to increasing nutrients in the water, recent experimental work has demonstrated that the average body size of catfish increases in response to the presence of certain 'contaminants' in the water supply.[131]

The catfish's omnivorous diet improved its odds of finding food and adapting to changing food webs,[132] increasing its presence within the local ecosystems. While the catfish's higher chances of surviving inhospitable micro-habitats may have given it an adaptive advantage,[133] its improved success may have resulted in it becoming, through simple probability, a larger part of the diet of the Egyptian population.[134] Noted was the significant number of catfish bones recorded in an archaeological context, a huge proportion compared to other types of fish.[135] This conclusion has already been drawn by Luff and Bailey, suggesting that the bone records in spoil heaps point to an increasing role of that particular species as a source of protein.[136] This evidence may corroborate the evidence of increasing catfish representations and/or vice versa. Some of the catfish species have been noted as 'air gulpers',[137] which allows them to break the surface of the water for some small amount of time: this behaviour may have been interesting enough to have resulted in its peculiar representation on the walls.[138] If the hypothesis under investigation is valid, then the consumption of catfish would have increased and should be more noticeable in the archaeological record. Interestingly, in modern day Botswana, in times of a low river, the catfish becomes the main source of fish protein.[139]

7.5.2. Catfish Depictions and A River In 'Drought'

Catfish representations increased perhaps due to their behaviour becoming more idiosyncratic and notable. A potential catfish-related indicator of a changing environment is described in the tomb report of Ankhires, dating to late Fifth Dynasty, from Abusir. Perhaps this explains why the instances of upside-down depictions increase over the time frame in question – the artists were simply drawing what they saw.[140] Another response of some catfish species to low levels of oxygen in waterways might be observable in their propensity to 'migrate', catfish being able to 'walk' on their pectoral fins.[141] Evans suggests that it is when they are observed in this act of locomotion, that the two eyes become obvious and may explain why they are represented in this manner.[142] Fortunately for catfish, their ability to survive out of water for at least twelve hours[143] may have given them the capability to escape from water becoming increasingly uninhabitable. If the A.R.I.D. hypothesis is valid, 'walking' behaviour would have become more frequently observed[144].

7.6. Summation: An Increasing Reliance upon Fishing?

The significant points arising from this chapter include:

- A river that retains its nutrients increases its biological potential, which encourages plant growth.
- This leads to an increase in organisms that use plants for food and/or shelter, which, in turn, results in an increase in these organisms' predators.

[124] Evans, *Animal Behaviour*, 103.
[125] Davies etal., *Saqqara Tombs*, pl. 25; Brewer & Friedman, *Fishing*, 67–69; Evans, *Animal Behaviour*, 163–166; Boulenger, *Fishes of the Nile*, 382; Chapman & Mackay, *Copeia*, fig. 2.
[126] Woods, *Marshes*, 380, Appendix 3: #175.
[127] Duell, *Mereruka*, pl. 43; updated by Kanawati & Abder-Raziq, *Waatetkhetor*, pl. 56.
[128] Chapman etal, *Swimming Upside Down*, 130–135.
[129] Chapman etal., *Limnological Observations*, 1821–1826.
[130] Habib & Samah, *Catfish Protein Synthesis*, 555; Sabri etal., *Catfish Infections*, 897. Awadan etal., *Parasitic Catfish Disease*, 20.
[131] Mona etal., *Catfish Supplementary Feed*, 5–6.
[132] Awadan etal., *Parasitic Catfish Diseases*, 17.
[133] Verreth etal., *Catfish Digestive System*, 296–297; Mona etal., *Catfish Supplementary Feed*, 1.
[134] Luff & Bailey, *Aquatic Basis Ancient Civilisations*, 100–113.
[135] Duliková, etal., *Ankhires*,27.
[136] Luff & Bailey, *Catfish Growth Structures*, 831–832.
[137] Brewer & Friedman, *Fishing*, 60; Evans, *Animal Behaviour*, 48.
[138] Brewer & Friedman, *Fishing*, 69; Woods, *Marshes*, 402–403; Evans, *Animal Behaviour*, 48, n.88–90.
[139] Steward etal., *River Runs Dry*, 204.
[140] Evans, *Animal Behaviour*, 103–104.
[141] Johnels, *Catfish Terrestrial Locomotion*, 122–129.
[142] Evans, *Animal Behaviour*, 48, n. 88.
[143] Stewart, *Fossil Fish*, 683.
[144] Stanley etal, *Nile Flow Failure*, 395–402; Brewer & Friedman, *Fishing*, 60.

- Over the time frame in question, a greater population of fish would develop, making the harvesting of these organisms a more rewarding endeavour.
- The greater availability of fish for harvest would have required fishing techniques to adapt to the new habitat and encouraged adaptation to the use of more suitable technologies such as smaller nets and traps.
- Progressively clogged river margins would have decreased the effective use of large nets that required many operators, such as seine and drag nets.
- The relative proportion of representations of smaller less human intensive technologies coincides with this time.
- The mounting abundance of fish bones, as an overall proportion, in refuse dumps in a number of archaeological sites suggests that fish as a ready source of protein had become progressively more important.
- This dietary change, as suggested in the archaeological record may be evidence of a response to changing ecological circumstances.

As large dragnetting endeavours had begun to decrease, perhaps the organisational skills or administrative apparatus needed to organise such activities would also diminish. Perhaps fishing had become a less 'industrialised' process with an increasing emphasis upon family or small-group-oriented actions. The data suggests that fishing required smaller groups of people, perhaps to avoid overcrowding the secondary channels that had developed.

8

Waterfowl and A River In 'Drought'

As a food resource, birds were a significant supply of protein. Previously, it was identified that the proportion of representations depicting birds and activities related to them declined slightly over the time frame in question. The data suggesting a decline in bird depictions seems to be inconsistent with the hypothesis under investigation and recognised the significant increase in waterfowl depictions. In this chapter, the distribution and abundance of those depictions relating to those birds that may have been directly affected by a changing riverine habitat will be studied in more detail to scrutinise the changing role of waterfowl as a food resource.

8.1. Avian and Associated Attestations

Birds have played an important role in ancient Egypt: in religious[1] and symbolic roles[2] as well as by other more pragmatic means,[3] with their economic role well described.[4] In a similar manner to the way in which depictions of the agricultural cycle appear to be an 'instruction manual', clap-netting seems to be 'instructional' in its portrayals; indeed, the overall process of clap-netting can be understood by amalgamating these types of scenes from many tombs.[5]

Earlier, it was noticed that the relative proportion of scenes produced during the Sixth Dynasty whose themes were linked by their associations to avians decreased by about three percent (3%) compared to the proportion attested in the previous dynasties, a statistically insignificant change. However, if we were to look at the changes across a few more boundaries, a clearer picture may emerge (see Table 8.1). From before the time of Niuserra compared to Niuserra onwards, there is a significant change, increasing by a third during the time of Niuserra and the latter stages of the Fifth Dynasty, then declining slightly during the Sixth Dynasty (see Figure 8.1). Before the time of Niuserra only two tombs contained images of the trapping of songbirds.[6]

During the time of Niuserra, the total attestations with the subject relating to birds increased. As well as a slight increase in overall proportion, new themes entered the repertoire with pigeons, for instance, joining the bird

Table 8.1: Proportion in avian representations: pre- and post Niuserra

Avian & associated scenes	Date		
	Pre-Niuserra	Niuserra → D6	D6 Onwards
Proportion of overall attestations	6.4%	8.4%	7.9%

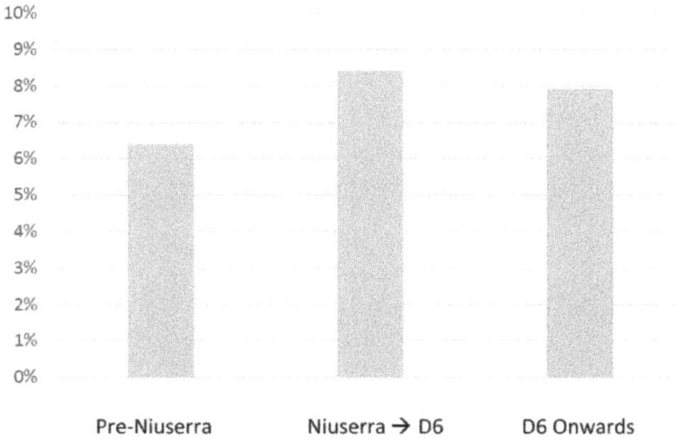

Figure 8.1: Avian representations: proportion change over time.

procession at this time.[7] From the time of Dynasty Six onwards, however, this proportion declined once more. Notably, the trapping of songbirds in orchards using hand-set traps began to be attested early in Dynasty Six, at Saqqara.[8]. It is worth recalling that the overall proportion of scenes depicting orchard scenes declined over this same time frame. Previously it was explained that, as some marshland plants become more numerous and larger, they provide a greater diversity of habitats for bird species. It would be expected, therefore, if decorations changed in response to changing cultural behaviours, that more depictions of birds and their capture should have been identified. This was not the case.

8.1.1. Poultry Proportions

Within the avian category, however, there appears to exist two distinctly different type of categories: poultry (the waterfowl) and others, each with different behavioural and habitat characteristics. Considering its different habitat settings and distinctive behaviour patterns in relation to the river, the poultry class may be seen to be an individual class of resources on its own. It was deemed worthwhile,

[1] Bailleul-LeSuer, *Avian Resources*, 1–14.

[2] Ciałowicz, *La naissance d'un royaume*, 217–218; Patch, *Egyptian Art*, 44–46; Hendrickx etal., *Early Falcons*, 129.

[3] Vernus & Yoyotte, *Bestiaire*, 342–427; Bailleul-LeSuer, *Avian Resources*, 2; Ikram, *Choice Cuts*, 24.

[4] Bailleul-LeSuer, *Avian Resources*.

[5] Mahmoud, *Vögel im Alten Reich*, 155–162.

[6] Sekhemkare: Hassan, *Giza* IV, 110–111, fig. 57 and Rawer: Hassan, *Giza* I, p. 33, pl. 34 [3].

[7] MMA, *Handbook*, 31, fig. 11(part).

[8] Mereruka: Harpur, *Miscellaneous Reliefs*, 35–36, fig. 5; Hesi: Kanawati & Abder-Raziq, *Hesi*, 34, pl. 27, 29, 56.

Table 8.2: Avian versus poultry attestations: pre- and post-Niuserra

Avian vs poultry	Date		
	Pre-Niuserra	Niuserra → D6	D6 Onwards
All avian incl. poultry: proportion all scenes	6.4%	8.4%	7.9%
Poultry only: proportion of all scenes	1.3%	2.2%	4.7%
Poultry versus avians: relative proportion	20%	26%	60%

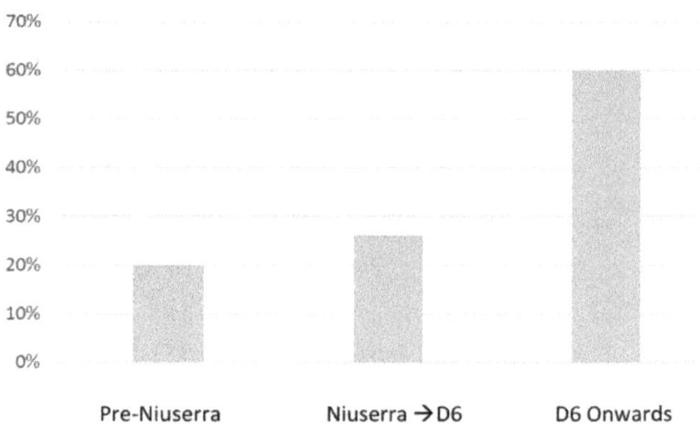

Figure 8.2: Poultry within avian attestations: proportion change over time.

therefore, to observe the changes in numbers of avian attestations with and without poultry (see Table 8.2). The overall proportion of scenes depicting an avian theme, without the presence of poultry, decreases by almost one-half during the Sixth Dynasty. At the same time, those scenes depicting poultry-related themes more than double, while those depicting the various netting activities occurring in garden and orchards, for example, decline.

8.2. The Parable of Proliferating Poultry

Chickens, the most commonly assumed member of the poultry family, are a late-comer to the archaeological record,[9] with domestic breeding of them becoming significant during later times.[10] The main members of the poultry family associated with Old Kingdom tomb decorations are not chickens,[11] but geese and ducks, along with similarly related waterfowl.[12] While poultry was always a significant part of the Egyptian diet, documented since Predynastic times,[13] some impetus must have spurred their increased presence in the elite tomb decoration programme in the Old Kingdom. The rapid increase in the numbers of tomb scenes depicting fowling *and* fishing in the same scene indicates the increasing importance attributed to fowling.[14] Of all attestations regarding avian resources, only one-fifth of the attestations before Niuserra represented the poultry subset. This proportion rose to more than one-quarter from the time of Niuserra to well over one-half by the time of Dynasty 6 (see Figure 8.2). In bird procession scenes in the tombs of Waatetkhethor[15] and Ibi[16], for instance, it is difficult to identify birds that are not aquatic.

8.2.1. Waterfowl Depictions and A River In 'Drought'

It has been shown that representations of orchards and gardens decreased over the time frame in question. Consequently, it seems reasonable that attestations of animals associated with orchards and gardens would also lessen in number over the same time frame, explaining the decrease in depictions of songbirds, for example. It remains important to investigate why a rapid increase in the apparent importance of poultry took place at that time.

Previously, it has been noted that, by the time of the Sixth Dynasty, the preponderant subject in tomb decorations with an avian theme seemed to be poultry. The rapid decline in non-poultry avian depictions, as well as the even more rapid increase in tomb decorations depicting poultry, can be explained in relation to the A.R.I.D. hypothesis. It is easily possible to relate the increase in poultry scenes to the ecological changes postulated: changing riparian properties affected the distribution and abundance of some plant species, changing the riparian habitats. If a change in the distribution and abundance of some plant species occurred, then the ecosystem would have experienced instability until a new natural equilibrium was achieved. An environment with characteristics now favouring papyrus and reeds over bulrushes and lilies would have altered the balance of the food webs associated with the river.

Poultry animals such as geese and ducks are happily omnivorous and can quite successfully forage for themselves.[17] Geese browse on grass,[18] so an increase in grasses and juvenile reeds at the river's edge would have enhanced their diet diversity and lessened the amount of feed they needed to source elsewhere. An increase in dense and moist vegetation on or around the river's edge would also benefit their survivability. Ducks enjoy a bug-heavy diet,[19] contentedly eating worms and bugs that are found as predators on important cereal crops, so inadvertently helping weed an area and act as living pesticides. An

[9] Coltherd, *Domestic Fowl*, 217–223; von den Driesch & Peters, *Störche über Elephantine*, 667, Taf. 5; Darby etal, *Gift of Osiris*, 1; Ikram, *Choice Cuts*, 26.
[10] MacDonald & Edwards, *Chickens in Africa*, 584–90; Bailleul-LeSuer, *Avian Resources*, 411, 419.
[11] Coltherd, *Domestic Fowl*, 218–219; Houlihan & Goodman, *Birds*, 54.
[12] Houlihan & Goodman, *Birds*, 54; Luff, *Ducks*, 519.
[13] Ikram, *Meat Processing*, 656.
[14] Harpur, *Decoration*, 257.
[15] Kanawati & Abder-Raziq, *Waatetkhethor*, pl. 62, 65–66.
[16] Kanawati, *Deir el-Gebrawi* II, 30, pl. 11–13, 47–58, 68.
[17] MacDonald & Blench, *Geese*, 529.
[18] https://poultrykeeper.com/keeping-geese (07/08/2017).
[19] https://poultrykeeper.com/keeping-ducks (07/08/2017).

increase in poultry numbers as an aid to agricultural production during times of pastoral stress seems once again perfectly feasible.

8.2.2. Poultry Becoming a More Significant Resource

The archaeological remains of geese and the relative largeness of an individual's size, compared to their keepers in artistic representations,[20] as shown in the tombs of Merefnebef and Kahai at Saqqara,[21] for example, suggest the animal would have represented a significant protein source.[22] Providing ready access to feed, or force- feeding them, would have made the animals more massive, reducing their ability to fly long distances.[23] A decline in available water for orchards and gardens would have led to a decrease in availability of the avian occupants of these regions. Songbirds and other birds attracted to the invertebrate pollinators of these plants would have diminished in numbers. On the unsolidified boundaries of the river, waterfowl would have been increasingly attracted to the expanding variety of aquatic food webs that would have been developing.[24] On the more solid banks of the river, geese and ducks would have found more opportunities to graze and forage as the margins between the land and river enlarged. Enhanced opportunities would have led to rapid population growth of the waterfowl, providing greater prospects for hunting, or harvesting them, and perhaps explaining the rapid increase in the number of representations depicting this activity.[25]

8.3. Avian Predators

An overgrown riverbank, with taller stands of reeds and denser overhanging thickets of papyrus, would have provided a greater refuge from airborne predators, perhaps allowing a boost in the overall numbers of the avifauna populations.[26] With an increase in avian numbers, it would be expected a commensurate increase in the populations of their predators would be noticeable. Common mammalian predators of riverbank and marshland bird species are the genet (*Genetta genetta)*[27] and the mongoose (*Herpestes ichneumon),*[28] which is slightly larger.[29] This similarity in size and structure has led to some confusion in the identification of the two types of predators by previous researchers: in distinguishing between these representations, this study followed the guidelines outlined out by Evans.[30]

8.3.1. The Genet and the Mongoose

Genets are abundant in vegetated areas,[31] especially beside watercourses,[32] and while they can climb,[33] most of their prey is caught at ground level.[34] An increase in plant cover would be expected to have benefited their behaviours by providing greater hiding places. Similarly, with an increased biomass extending protection to nesting and burrowing birds, avian population growth would have provided the genet with more opportunities to harvest eggs and young birds. The mongoose, a terrestrial hunter, preferred covered waterways;[35] however, it displays an aptitude to adapt to many varieties of wetland conditions.[36] The mongoose hunts mainly during the day,[37] as opposed to the nocturnal genet,[38] but still prefers to remain in cover.[39] It is unable to climb,[40] despite some depictions of it doing so,[41] so relies on food on the ground or within its reach.[42] Food items are usually taken into dense cover to be eaten.[43] An increase in plant cover would have aided the hunting activities of mongooses, as well as protecting them from larger predators.[44] Does the large number of decorations depicting this indicates that mortality to waterbirds may have more frequently arrived on four legs rather than two?

8.3.2. Avian Predator Depictions and A River In 'Drought'

Evans identified the genet in a number of marsh scenes in tombs dating from the late Fourth to the late Sixth Dynasty,[45] with all but one dating from the time of Niuserra. Similarly, of Old Kingdom tombs depicting the mongoose, only one dates to before Niuserra.[46] Both these predators would have been in direct competition with human hunters for waterfowl.[47] This may explain why the depictions of these animals focus primarily on their hunting of waterfowl,[48] an (increasingly) important source of food for the local population at the time. However, in depicting a potential rival, the artists represented both mongooses and genets as able to climb, a physical unlikelihood in the

---

[20] Boessneck, *Riesige Hausgänse*, 105–10.
[21] Myśliwiec, etal., *Merefnebef*, 151, pl. 22, 70d; Lashien, *Kahai*, pl. 39a, b;Vandier, *Mo'alla*, fig. 14.
[22] Kiple & Ornelas, *Geese*, 529–531.
[23] Zeuner, *Domesticated Animals*, 467–468; MacDonald & Blench, *Geese*, 530; Mannermaa, *Goose Domestication*, 3096–3097.
[24] Luff, *Ducks*, 519.
[25] Harpur, *Decorations*, 257.
[26] Guisan & Thuiller, *Species Distribution*, 993–1009.
[27] Gaubert etal., *Genetta genetta*, 1.
[28] Do Linh San etal, *Herpestes ichneumon*, 1–2.
[29] Hoath, *Field Guide*, 88–90; Estes, *Behavior Guide*, 298–302.
[30] Evans, *Animal Behaviour*, 36, n.43, 208–214.

[31] Haltenorth-Diller, *Field Guide*, 188; Larivière & Calzada, *Genetta genetta*, 3.
[32] Virgós & Casanovas, *Genet Habitat Selection*, 169–177; Gaubert, etal, *Genetta genetta*, 4.
[33] Estes, *Behavior Guide*, 281; Wemmer, *Comparative Ethology*, 13.
[34] Schliemann, *Viverrids*, 518.
[35] Özkurt, *Egyptian Mongoose*, 486.
[36] Barros etal., *Egyptian Mongoose Expansion*, 5–6.
[37] Haltenorth & Diller, *Field Guide*, 201–202; Estes, *Behavior Guide*, 298.
[38] Lariviére & Calzada, *Genetta genetta*, 3.
[39] Estes, *Behavior Guide*, 299; Özkurt, *Egyptian Mongoose*, 484; Barros etal., *Egyptian Mongoose Expansion*, 11.
[40] Delibes, *Herpestes ichneumon*, 356–357.
[41] Evans, *Animal Behaviour*, 43–45.
[42] Estes, *Behavior Guide*, 299; Kingdon, *East African Mammals*, 180.
[43] Ben-Yaacov & Yom-Tov, *Egyptian Mongoose*, 39, 48.
[44] Lariviére & Calzada, *Genetta genetta*, 3; Galantinho & Mira, *Occurrence of Genet*, 680–681; Martínez-Jauregui etal, *People & Predators*, 241–242.
[45] Evans, *Animal Behaviour*, 41.
[46] Evans, *Animal Behaviour*, 41–42.
[47] Admasu etal, *Genetta in Farmlands*, 160–162; Delibes, *Herpestes ichneumon*, 357.
[48] Evans, *Animal Behaviour*, 43.

case of the mongoose that has been noted by a number of scholars.[49] Evans points out that, while mongooses are not thought to climb at all,[50] she posits that perhaps scholars are misinterpreting these representations and the portrayals of the animals on papyrus stems is not depicting them 'climbing' but perhaps moving away through the vegetation, deeper into it.[51] If the images of these animals on single stalks is to be taken more literally, they could represent the creatures walking on flattened papyrus stalks, a consequence of papyrus that has grown to larger-than-normal heights, has fallen over and acts as a floating ramp on the water.[52] This mat would act as excellent cover for birds, increasing the hunting potentialities for the mongooses present. Interestingly, Evans now believes that during this time, Egyptians began to use these animals in hunting activities, with these animals partly domesticated; she suggests that the representations of these animals with straps around them suggests they have been trained.[53]

8.4. Changes in Art: Fowling in the Marshes

As noted earlier, from the time of Niuserra onwards some decorations depicted fowling activities in front of a very crowded papyrus thicket with a congested composition. Before that, the arrangement of the thicket seemed rigid and formulaic.[54] In the Fourth Dynasty tomb of Nebemakhet, the papyrus stems are regularly spaced and the birds atop lined up as if in a queue.[55] Also increasingly prevalent over this time frame is the depiction of the thicket with fewer stems represented than umbels:[56] this presentation enters the artistic repertoire mid-Fifth Dynasty and continues through the Old Kingdom;[57] with the tomb of Khunes at Zawiyet el-Maiyetin, for example, depicting many fewer stems than umbels, allowing the placement of more bird images within the thicket.[58]

8.4.1. Fowling Scenes and a Cluttered Papyrus Thicket

In some instances, the artist has represented the papyrus thicket with stems separated and unevenly spaced, rather than touching and forming a wall or backdrop.[59] Woods notices this as unusual and associates this characteristic more with provincial art,[60] observing that most of these compositions date to the latter half of the Old Kingdom. While this may seem illogical within an argument for more congested papyrus thickets, the opening of spaces within the thicket has now allowed the artist to fill the gaps with birds and their land-based predators. The fowling scene spreads more widely along the wall, presenting to the viewer an impression of vast width and great depth, emphasising the resources available *within* the papyrus thicket. In the Sixth Dynasty tombs of at Saqqara[61] and Meir,[62] for example, the birds are represented within the thicket, showing the viewer the varied offerings available within that habitat. Similarly, the tomb of Merefnebef, at Saqqara, for example, depicts the tomb owner preparing to throw a stick into a very chaotic scene of high papyrus thickets, staggered rows of umbels and birds represented haphazardly within.[63] The large sturdy vessel from which Ibi, at Deir el Gebrâwi, is fowling and fishing may be another indication of the growing significance of these activities.[64]

Therefore, it is suggested that the representations of fewer and wider papyrus stalks allow the viewer to better see the bounty offered inside the thicket.[65] These images contrast well with those spearfishing scenes of the latter Old Kingdom where the papyrus thicket is not represented at all, suggesting perhaps that these two activities occurred at different places within the marshes. While Iyenefert at Giza does not have an extant spearfishing scene, his fowling activities appear in a crowded papyrus thicket, the pleasure cruise scene depicted nearby in more open waters.[66]

8.4.2. Poultry Farms

Poultry farms became a significant decorative feature in the tombs of the late Fifth Dynasty, starting with the tomb of Ty,[67] and continuing into the Sixth Dynasty. Very few examples are found in the provinces: Pepyankh the Black at Meir[68] depicts one, as does Ibi at Deir el-Gebrawi.[69] Kanawati suggests this theme is mostly encountered at Saqqara in the Teti Cemetery,[70] however, the tomb of Ptahshepses at Abusir also contains this depiction.[71] The development of poultry farms from the time of Niuserra onwards may be explained as another cultural adaptation to take advantage of the changing environmental situation. As the growing marshland provided habitats more favourable to waterfowl, the pragmatic response would have been to organise and exploit this increasing resource. Scenes depicting the force-feeding of poultry began

[49] Houlihan, *Ti Swampland Scene*, 20; Evans, *Animal Behaviour*, 42–43.
[50] Evans, *Animal Behaviour*, 42–43.
[51] Evans, *Animal Behaviour*, 44, n. 68–69.
[52] Britton, *Papyrus Swamps*, 450; Thompson & Hamilton, *Peatlands*, 334–335.
[53] Pers. comm. 1508/2019; (in preparation).
[54] Harpur, *Decoration*, 199–201.
[55] Hassan, *Giza* IV, 138, fig. 79.
[56] Woods, *Marshes*, 175, n.226; appendix 3, n.136.
[57] Woods, *Marshes*, 361.
[58] Lepsius, *Denkmaler* II, Band IV, pl. 106a.
[59] Harpur, *Decoration*, 265 [8]; GAP: Photo #C12741_NS ©Mohammedani Ibrahim, Jan. 20, 1931.
[60] Woods, *Marshes*, 175.
[61] Seankhuiptah: Kanawati & Abder-Raziq, *Teti III*, pl. 69, 76; Nikauisesi: Kanawati & Abder-Raziq, *Nikauisesi*, pl. 50; and Hesi: Kanawati & Abder-Raziq, *Hesi*, pl. 54.
[62] Pepyankh the Black: Blackman & Apted, *Meir* V, pl. 28; Kanawati & Evans, *Meir* II, pl. 88.
[63] Myśliwiec etal., *Merefnebef*, pl. 21, 63–65.
[64] Kanawati, *Deir el-Gebrawi* II, 27, pl. 7–10, 45–46.
[65] Evans, *Animal Behaviour*, 73–76; Woods, *Marshes*, 227.
[66] Schürmann, *Ii-nefret*, pl. 6 (fowling), pl. 21 (pleasure cruise).
[67] Harpur, *Decoration*, 114; Wild, *Ti*, II, pl. 6–8.
[68] Blackman & Apted, *Meir* V, 31, pl. 22 [2]; Kanawati & Evans, *Meir* II, 36–37, pl. 83a.
[69] Davies, *Deir el-Gebrâwi* I, p. 21, pl. 16; Kanawati, *Deir el-Gebrawi* II, 51, pl. 53, 72.
[70] Kanawati & Abder-Raziq, *Hesi*, pl. 27, 56; Kanawati & Abder-Raziq, *Nikauisesi*, pl. 49; Kanawati & Abder-Raziq, *Mereruka*, pl. 48; Kanawati & Abder-Raziq, *Meryteti*, pls, 14, 48; Harpur & Scremin, *Kagemni*, figs. 13, 14.
[71] Vachala, *Ptahshepses*, 164–165, pl. 13.

during the reign of Niuserra and continued throughout the Sixth Dynasty.[72] Force-feeding of waterfowl would have increased their mass more rapidly and this depiction coincided with their increasingly frequent appearance in offering procession scenes. It would have had the secondary benefit of making the heavier waterfowl less likely (or able) to escape.

8.5. Summation: A Perception of Predominating Poultry?

The main points developed in this chapter include:

- Over the time frame in question, as representations of orchards and gardens decreased a corresponding decrease in depictions of those birds that were exploited from these resources was observed.
- Over the same time frame in question, poultry/ waterfowl was depicted with increasing regularity and became significant part of the elite tomb decoration programme.
- An increase in the marshlands led to an apparent increase in the relative importance of the waterfowl.
- New scene types, such as the poultry yard and the poultry procession were added to the lexicon.
- Representations of the major predators of the waterfowl entered the tomb scene repertoire at the same time.

Birds better suited to moist riverine boundaries and whose nesting sites were protected from aerial predators by overhangs, and who benefited from an increase in fish numbers, would have taken advantage of the environmental changes proposed, explaining the increased attestation rates of waterfowl species such as duck and geese. This led to their habituation and eventual domestication through the use of poultry yards, even to the practice of force-feeding. This same stimulus saw the rise to prominence of the fowling scene, with its associated significance most likely drawn from the underlying ecological need to hunt waterfowl.

[72] Vachala, *Ptahshepses*, 164–165 [C262 (1208)].

9

The Rise of Cattle and A River In 'Drought'

According to the A.R.I.D hypothesis, excess riverine nutrients would have led to a river becoming more congested with an abundance of grasses and reeds. As the riverbank habitat flourished, cattle and other herbivores would have found huge amounts of riverside nutrients available, with these conditions providing increased opportunities for cattle grazing. The surroundings should have become more agreeable to those individuals within society that relied on pastoralism as a major source of wealth. The rapid increase in depictions of cattle could be explained by the society taking advantage of this bloom in resources. In this chapter, the increase in pastureland attestations will be investigated in greater depth. An explanation of why reeds and papyrus grasses would encourage pastoralist activity will be presented. An explanation of why pigs appeared to have declined in relative importance compared to Pre- and Early dynastic times will be suggested. The increasing presence of goats and their role in a changing ecology will be proposed and the idea of an increasing reliance on 'mobile wealth' among the elite will be advanced.

9.1. The Developing Importance of Cattle

Perhaps the ready availability of wild grains to the early users of the Nile valley[1] was one of the reasons behind the late adoption of agriculture in Egypt compared to other Middle Eastern civilisations.[2] It seems this adoption was not undertaken in response to a changing climate but an apparently conscious decision to take advantage of the regular flood.[3] Similarly, the first significant addition of cattle to the resource mix was apparently made as a result of society's adaptation to the seasonal offerings of the river.[4]

As the inundation became more unreliable, especially in the south, cattle rearing seemingly was adopted in response to climate change.[5] The rise of cattle as an adaptation to environmental change seems a valid response to an increasingly unreliable river system. Cattle can move to where the feed is. Cattle herds as mobile wealth storage systems may have been a natural response to changing ecosystems that necessitated a lesser reliance on agriculture. From the reign of Niuserra, themes related to cattle-raising and pastureland activities increased significantly, continuing into and throughout the First Intermediate Period[6] (see Table 9.1). It appears that pastoral activities became increasingly important to the tomb owners as the Old Kingdom progressed. The proportion of images representing pastureland-related decorations increased to more than one-sixth of all attested scenes by the end of the Old Kingdom (see Figure 9.1).

Table 9.1: Cattle and pastureland scenes: pre- and post Niuserra

Cattle & pastureland scenes	Date		
	Pre-Niuserra	Niuserra → D6	D6 Onwards
Proportion of overall attestations	7.7%	14.7%	16.8%

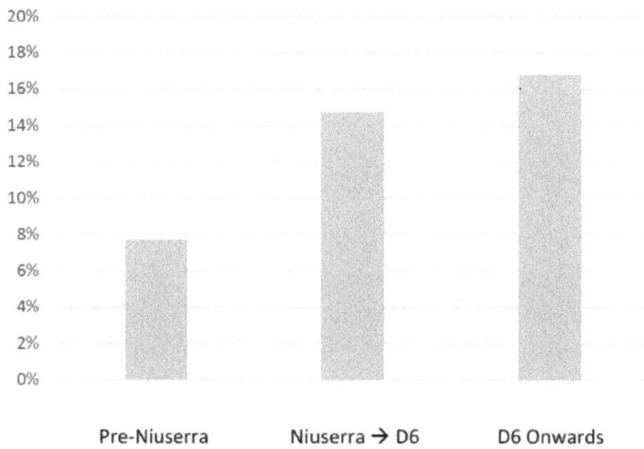

Figure 9.1: Cattle and pastureland scenes: proportion change over time.

As a consequence of an increasing role for cattle in the food chain, it could be expected that the products of cattle would also be depicted in increasing frequency (see Table 9.2). Depictions of the preparation and cooking of red meat increased over the time frame under investigation. This appears to be the case and coincides with increasing archaeological evidence of cattle products in a number of archaeological settings,[7] with analyses suggesting an increasing relationship between the consumption of beef and social status.[8] Despite the small absolute number of scenes, the fact that the proportions of these representations had almost doubled by the end of the Old Kingdom should be seen as an important indicator (see

[1] Midant-Reynes, *Prehistory*, 141.
[2] Hartung, *Livestock Production*, 22–23.
[3] Midant-Reynes, *Prehistory*, 138.
[4] Midant-Reynes, *Prehistory*, 137.
[5] Barich, *People, Water & Grain*, 128; Wright, *Human Holocene Termination*, 7–9.
[6] Hagseth, *Livestock Transport*, 51.
[7] Helck, *Pork Consumption*, table 1; Ikram, *Choice Cuts*, 35–36; Willems etal., *Wadi Zabayda*, 308–325; Vymazalová and Arias Kytnarová, *Sheretnebty*, 41; Duliková, etal., *Ankhires*,27.
[8] Bárta etal., *Hetepi*, 159; Ikram, *Choice Cuts*, 14.

Table 9.2: Meat depictions: pre- and post Niuserra

Meat Depictions	Date		
	Pre-Niuserra	Niuserra → D6	D6 Onwards
Proportion of overall attestations	1.25%	1.30%	2.45%

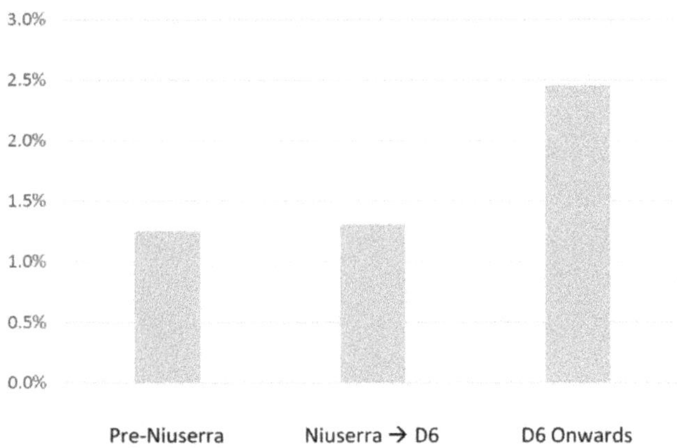

Figure 9.2: Meat depictions: proportion change over time.

Figure 9.2). As bio-archaeological dating methods become more precise, it should be possible to trace these changes in the diet more precisely.[9] Increasingly sophisticated testing regimes within the field of settlement archaeology should be able to note if these proportional changes to the visual representational record were manifested in a similar change in the average diet, as evidenced by the apparent increase of fish and fowl in the average diet.[10]

9.1.1. Cattle Depictions and A River In 'Drought'

Cattle appear to have become a more prized commodity in the latter years of the Old Kingdom: this is attested by the proximity of the tomb owner's image to representations of fighting bulls.[11] The role of bull fighting may have been in order to identify the strongest bull with which to breed the next generation.[12] Cattle depiction numbers relate to the wall scenes of the tombs of the elite class, so this may skew the data when comparing the proportion of meat in the diet of the entire population. However, the investigation is looking for evidence of an adjustment as a result changing environmental conditions. Despite developing into a successful symbiosis between humans and those animals they exploited for various resources,[13] pastoralism, can sometimes lead to negative modifications to the surrounding landscape, usually with land clearing a major concern. [14]

According to the A.R.I.D. hypothesis, the rising nutrient load remaining within the waterway would have led to a substantial growth of papyrus and reeds, from which the animals that fed on these plants would benefit. While herbivores, including cattle, are not particularly favourably disposed towards eating mature reeds and grasses; immature plants are more palatable;[15] with both papyrus and reeds responding favourably to regular cropping, which seems to encourage regrowth of new tender shoots. Typhae, are more palatable than both of these plants,[16] so would have been eaten more rapidly; perhaps this is another reason for its suggested decline at this time. Since large herbivores can have an important impact upon the quality of a local habitat,[17] for example, changing the floral mix and causing erosion, one could expect that if the A.R.I.D. hypothesis is valid, then a greater proportion of pastoral representations would have been produced. As the environmental situation changed, the increased adoption of pastoralism as a strategy seems an obvious pragmatic solution.[18] A similar pattern has been identified as having developed in the Sahara.[19]

9.2. Fording Scenes and A River In 'Drought'

Cattle crossing scenes have been interpreted as fording a canal or stream to, one would assume, move to a source of food.[20] The overall proportion of the many representations of this theme increased over the time frame under investigation, suggesting that depictions of moving cattle between varied food sources alongside the river became more important to tomb owners over time (see Table 9.3).

9.2.1. Fording Impacts

Perhaps, since food was more readily available, the cattle were moved more regularly to take advantage of it. Since the younger parts of rapidly growing plants are more toothsome, the cattle may have needed to be moved more often than previously, in order to take advantage of the high-quality nutrition in newly growing shoots.[21] The foraging pathways would have remained open more consistently, becoming temporary trails for the duration of the season,[22] similar to the fording impacts of the hippopotamus (see Figure 9.3). It is also interesting, that scenes depicting the threat of crocodiles upon fording cattle increase in number at this stage;[23] with some of the where the crocodiles are depicted in pairs could easily represent the situation where each on is one his 'territory' on either side of a narrow secondary channel.

Producing the trail may have helped open waterways between different pools, enabling water to move more

[9] Wing, *Animals as Food,* 57.
[10] Sterner, *Herbivores Effect on Algae*, 605–607.
[11] Kanawati, *El-Hawawish* IV, 18–19; Galan, *Bullfight Scenes*, 81, 93.
[12] Kanawati, *Tomb & Beyond,* 90.
[13] Hartung, *Livestock Production*, 20.
[14] Hughes, *Deforestation Views*, 439.

[15] Evans, *Animal Behaviour*, 87.
[16] Zahran & Willis, *Vegetation of Egypt*, 298; Shaltout & Ahmed, *Plant Life*, 117.
[17] Ashley & Liutkus, *Tracks, Trails & Trampling*, 31.
[18] Hassan, *Climate & Cattle*, 61–86.
[19] Jelinek, *Pastoralism & Social Stratification*, 41–44.
[20] Harpur, *Decoration*, 199.
[21] Evans, *Animal Behaviour*, figs. 7-5, 7-6, 7-7.
[22] McCarthy etal., *Hippo Impact*, 44; Deocampo, *Hippo Structures*, 217.
[23] Evans, *Animal Behaviour,* 127–128.

Table 9.3: Cattle fording and feeding scenes: pre- and post Niuserra

Fording & Feeding	Date		
	Pre-Niuserra	Niuserra → D6	D6 Onwards
Proportion of overall attestations	2.1%	2.8%	3.5%

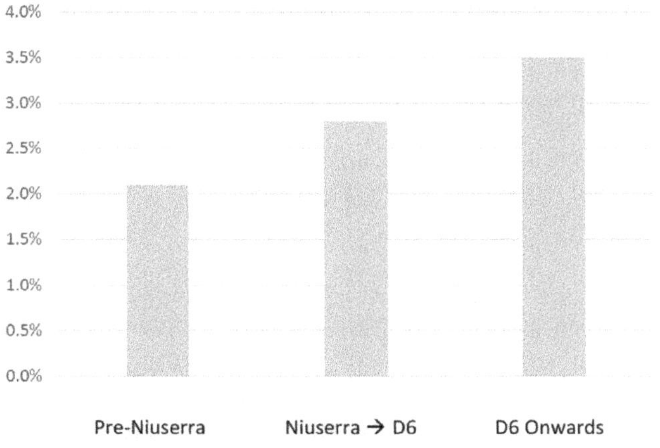

Figure 9.3: Cattle fording and feeding scenes: proportion change over time.

Table 9.4: Crossing cattle immersion levels: pre- and post Niuserra

Fully immersed	Date		
	Pre-Niuserra	Niuserra → D6	D6 Onwards
Proportion of all cattle crossing scenes	79%	75%	69%

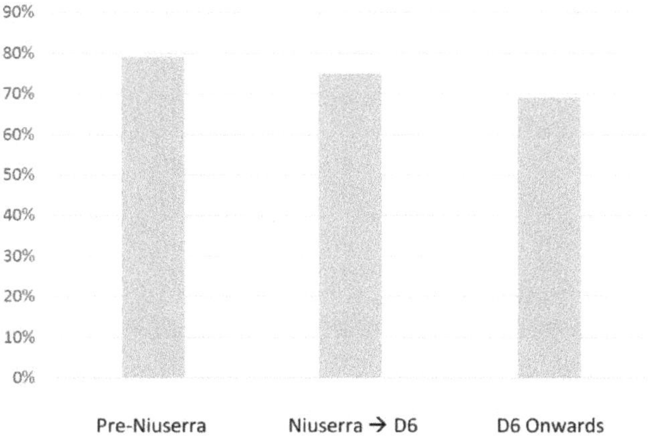

Figure 9.4: Crossing cattle immersion levels: proportion change over time.

freely and avoid localised stagnation.[24] While cattle do not live within waterways as hippopotamus do, their herd-size is larger, leading to similar impacts on the riverine ecosystem overall.[25] Trampling of plants breaks up the vegetation's physical structure,[26] as well as increasing the rate of decomposition and facilitating more rapid microbial attack;[27] providing further impetus for plant diversity as other faster growing species may be able to take a hold before the damaged species recover.[28] It may also suggest that cattle were undertaking more fording activities due to an increase in secondary channel formation as a consequence of a weaker-than-normal river.

9.2.2. Immersion Levels

It is possible to observe other changes within the cattle crossing scene that can be related to the A.R.I.D. hypothesis. Of those depictions of cattle crossing before the time of Niuserra, most depict the animal immersed above the shoulder, with only one depiction displaying parts of the animal below the body.[29] From the time of Niuserra onwards, cattle crossing scenes begin to depict the lower part of the animal in more detail,[30] revealing most of the animal and often the knees.[31] A survey of the amount of the animal covered by the water was performed upon these depictions. Boundaries were marked at the neck and shoulders ('fully immersed'), the belly and the knees (see Table 9.4).

From the time of Niuserra onwards, it is increasingly common for a larger proportion of the animal to be on show above the waterline (see Figure 9.4). That may mean that the water is so shallow, that in real life you would see the cattle's legs, and therefore the artists have shown them. When the animal is depicted deeply immersed, it is sometimes accompanied by an indication that it is entering the deeper part of the river – for example, in the tomb of Iteti/Shedu,[32] and that of Senedjemib Inti, where the cattle are accompanied by men in boats.[33] Increasing visibility of the lower part of the animal suggests secondary channels becoming shallower. Also, the more regular transfers between the many secondary channels, as proposed, would give the artist greater opportunities to witness the animals moving between channels. However, it should not be assumed that the water level was so low that cattle could cross the Nile itself, despite this being suggested

[24] McCarthy etal., *Hippo Impact*, 54–55.
[25] McCarthy etal., *Hippo Impact*, 49; Ashley & Liutkus, *Tracks, Trails & Trampling*, 31.
[26] Silliman etal., *Livestock as Control Agents*, 10; Brundage, *Grazing as Management*, 61.
[27] Brundage, *Grazing as Management*, 51–52.
[28] Steward etal., *River Runs Dry*, 204, 207; Brundage, *Grazing as Management*, 66.
[29] See Harpur & Scremin, *Niankhkhnum & Khnumhotep*, fig.8.
[30] Harpur, *Decoration*, 195; Lashien, *Kahai*, 30.
[31] Moussa & Altenmüller, *Nefer and Ka-hay*, pl. 1, 5; Lashien, *Kahai*, pl. 9b, 13d; Vachala, *Ptahshepses*, 96–97, A815 (1102), C173 (1189); Moussa & Altenmüller, *Nianchchnum und Chnumhotep*, Taf. 76–77; Steindorf, *Ty*, pl. 112, 118; Roth, *Palace Attendants*, pl. 73a–b, 74a, 181, 182; McFarlane, *Irukaptah*, pl. 16, 41, 46: Schürmann, *Ii-nefret*, Abb. 7 [a, b], 21; Davies, *Ptahhetep & Akhethetep* II, pl. 14; Hassan, *Gîza* II, fig. 236; Borchardt, *Cat. Caire* II, Bl. 109; MMA, *Egyptian Art*, 468–471; Davies, *Deir el-Gebrâwi* II, pl. 20; Davies, *Deir el-Gebrâwi* II, pl. 20; Kanawati, *El-Hawawish* VII, fig. 18.
[32] Kanawati & McFarlane, *Deshasha*, 50–51, pl. 14, 46.
[33] Brovarski, *Senedjemib* I, 42, pl. 18, figs. 29–30 (Scene 2).

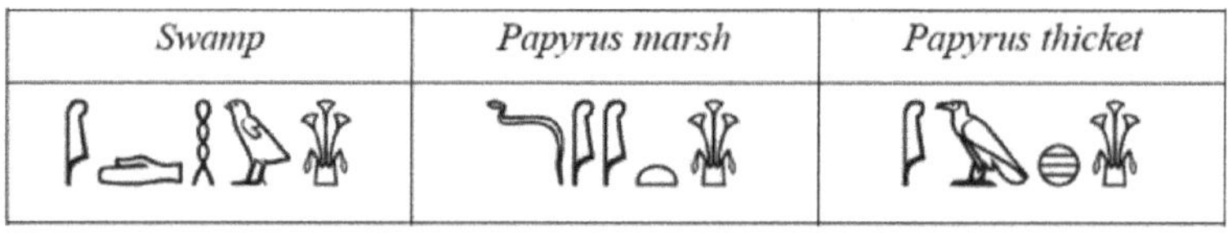

Figure 9.5: Waterway Vocabulary.

in commentary on the tomb of Hetepherakhti, where the caption reads "*...crossing the Nile...*".[34]

A number of depictions display the animal with just the leg below the knee covered.[35] All these examples date to the latter half of the Fifth Dynasty and beyond. Some may suggest that these depictions represent cattle as they are leaving the ford, but those depictions portraying an animal with a sloping back give the impression of climbing out onto the land more effectively; it seems these 'knee-deep' scenes represent entering and leaving shallow water. Tombs depicting cattle actively climbing out of the marshlands all date from the time of Niuserra and beyond:[36] this could indicate that an increasingly shallower river inspired representations of animals with less of their bodies immersed.

9.2.3. Swamps, Marshes and Channels

In the mid-Fifth Dynasty tombs of Niankh-khnum and Khnumhotep and Akhethotep, the cattle are specifically described as coming out of the 'swamp',[37] as opposed to the 'marshlands' referred to in many other tombs. The ancient Egyptians had different words for each,[38] (see Figure 9.5) so would be expected to use them appropriately, so the use of swamp must have particular significance. The word for swamp appears specifically related to the delta, where water flows less powerfully, whereas the word for papyrus marshes and papyrus thickets are significantly different. Some of the river trails may have been improved through human activity: for example, the text...

w3ḥ š iḥw ḫsf mt ḫsf mzḥ in mniw

"*the channel has been prepared for the cattle*"[39]

in the tomb of Senedjemib Inti suggests that existing pathways were improved upon rather than new ones constructed. That phrase or something similar is used in the Senedjemib complex at least three times.[40] Also relevant to the A.R.I.D. hypothesis is the identification of canals or pathways within the marshlands, such as in the tomb of Akhmerutnesut.[41] Some of these 'channels' appear to be a part of the geographical arrangement at that time. While not overwhelming in statistical terms, the fact that scenes suggesting lower-than-normal levels of water all appeared during and after the reign of Niuserra does seem noteworthy. With the significant increase in representations of cattle crossing between marshland feeding sites, it seems feasible that the local people were taking advantage of the burgeoning floral resources on offer.

The potential for semi-permanent pathways between sources of fodder may have led to unexpected fortunate side effects. The pathways also become a channel for fish to traverse, forming a potential site for the laying of traps. These short-lived pathways would have provided easier access to the river's edge at which to lay larger traps or from which to launch boats. The constant use of the pathways along the edges of the river would have pressed down the sedimentary layers, leading to less disturbance of the river floor, making the water less cloudy, which would have made it easier for fish to be seen. Those smaller channels that formed fish pathways needed smaller nets and traps to lay within them, perhaps occasioning the increase in the tomb scene representations of small nets observed at the same time as the increase in fording scenes. This permits speculation that some form of rudimentary aquaculture may have been practised during this time.[42]

9.2.4. Travelling North to the Delta

With the deposition or more sediment in the delta, it is expected that the delta would have grown during this time. While the Delta has been recognised as a desirable location for the grazing of cattle,[43] the idea that ALL cattle travelled to the Delta to graze, however, deserves to be challenged.[44] Travelling to marshlands to provide fodder for cattle seems a sensible response to the problem of ensuring adequate nutrition. However, it is a long way from Elephantine, to Kom el-Hisn, in the north.[45] Cattle can travel at speeds that vary from less than three

[34] Mohr, Hetep-her-akhti, 63.

[35] Vachala, *Ptahshepses*, 96–97, A815 (1102), 98–99, C173 (1189); Moussa & Altenmüller, *Nianchchnum und Chnumhotep*, 155–156, Taf. 76–77; Roth, *Palace Attendants*, 111, pl. 73a–b, 74a, 181, 182; Davies, *Ptahhetep & Akhethetep* II, 15, pl. 14; Schürmann, *Ii-nefret*, 33, Abb. 7 [a, b], 21.

[36] Brovarski, *Senedjemib* I, pl. 107, 110–111, figs. 96–98, 104–105; Davies, *Ptahhetep & Akhethetep* I, 9, pl. 3, 21, 23; Hassan, *Gîza* II, 126, fig. 140; HESPOK, 347–348, fig. 229; Davies, *Deir el-Gebrâwi* I, 4, pl. 5–6; Davies, *Deir el-Gebrâwi* II, 6, pl. 5; Kanawati, *Deir el-Gebrâwi* III, 30, n.160.

[37] Moussa & Altenmüller, *Nianchchnum und Chnumhotep* ,155–156, Taf. 76–77; Ziegler, *Akhethetep*, 81, 133–134, 154, fig. 37.

[38] Gardner, *Grammar*, 481.

[39] Brovarski, *Senedjemib* I, figs. 104–5 = Scene 2.

[40] Brovarski, *Senedjemib* I, 38, 134–137; Brovarski, *Senedjemib* I, 137, pl. 110–111, figs. 104–5.

[41] HESPOK, 347, 348, fig. 229.

[42] See Burn Hippos, Cattle, theARID Project and Incidental Aqualture (in prep.).

[43] Klebs, *Die Reliefs*, 60.

[44] Thanks to Dr L Evans, Macquarie (11/07/2013).

[45] ~920km: https://www.distancecalculator.net/ (11/11/2017).

kilometres per hour[46] up to a breath-taking four kilometres per hour;[47] dropping to about one kilometre per hour while grazing.[48] Having to traverse marshes, cultivated lands and settlements would have added substantially to the actual distance travelled. Allowing twelve kilometres per day,[49] cattle from Elephantine would take at least eighty days to complete the almost one thousand–kilometre trip.

Is this journey feasible? The logistics seem more than problematic. Did each nome send its cattle north to the Delta at the same time? Did they depart at a set time, presumably following the river's rise? What did the subsequent herds eat if they are following preceding herds? How did the communities along the river cope with successive herds of foraging cattle? How long did they stay in the Delta and what did they eat upon their return? While resting rates would have been lower than what is considered acceptable nowadays, and mortality rates higher,[50] there must have been some layover days for those accompanying the cattle to repair and recover.

Perhaps the phrase 'travelling north to the marshlands'[51] did not necessarily mean the Delta for all cattle, but for those from the Memphite region alone. There are areas of marshland identified as extant in Old Kingdom times before one reaches the Delta: the Fayum and, in Middle Egypt, areas near modern day el-Minya and Assyut.[52] Cattle travelling from Upper Egypt would have had to travel north to reach any of these sites. In the tomb of Nesutpuneter in Giza, the cattle crossing scene is accompanied by text stating the cattle are being "guided to the south".[53] It seems feasible that Minya or Assyut could be one of those southern locations. The tomb of Sopedhotep at Saqqara, describes cattle as… "*Proceeding out of the Lower Egyptian marsh towards the chief of this herdsman….*"[54]

9.2.5. Foraging Impacts

Another consideration of increased cattle population along the river's edges would be the changes to localised food webs as a result of their grazing impact.[55] Overgrazing of one component in a habitat may lead to a spread of others.[56] While grazing sometimes encourages such benefits as secondary regrowth, however, overgrazing leads to more significant negative impacts from which recovery becomes more difficult.[57] Foraging in a region has some beneficial side effects, with cattle grazing and goat browsing found to limit unwanted plant spread.[58] In modern-day times, cattle are brought in to feed upon crops that have failed, or been damaged by either pests or too little or too much water.[59] Similar pragmatic decisions must have been made in the past; flood damage would have spoiled crops as readily as it does today.

Waste products add to the fertility of the region, and some seeds deposited in a small pile of dung experience greater germination rates than those deposited without their very own compost pile.[60] Some grasses, however, are cooked by the high temperatures of the decomposing manure.[61] Papyrus is not mentioned as a plant that uses animal dung as a vector for germination, but the seed has been observed as being spread by waterfowl.[62] Cattle dung is not reported as a transmission for *Phragmites*, *Typhae* or the *Nymphaea*, either. Perhaps this relationship is one that needs to be tested. The presence of dung in the water adds to further chemical changes in the local habitat,[63] and the decomposition of dung in the water is another factor in reducing its oxygen content.[64]

9.3. Small Cattle: Goats

The evolutionary story of Egyptian goats is not fully complete,[65] though the narrative of domestication has been outlined, as has a survey of the artistic representations of these animals for corroboration of the morphological changes with those noted in archaeological findings.[66] Notwithstanding they have very little recognition in the iconographic record, goats were among the earliest, if not actually the first, species to be domesticated as livestock in the Middle East,[67] providing many useful materials.[68] It is difficult to imagine domesticating larger animals without first having some experience at controlling the smaller versions.[69] This verification is incomplete because the

[46] https://www.dairynz.co.nz/media/214237/Understanding-cow-movement.pdf (11/11/2017).
[47] Di Marco & Aello, *Energy Expenditure*, 105–110.
[48] Di Marco & Aello, *Cattle Walking*, https://journals.uair.arizona.edu/index.php/jrm/article/viewFile/9272/8884 (11/11/2017).
[49] This is 140% of the average speed recorded in a recent cattle drive in Australia: http://www.abc.net.au/local/stories/2013/08/07/3819912.htm (09/11/2017), and twice that quoted in McLean, A. 1982. *History of the cattle industry in British Columbia*, 130–134, and Wagoner, J.J., *The history of the cattle industry in Southern Arizona, 1540-1940*.
[50] Miranda-de la Lama etal., *Livestock Transport*, 1–2, 10–11; Swanson & Morrow-Tesch, *Cattle Transport*, 103.
[51] Roth, *Palace Attendants*,130, pl. 89, 185; Kanawati & Abder-Raziq, *Unis II*, 46, pl. 16 [b].
[52] Köln, *Project A5*, Picture 6.
[53] HESPOK, 191; Junker, *Giza* III, 51.
[54] Borchardt, *Cat Caire* II, 16, 18, 165.
[55] Turner, *Grazing Effects*, 54–60.
[56] Frank etal., *Grazing History*, 1–3; Brundage, *Grazing as Management*, 43–47.
[57] Fernandez etal., *Degradation & Recovery*, 297–323; Tal, *Grazing Regulations*, 455; Ash, *North Australia Food Supply*, 19; Steward etal., *River Runs Dry*, 207.
[58] Brundage's *Grazing as Management*; Silliman etal., *Livestock as Control Agents*, 6–9.
[59] Ash, *North Australia Food Supply*, 18.
[60] Milotić & Hoffmann, *Dung & Competition*, 774; Brundage, *Grazing as Management*, 49.
[61] Milotić & Hoffmann, *Reduced Germination Success*, 1045.
[62] *'Cyperus papyrus'*, https://www.cabi.org/isc/datasheet/17503 (16/11/2017).
[63] Frank etal., *Grazing History*, 2; Brundage, *Grazing as Management*, 53.
[64] Brundage, *Grazing as Management*, 4–8.
[65] Riemer, *Risks & Resources*, fig. 5.15, depicts a most logical pathway.
[66] Boessneck, *Die Haustiere in Altägypten*.
[67] Vermeersch, etal., *Sodmein Cave*, 486–487; Zeuner, *Domesticated Animals*, 138–139; Gautier, *Domesticated Fauna*, 304; Clutton-Brock, *Domestic Animals*, 68–69. Hatziminaoglou & Boyazoglu, *Ancient Goats*, 123–124.
[68] Hatziminaoglou & Boyazoglu, *Ancient Goats*, 126–128; Iñiguez, *Goats in Dry Environments*, 147; Hauschteck, *Goats in Houses*, 59–70.
[69] Brass, *Domestication Reassessment*, 107.

overall genetic relationships among breeds have not been fully evaluated, leaving an unclear grasp of the progression of the different breeds.[70]

Goats performed a sizeable role in the economy of ancient Egypt,[71] especially for those inhabitants of the poorer-quality pasturelands on the margin of the Nile valley and in vulnerable ecosystems and resource-poor environments.[72] Despite being less efficient than cattle at converting food to meat, and producing more offal per kilogram,[73] goats are thought to have been the most significant source of red meat in the diet of the majority of the non-elite ancient Egyptians.[74] The proportion of attestations depicting goats is quite consistent throughout the early and middle Old Kingdom,[75] only changing during the 6th Dynasty (see Table 9.5).

While it seems likely that cattle figured more predominantly in tomb decorations due to their important role in elite ritual,[76] the changes in abundance of goat representations suggest a relative importance almost as significant (see Figure 9.6). From Dynasty Six onwards, the absolute number of representations almost doubles; more than half of the attestations depicting goats in trees over the time frame under investigation date to the Sixth Dynasty and beyond.[77] By the end of the Sixth Dynasty, goats had become a significant feature of the animal procession in the provincial tombs,[78] and were an integral part of the basic food supply by the Middle Kingdom.[79] Certainly, by the end of the Old Kingdom, the emphasis had changed from husbandry of the individual animal to the acquisition of larger herds.[80]

There may have been some areas of conflict between goats and other means of sustenance: because cereal cultivation is more productive per unit area, it would be expected that agriculture would take priority.[81] Therefore, animal grazing would be expected to occur within regions less conducive to cultivation, such as the Delta, or along the wetland margins of the floodplain,[82] as well as in the fields after the harvest,[83] where the animals would add the nutrient load of their manure to the soil.

Table 9.5: The narrative of the goat: pre- and post-Niuserra

Goats	Date		
	Pre-Niuserra	Niuserra → D6	D6 Onwards
Proportion of overall attestations	0.9%	0.9%	1.7%

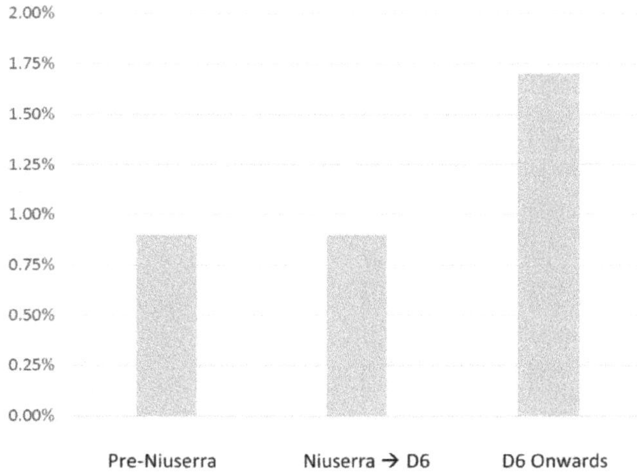

Figure 9.6: The narrative of the goat: proportion change over time.

9.3.1. Goats in Trees: an Indicator of Environmental Deterioration?

Goats have long been used as 'proof' that an environment is in ecological decline. It is important to recognise the potential for damage by goats when added to poor-quality land.[84] Once sheep and/or cattle have denuded a landscape, goats are often introduced to eat the trees and remaining scrubby material[85] (see Figure 9.7).[86] Goats in trees may represent the ultimate last resort in environmental management, for there is very little likelihood of recovery once the trees have been removed from their habitat. From this time onwards, the environment struggles to recover.[87] The iconography depicting goats in or around trees is first attested in the mid-Fifth Dynasty,[88] becoming a predominant scene type in the provincial tombs prepared after the time of Niuserra, with more than half of the attested scenes found in provincial tombs and all but two produced before the time of Niuserra. Scenes depicting goats rearing up on their hind legs to eat leaves from (presumably, living) trees may perhaps be seen as an

[70] Kim etal., *Arid Adaptations*, 255–256, 263.
[71] Gade, *Goats*, 531; Wenke, *Egyptian Civilisation Evolution*, 292.
[72] Iñiguez, *Goats in Dry Environments*, 137; Gade, *Goats*, 532, 534; Hatziminaoglou & Boyazoglu, *Ancient Goats*, 126.
[73] McDowell & Woodward, *Goat Adaptation*, 387–91; Brass, *Domestication Reassessment*, 104; Shirai, *First Farmer-Herders*, 312.
[74] Morenó García, *Production alimentaire*, 73–90; Morenó García, *Village*, 4; Gautier, *Domesticated Fauna*, 304. Contra Swinton, *Resources*, 102, 113.
[75] Swinton, *Resources*, 110–111.
[76] Eyre, *Cannibal Hymns*, 191; Warden, *Old Kingdom Economy*, 57.
[77] 90% date to the time of Niuserra onwards.
[78] Swinton, *Resources*, 113; Clarke, *Upper Egypt Overseer*, 127.
[79] Hauschteck, *Goats in Elephantine*, 59–70.
[80] Swinton, *Resources*, 111.
[81] Swinton, *Resources*, 100; Hassan, *Riverine Civilization*, 51–55.
[82] Redding, *Animal Use Patterns*, 15; Redding, *Pig vs. Sheep/Goats*, 171–178.
[83] Hassan, *Riverine Civilization*, 56; Redding, *Animal Use Patterns*, 99–107.

[84] Iñiguez, *Goats in Dry Environments*, 143; Macias, *Degradación de los Suelos*, 3–18.
[85] Silanikove, *Adaptation in Goats*, 184–185; Mellado, *Goats Dietary Selection*331–333.
[86] http://i.huffpost.com/gadgets/slideshows/349084/slide_349084_3726705 (03/08/2017).
[87] Evans, *Animal Behaviour*, 191; Mellado, *Goats Dietary Selection*, 331; Shaker etal., *Salt Tolerant Fodder*, 66–68.
[88] Harpur, *Decoration*, 105; Akhmerutnesut, in HESPOK, fig. 239; Moussa & Altenmüller; *Nefer & Ka-hay*, pl. 18; Lashien, *Kahai*, pl. 6a, 6c, 28, 30.

Figure 9.7: Goats in trees.

indicator of drought; all such scenes date from the time of Niuserra and beyond.[89]

9.3.2. Goats AND Trees

At el-Hawawish,[90] el-Hagarsa[91] and el-Khoka,[92] many tombs depict goats rearing up to eat. Not all images represent goats looking for sustenance in trees; some depict goats enjoying fresh fruit, the seeds of which are then deposited with their own pile of dung, ready for re-growth. It is important to remember that art is to be interpreted from the perspective of the producer, and not the viewer.[93] Some of the feeding goats appear to be reclining; for example, in the tomb of Khunes at Zawyet el-Maiyetin, the uppermost register shows goats on their haunches,[94] lying down to browse, something difficult to do halfway up a tree, suggesting that the artist is attempting to render depth and distance rather than height.[95]

Some of the scenes[96] may represent goats eating the leaves of trees that have been purposefully felled[97] for their consumption, suggesting a need to broaden the resources available to the goats.[98] Notably, some of the scenes depicting goats feeding in trees are nearby those depicting the felling of trees and boatbuilding. As agriculture's role in supplying food and resources dwindled, the need to exploit the river in different ways grew. In the nearby tomb of Niankhpepy, goats are depicted in (more likely, on) a (felled) tree beside images of workmen cutting down trees and above scenes of workers removing branches. Presumably, the tree has been felled to begin the process of making a boat to do so,[99] and feeding the unnecessary branches to the goats seems an eminently pragmatic use of resources[100]

9.3.3. Goat Attestations and A River In 'Drought'

Goat scenes may indicate a response to fluctuating ecological conditions, for several reasons. A decline in excess water would have resulted in a decline in excess grain, normally fed to sheep. Goats, in particular, are well adapted to take advantage of changing environmental situations with their wider feeding range.[101] Goats are a good option for a region experiencing ecological stress, due to their ability to reproduce rapidly in fertile times,[102] with goats displaying a greater proclivity for producing twins compared to sheep.[103] Also, there is less competition for food between goats and cattle, compared to sheep versus cattle,[104] allowing the herder to best utilise the local habitat.

However, the adaptation of goats to the prevailing harsh conditions, with respect to the feeding responses of goats to that of sheep, cattle and other animals, has not been studied as deeply as has that of other grazing animals.[105] As increasingly sophisticated techniques to correctly distinguish between goats and sheep are utilised,[106] a clearer understanding of the relationship between these two species will develop. A more detailed study investigating the relationship between the zoology of the ancient species to the reported archaeology and the artistic representations may be in order.[107] According to the pictorial evidence, goat husbandry had become a significant part of the culture at the end of the Old Kingdom, so much so that goats fighting, in imitation, perhaps to scenes of bull fighting, entered the artistic repertoire.[108]

9.4. But ... Where are the Pigs?

While explanations about changes in the distribution and abundance of decorations in terms of changing environmental circumstances have been proposed, the

[89] Iteti: Petrie, *Deshasheh*, pl. 15, updated Kanawati & McFarlane, pl. 46; Ibi: Davies, *Deir el-Gebrâwi* I, pl. 11, updated by Kanawati, *Deir el-Gebrawi* II, pl. 52; Pepyankh the Middle: Blackman, *Meir* IV, pl. 14, updated Kanawati, *Meir* I, pl. 84; and Pepyankh the Black: Blackman & Apted, *Meir* V, pl. 32, updated Kanawati & Evans, *Meir* II, pl. 84.
[90] Kanawati, *El-Hawawish* I, fig. 15; *El-Hawawish* VII, fig. 3 [c] and fig. 30.
[91] Kanawati, *El-Hagarsa* III, pl. 22–23 for the tomb of Wahi and pl. 37, 40 for the tomb of Mery-aa.
[92] Saleh, *OK Theban Tombs*, pl. 18.
[93] Evans, *Animal Behaviour*, 8 (n. 24).
[94] Lepsius, *Denkmäler*, II, B.IV, Bl. 108; Hassan & Iskander, *Neb-Kaw-Her*, figs. 12–13.
[95] Evans, *Animal Behaviour*, 8, 48, 90–92, 192–193.
[96] Hassan & Iskander, *Neb-Kaw-Her*, figs. 12–13; Klebs, *Reliefs*, 24.
[97] Evans, *Animal Behaviour*, 41.
[98] Swinton, *Resources*, 107.
[99] Harpur, *Decoration*, 110; contra Evans, *Animal Behaviour*, 90–91, n.26.
[100] Lashien, *Kahai*, pl. 28, 30, 34; Evans, *Animal Behaviour*, 91; contra Schäfer, *Principles*, 253–255.
[101] Gade, *Goats*, 531–532; Shaker, *Salt Tolerant Goats*, 66–77.
[102] McGregor, *Goat Notes*, 81–83.
[103] Evans, *Animal Behaviour*, 172–173; Swinton, *Resources*, 102.
[104] Redding, *Animal Use Patterns*, 102; fig. 3.
[105] Kay, *Responses to Drought*, 683–694; Kim etal., *Arid Adaptations*, 255–264; Bryant etal, *Nutritive Content*, 410–414; Silanikove, *Adaptation in Goats*, 182–183.
[106] Copley etal, *Foraging & Foddering*, 1274; Payne, *Morphological Distinctions*, 139–142.
[107] see Bárta etal., *Hetepi*, 159, 179, Table 3.2. Evans, *Animal Behaviour*, 19, recommending Porter, *Wild Goats*, 295–315, as a suitable scaffold for similar studies.
[108] Evans, *Animal Behaviour*, 142, figs 9-19.

elephant in the room is the lack of porcine representations.[109] The explanation of 'taboo' against eating pork does not hold based on the archaeological evidence:[110] and, with increasing refinements of technology, pigs' presence is being more readily identified.[111] Notwithstanding the claim that pigs were not a part of the funerary cycle,[112] pig bones have been found in many different offering contexts in recently uncovered tombs at Abusir.[113] Despite this, they are noticeably rare in the cultural context of Old Kingdom tomb decoration.[114] What pig representations there are, mostly depict naturally occurring wild boars.[115] The example in Kagemni's tomb at Saqqara is undoubtedly a dog,[116] despite some scholars re-drawing the image to show hooves in place of paws,[117] presumably correcting the artist's original 'mistake'!

9.4.1. Pigs and Early Egypt

Although it is difficult to assess archaeologically whether early remains came from hunted or domestic animals, skeletal remains of wild pigs are marginally smaller, but more robust, than those belonging to truly domesticated pigs.[118] Changes thought to be the result of the domestication process include decrease in size and the loss of hair and pigmentation, which corresponded to a decrease in brain size.[119] During the late Palaeolithic and Epipalaeolithic in Egypt, the people of Helwan ate pork;[120] with pig bones, mostly of wild stock, also recovered from these times at sites in the Fayum and from most Lower Egyptian Predynastic sites.[121] This suggest pig as the main source of domestic animal protein,[122] probably reflecting the advantage of a damper habitat.[123] The evidence is less precise for Upper Egypt, though the presence of swine at Hierakonpolis is confirmed,[124] and early settlers of Merimde consumed such quantities of pork[125] that some archaeologists concluded that pork consumption was a characteristic of early Delta culture.[126] The large number of bones recovered and their relative frequency compared to sheep/goat and cattle bones strongly suggests that they were kept animals.[127] Despite this, it appears that pigs lost their relative importance. Brewer suggests a cultural preference for less pork as well as an environmental one;[128] as cultivation rates increased, pig populations declined. Most likely, pig raising flourished where cultivation did not.[129]

9.4.2. Pigs and Old Kingdom Society

With its omnivorous habits,[130] it is not difficult to get a pig accustomed to whatever food is available in village settlements. Feeding and sleeping are the two activities that dominate a pig's life.[131] Ruminants will feed for a short time, ruminate, perhaps sleep for a while and then feed again; pigs will feed continuously for many hours and then sleep for many hours. This means that in captivity they do not require food during the night and their sleeping and feeding programs can be easily be accommodated to that of humans.[132] In addition, even wild pigs have social tendencies, with the basic family unit making them suitable for domestication.[133] Pigs have a rapid reproduction cycle,[134] they mature quickly and produce harvestable protein more rapidly than sheep and goats.[135]

The physical stature of the pig, its preferred habitat, and its innately stubborn personality (perhaps pig-headedness, even?) make it an unlikely selection for herding over great distances.[136] It is difficult to visualize nomadic or semi-sedentary peoples being accompanied by pigs as we know them:[137] pigs seeming more suitable to settled, agricultural societies. Because pigs are beneficial to village farmers and less so to more mobile groups,[138] one would not expect to find domestic pigs until such agriculture-based communities developed. It seems unlikely, therefore, that the pig was domesticated before the establishment of permanent settlements, yet the presence of domesticated pigs in the Middle East has been verified for the last eight to ten thousand years, with some parts of ancient Turkey providing evidence suggesting that pig herding had developed well before the adoption of agriculture,[139] making it the oldest known domesticated creature apart from the dog.[140] Perhaps, their ability to scavenge among human refuse[141] led to, at first, 'accidental' domestication,

[109] Keimer, *Remarques sur le porc*, 147–156; Evans, *Pig Overboard*, 153–158.
[110] Evans, *Pig Overboard*, 159; Faust, *Pigs in Space*, 279–280; Blench, *Pigs in Africa*, 358–359; Amills etal, *Domestic Pigs*, 76–77; Hecker, *Pork Consumption*, table 1.
[111] Hecker, *Pork Consumption*, 61–64.
[112] Hecker, *Pork Consumption*, 59.
[113] Duliková, etal., *Ankhires*,27–30; Bárta etal, *Hetepi*, 77, 80, 159–161, 169–71, 179, Table 3.2; and tombs AS34–35, 50–53 and 57.
[114] Blench, *Pigs in Africa*, 357.
[115] For example, Keimer, *Le Sanglier*, 26–33; Manlius & Gautier, *Sanglier en Egypte*, 573–574.
[116] Harpur, *Decoration*, 276; Harpur & Scremin, *Kagemni*, fig. 7; Miles, *Enigmatic Scenes*, 71–88.
[117] Evans, *Animal Behaviour*, 107–108.
[118] Brewer etal., *Domestic Plants & Animals*, 94–95.
[119] Zeder, *Animal Domestication*, 167; Price, *Domestication & Behavior*, 89.
[120] Köhler, *Of Pots and Myths*, 175.
[121] Brewer & Wenke, *Mendes*, 191–197.
[122] Brewer etal., *Domestic Plants & Animals*, 96, table 7.1.
[123] Köhler, *Of Pots and Myths*, 175.
[124] McArdle, *Hierakonpolis Fauna*, 116–121.
[125] Brewer etal., *Domestic Plants & Animals*, 97.
[126] Redding, *Animal Use Patterns*, 106; Wenke etal., *Kom el-Hisn*, 5–34; Brewer etal., *Domestic Plants & Animals*, 96, table 7.1; Hecker, *Pork Consumption*, 62.
[127] Wenke etal., *Kom el-Hsin*, 18–21.
[128] Brewer etal., *Domestic Plants & Animals*, 95, table 7.1.
[129] Redding, *Animal Use Patterns*, 103.
[130] Edwards, *Pigs*, 253.
[131] Edwards, *Pigs*, 253.
[132] Edwards, *Pigs*, 254–255.
[133] Edwards, *Pigs*, 254; Gades, *Hogs*, 536–537.
[134] Williamson & Payne, *Animal Husbandry*, 552–554.
[135] Redding, *Animal Use Patterns*, 104.
[136] Edwards, *Pigs*, 253; Brewer etal, *Domestic Plants & Animals*, 94.
[137] Fang etal., *Domesticated Coat Colours*, 4–5.
[138] Edwards, *Pigs*, 254–255.
[139] Gades, *Hogs*, 537.
[140] Fang etal., *Domesticated Coat Colours*, 4–5; Blench, *Pigs in Africa*, 357; Amills etal, *Domestic Pigs*, 73–74.
[141] Zeder, *Animal Domestication*, 173.

which might explain why domestication occurred simultaneously at multiple locations.[142]

Redding has shown that pigs were an important resource,[143] primarily in rural areas, and therefore out of sight of those artists decorating the tomb. This may perhaps explain the persistent erroneous belief that pigs played a minimal role in the provision of protein. The primary role of the pig was as a source of meat and fat for the community, complementing that provided by cattle, sheep and goats, and further complementing sheep/goats/cattle by eating what those animals did not.[144] As well, pigs' foraging behaviour would have helped turn the soil, producing favourable conditions for cultivation. Their adaptability allows them to freely range and root in the village or to live in a confined space. Their versatility offers a unique advantage over other domesticates, particularly in densely populated areas, yet the question remains … why are there so few depictions of this valuable resource?[145]

9.4.3. Pigs and A River In 'Drought'

As the river weakened and cultivation production diminished, the flexible eating habits of pigs, like goats, should have ensured them a key role in the food resource base. This doesn't seem to be evidenced in the decoration programme. While a taboo has been discounted, it has been suggested that the absence of pigs in the elite tomb decoration programme may be explained as an example of elitist snobbery.[146]

Pigs would be expected to survive handily foraging on the edge of newly flourishing papyrus and *Phragmites* plants. Despite their potential to survive readily, the extra distances pigs needed to travel while foraging may have meant that they were less suitable at this time than goats and cattle.[147] Perhaps the competition with cattle for the same feed sources may have led to a disinclination of the society to take advantage of pigs. Also, since pigs enjoy a similar diet to humans,[148] possibly starting out as competitors,[149] maybe it is reasonable to suggest that, as food resources declined, so, too, did their relative importance. As fish, fowl and cattle took advantage of changing environmental conditions and foraged in the flourishing flora, pigs may have become less desirable as sources of food. Perhaps, due to an increasing need for mobility in times of less reliable water supply, pig herds were too cumbersome at this time in Egyptian history. Waterfowl and fish were not domesticated, so needed no care, and cattle could forage for themselves.

[142] Larson etal., *Pig Domestication Multiplicity*, 1616–1621.
[143] Redding, *Pyramids & Protein* (20/03/2017).
[144] Redding, *Animal Use Patterns*, 103–104, fig. 3; Brewer etal., *Domestic Plants & Animals*, 95.
[145] Moreno Garcia, *J'ai rempli les pâturages*, 251–254.
[146] Evans, *Pig Overboard*, 159; Rossel, *Food for the Dead*, 198–202; Rossel, *Tale of the Bones*; Redding, *Pyramids & Protein* (20/03/2017).
[147] Bárta etal., *Hetepi*, 159.
[148] Miller, *Hogs & Hygiene*, 125; Gade, *Hogs*, 537.
[149] Porada, *Chronologies*, 280; Redding, *Role of the Pig*, 22–23.

9.5. Increasing Reliance on a 'Mobile' Economy?

As mentioned earlier, it has been suggested that the marshlands could be expected to have responded differently to the floodplains A.R.I.D. times. A weaker river would have changed habitats, leading to circumstances where reeds and papyrus grew more widely, and were more readily available to be easily cropped by cattle. If the produce of cultivation was declining, then perhaps the local population relied more on resources 'on the hoof'.

9.5.1. The Rise of the 'Cattle Barons'

Landscape research points to several micro-regions within the Nile Valley; some of these appeared to be more conducive to the successful rearing of cattle.[150] Some authors have suggested influence from the west and south. Since the first examples of Saharan dairying, for instance, are recorded at this time,[151] Midant-Reynes suggests a strong influence from Saharan culture.[152] Recent research by Brass has shown that the idea of an independent emergence of cattle domestication from the south also seems unlikely, with the Euphrates the most likely inspiration for domestication.[153]

Before domestication of cattle could develop, it seems more likely that domestication of smaller animals occurred.[154] The time for selective breeding to decrease the size of the wild cattle to a more manageable size may be a reason for the delay is domesticating cattle, as opposed to the caprines. The increasing proportion of cattle bones, in comparison to other ungulates like sheep and goats, as well as pigs' remains, suggest that cattle were becoming an increasingly significant part of the resource mixture.[155] Evidence from the industrial site of Wadi Zabayda suggests that the food provided to the workers was organised in a very controlled manner, with bone fragments suggesting entire self-replicating herds onsite, not groups delivered at regular intervals.[156]

A greater understanding of the role pastoralism plays in protecting an environment is developing.[157] The ability of cattle to rapidly recycle nutrients may have provided added material, in the form of fertiliser, to areas of cultivation.[158] In some parts of Africa the environment relied so much upon the nutrients released by cattle that some herbaceous

[150] Manning & Timpson, *Saharan Holocene Demographics*, 28–35; Moreno García, *Social & Economic History*, 231; Brass, *Cattle Domestication Reassessment*, 105.
[151] Dunne etal., *Green Saharan Dairying*, 390–394.
[152] Midant-Reynes, *Prehistory*, 148, 164.
[153] Brass, *Cattle Domestication Reassessment*, 109.
[154] Brass, *Cattle Domestication Reassessment*, 107.
[155] For example, Dulíková, *Ankhires*, 27.
[156] Willems etal., *Wadi Zabayda*, 318.
[157] Homewood & Rodgers, *Pastoralism & Conservation*; Notenbaert etal., *Pastoralism & Biodiversity*; Niamir-Fuller, *Legitimization of Transhumance*; Hart, *Plant Biodiversity*; Neely etal., *Drylands Pastoral Systems*.
[158] McNaughton, *Grazers Nutrient Cycling*, 1798–1800.

plants evolved to take advantage of these offerings,[159] so it is possible that the same response could have developed in the regions alongside the river. In some areas, perhaps the importance of cattle was such that it precluded the exploitation of another resource, such as fishing. In order to minimise the contamination of livestock water supply by waste from the cleaning and preparation of fish, some herders in northern Sudan at this time appeared to actively chose *not* to exploit local fish stocks,[160] perhaps responding to an understanding of the need to minimise biological pollution in order to maintain the ecological balance.

The rise of local chieftainships in Upper Egypt during this time[161] may have been aided by a declining reliance on food distributed from the capital and the increasing independence that mobile wealth delivered: a 'livestock economy'.[162] In southern Africa, cattle are regarded as 'portable wealth' as an adaptation to unreliable water supply.[163] Some of the early cattle cults developed as a social response to environmental changes in the Holocene, with an increasing emphasis on portable wealth, or a "walking larder",[164] and this same thinking may have established itself among land holders in Upper Egypt as a comparable response to a similar situation .[165] The fact that societies practising pastoralism persisted in arid regions for so long demonstrates an inherent flexibility in traditional African pastoralist strategies that enables them to make the most efficient use of an increasingly fragile environment.[166] With regard to the exploitation of particular environments, the increase in the relative importance of cattle in Upper Egypt may be linked to the increasing exploitation of a particular environment – wadis.[167] This linkage requires further investigation.

9.5.2. Move your Ass: Donkeys and their Increasing Role in Society

Despite being domesticated relatively early in Egyptian history, donkeys' representation upon tomb walls appeared much later.[168] While it is difficult to distinguish, archaeologically, between wild asses and early domesticated donkeys,[169] domesticated donkeys seemed to have become significant features of Early Dynastic long-distance trade.[170] The donkey was a relatively insignificant subject in Old Kingdom tomb scenes,[171] but donkey meat had begun to be left as offerings in some Dynasty Five tombs.[172] Evans mentions that, by the time of the Middle Kingdom, donkeys were often depicted performing actions that were usually performed by sheep in the Old Kingdom; for example, treading in the seed.[173]

Representation of donkeys in tomb scenes increased from the middle of the Old Kingdom onwards,[174] at the same time that organised way-stations for donkey caravans were being introduced through the desert.[175] Donkeys were important to the economy in many ways other than simple agricultural.[176] Harkhuf reported to King Merenre to have returned from one of his expeditions with 300 asses full of exotic goods.[177] Perhaps the need for such large numbers of these hardy animals may have led to the development of donkey breeding programmes at this time.[178] The physiological features of the donkey made it suitable for long-distance hauling. Donkeys are hardy animals and well adapted to working in arid environments.[179] Their dietary preferences lie towards the consumption of plants that did not flourish alongside the banks of the river,[180] meaning they did not become a source of competition to the cattle and other large herbivores.[181] This also meant that donkeys were able to travel long distances away from the river, enabling the development of overland trade networks.[182]

The apparent rise in the significance of donkeys in tomb scenes can be readily interpreted by the A.R.I.D hypothesis. As the river became increasingly congested, this must have required an increasing frequency of unpacking and re-packing the vessels plying their trade along the river.[183] Perhaps it had become more efficient to adopt donkey caravans to bypass these blockages. It may also help explain why the Abu Ballas Trail seems to have lost its importance after Dynasty Eight:[184] perhaps a more freely flowing river meant long-distance trails as described by Förster may have become less necessary.

---

[159] Stebbins, *Grass Herbivore Coevolution*, 75–86; Kuper & Kröpelin etal, *Saharan Succession*, 803–807.
[160] Linseele & Zerboni, *Done with Fish*, 237.
[161] Seidylmayer, *FIP*, 108–136.
[162] Wright, *Human Holocene Termination*, 5–8; Makarewicz, *Pastoralist Manifesto*, 159–174.
[163] Brierly etal., *Green Saharan Pastoralism*, 2–7; Wengrow, *Cattle Cults*, 91–104.
[164] Clutton-Brock, *The Walking Larder*, 115–118.
[165] di Lernia, *Cattle Cults*, 50–62; Jelinek, *Pastoralism & Social Stratification*, 41–44; di Lernia etal., *African Cattle Complex*, 25–26.
[166] Brierly etal., *Green Saharan Pastoralism*, 6–7; Brass, *Cattle Cult Origins*, 101; Kay, *Responses to Drought*, 683–694.
[167] This insight came from J. C. Morenó García; see also Embabi, *Landscapes & Landforms*, 5–10.
[168] Gifford-Gonzalez & Hanotte, *Domesticating Animals in Africa*, 13–14; Rossel etal., *Donkey Domestication*, 3715; Peters etal., *Animal Domestication*, 96–124; von den Driesch, *Buto im Nildelta*, 23–39; Kimura etal., *Donkey Ancestry*, 50–57.
[169] Rossel etal., *Donkey Domestication*, 3716–3717.
[170] Arnold etal., *Early Animal Trade*, 2, 8; Rossel etal., *Donkey Domestication*, 3718–3719; Hassan, *Food Metals & Towns*, 551–569; Förster, *Abu Ballas Trail*, 2; Osborn & Osbornová, *The Mammals*, 132–136.
[171] Power, *Donkey Drawing Deconstruction*, 131–151.
[172] Dulikova, *Ankhires*,30; Bárta etal., *Hetepi*, 77, 80, 159–171, 179-181, 334–335, 345–349, 368.
[173] L. Evans, Macquarie, pers. comm.
[174] Evans, *Animal Behaviour*, 63; Lashien, *Donkeys in Egyptian Art*.
[175] Förster, *Abu Ballas Trail*, 3–5; Yilmaz etal., *Domesticated Donkey*, 339–352.
[176] Swinton, *Resources*, 122–123.
[177] Lichtheim, *Literature* I, 23–27.
[178] Brewer etal., *Domestic Plants & Animals*, 100.
[179] Arnold etal., *Early Animal Trade*, 5; Rossel etal., *Donkey Domestication*, 3715; Förster, *Abu Ballas Trail*, 4–6; Dill, *Life, Heat & Altitude*, 104–109.
[180] Evans, *Animal Behaviour*, 85; Vogel etal., *Grasses in the Deserts*, 258–265; Bailey etal., *Bedouin Plant Utilization*, 145–162.
[181] Yilmaz etal., *Domesticated Donkey*, table 2.
[182] Rossel etal., *Donkey Domestication*, 3715; Vandier, *Mo'alla*, figs. 44, 48, pl. 36.
[183] Marshall, *Donkey Domestication*, 371–407.
[184] Förster, *Abu Ballas Trail*, 9.

9.6. Summation: An Emerging Status of Cattle, Goats and Donkeys?

The major points extracted from this chapter include:

- A rapid growth in reed and papyrus would have resulted in an increase of feed available for cattle.
- The rapid growth in cattle herd size coincides with the timeframe suggested in this investigation for an increase in plant growth on the riverbank.
- The increasing depictions of cattle crossing river channels suggest a regular moving the herds from one fresh source of nutrition to another.
- Despite the many advantages of pigs as a food source, their relative immobility may have led to their decline in relative importance during A.R.I.D. times.
- On the contrary, goats, with their ability to subsist well on marginal lands may be the reason why depiction of these animals increased in significant proportion over the time frame of this investigation.
- Goat representation changed from depictions in small groups to large herds over this time, suggesting their role in society changed.
- Similarly, donkey representations increased, and archaeological evidence suggests they had become a more significant part of the society.

Cattle became an increasingly important feature of society at this time. With the goat often identified as a 'marker' species in regions experiencing ecological stress, perhaps it is not surprising, therefore, that the goat first appeared in significant numbers in the tomb decoration programme at this time. Goats have a wide and varied diet and rarely compete for food with sheep and cattle, making them a good supplementary option in an environment that is experiencing limited availability of resources. Pigs may have been too difficult to organise in times of more necessary mobility, with fish and waterfowl perhaps providing an easier source of alternative protein. Donkeys, with their ability to carry significant mass, and an ability to survive harsher environments, became another important factor of the society newly adapting to changing ecological conditions.

10

The Desert and A River In 'Drought'

According to the artistic record, desert-based mammals had become an increasingly important aspect of the offering ritual by the end of the Old Kingdom. In light of the unfolding narrative of the A.R.I.D. hypothesis, it appears significant that the desert hunt motif was to re-establish itself as a significant element of tomb scenes around the time of Niuserra.[1] Even though some religious connotations are possible,[2] associating the desert with chaos and casting the desert hunt as a symbolic representation of victory over anarchy, with the cult of Re missing in response to a changing environment; the explanation for this change could be more prosaic: one of a reinvigorated artistic expression of a religious concept or outlook. Instead of viewing the surge in desert scenes as a reaction to metaphysical influences, it may be that a more physical influence led to these decorative changes, or a combination of these factors. In this chapter, desert hunt depictions will be re-examined in light of a possible change in Nile levels and the need to exploit desert-sourced protein.

10.1. The Desert Hunt

During the time frame under investigation, a number of new artistic themes arose, including those themes related to the desert hunt itself, with an increasing proportion of such scenes attested from the late Fifth Dynasty into the First Intermediate Period (see Table 10.1). The pre-Niuserra proportion of desert scenes at 2.6% may seem substantial, but most of these attestations can be accounted for in tombs of the early Fourth Dynasty. At this time, a greater proportion of decorated tombs were owned by members of the royal family, which perhaps points to a metaphysical role for the desert scene. Compared to the late Fourth, early Fifth Dynasty the absolute number and relative proportion increased significantly from the time of Niuserra onwards. It is likely that the re-adoption of the desert hunt scene is a manifestation of the same motivation. As noted previously, during a riverine 'drought', a 'blossoming' of plant life would be expected within the river, and as the drought extended in duration, increasing aridity would draw the desert closer to the river. Both circumstances might prompt a closer interaction between river and desert organisms. Knowing that tomb depictions of various resources changed over the time frame under investigation,[3] it may be possible to identify hints that the artists were sufficiently aware of their surroundings to notice changes within it. If the 'drought' led to increasing desertification, then it should be possible to identify a reflected trend in those scenes with themes related to the desert hunt and its associated activities (see Figure 10.1). Animals of the desert were drawn so precisely that individual species can be identified, and different variants of the same species recognisable.[4] In provincial scenes depicting the presentation of cattle and other large creatures, desert animals become a regular part of the display, suggesting an increasingly important role for them in the food chain.[5]

10.2. The Desert Hunt as a Tomb Decoration Theme

There are many different interpretations of the desert hunt scene: depictions of sport, the reality of hunting or the religious symbolism of a moral victory over the forces of chaos.[6] While religious symbolism became increasingly important, it is possible that a more physical influence led to these decorative changes as the increasingly realistic representation of the process suggests that this activity had a more pragmatic function, over time.

Table 10.1: Desert hunt scenes: pre- and post Niuserra

Desert hunt scenes	Date		
	Pre-Niuserra	Niuserra → D6	D6 Onwards
Proportion of overall attestations	2.6%	1.7%	2.4%

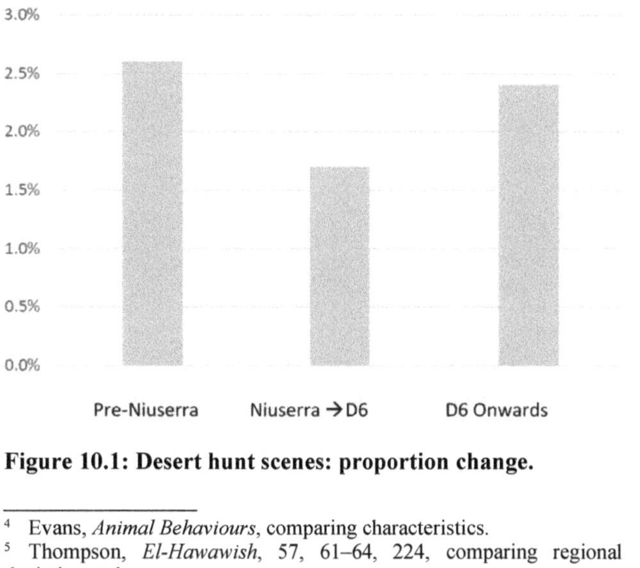

Figure 10.1: Desert hunt scenes: proportion change.

[1] Espinel, *Desert Hunters Iconography*, 88–90.
[2] Vischak, *Agency*; Allen, *Reading a Pyramid.* Contra Hays, *Unreading the Pyramid Texts*; Burn, *Pyramid Texts.*
[3] Swinton, *Resources*, 27.
[4] Evans, *Animal Behaviours*, comparing characteristics.
[5] Thompson, *El-Hawawish*, 57, 61–64, 224, comparing regional depiction styles.
[6] Säve-Soderbergh, *Hippopotamus Hunting*; Hendrickx, *Order over Chaos*, 723–749; Swinton, *Resources*, 245; Montet, *Les scenes*, 76.

10.2.1. Scene Development

The evolution of the desert hunt motif can be traced from Predynastic times.[7] On the Hunter's Palette (pre-dynastic Naqada III), numerous desert animals are depicted fleeing a group of hunters, while Hierakonpolis Tomb 100 depicts a desert hunt with archers and corralled animals.[8] In scenes found upon the mortuary temple of Sahura at Abusir, the king is depicted firing arrows into desert animals in an enclosure, similar to Hierakonpolis Tomb 100.[9] Strandberg has compared these scenes to those desert hunt scenes depicted in the Chamber of Seasons in Niuserra's Sun Temple at Abu Ghurob, noting Niuserra is not shown in an active role.[10] An interesting addition to the iconography in this sun temple are mating and birthing scenes,[11] implying fertility and regeneration: Abu Ghurob emphasises harvest, as opposed to that at Abusir, which seems to be a "veneration of slaughter."[12]

10.2.2. Non-Royal Depictions of the Desert Hunt: Dynasty 4 → 6

Early in Dynasty 4, beginning with the reign of Snefru, individuals other than the king had begun to adopt images of the desert hunt upon their tomb walls.[13] While all depict dogs hunting, it is only the tomb of *Itet* that depicts images of hunters returning,[14] as well as being the only tomb from this era that depicts a major figure holding the leashes of hunting dogs.[15] Interestingly, Itet is the wife of Nefermaat and it is worth wondering why her husband did not depict these scenes in his own tomb, especially considering that his also contains images of desert-related hunting with dogs.[16] Since no other tombs from this era contained the depiction of major figures actively hunting, it is conceivable the royal prerogative of depicting only the king hunting was still being observed.

After a long hiatus from the early Fourth Dynasty, desert hunt scenes made a significant re-appearance around the time of Niuserra (V.6),[17] with depictions of the desert hunt found in under every subsequent king from then to the end of the Old Kingdom. Thereafter, though the scene composition varied, the desert hunt scene appears regularly in elite tombs up until the Middle Kingdom and beyond.[18] Note that lions are not represented in desert hunt scenes of elite tombs until the reign of Niuserra: by this time, the lion motif was no longer restricted to the royal prerogative.[19] By the early Middle Kingdom, for example, in the tomb of Khnumhotep II, representations of felines appear in a way to suggest that the tomb is highlighting his power, with the individual emphasising his prerogative to use this motif. Khnumhotep II's biography[20] suggests that his governorship was not appointed but perhaps inherited – a 'fiefdom', perhaps.[21] From the distribution of the data, it appears that the reign of Niuserra was a significant time with respect to a change in the proportion of desert hunt representations. Many new scene types within the desert hunt theme appear in the iconographic record at this time (for an overview of this progression through the Old Kingdom, (see Table 10.2).

10.3. Changing Decoration Variants

While the wavy desert baseline is a feature common throughout the entire era in question, none of the Dynasty 4 tombs depicted desert foliage. However, over the time frame under investigation, the components making up the desert hunt scene type displayed changes.[22] By the time of the Old Kingdom, the desert hunt motif had been well developed.

10.3.1. Increasing Complexity of Desert Hunt Scenes

While the proportion of desert hunt scenes did not increase over the time frame of the investigation as perhaps would be expected from the A.R.I.D. hypothesis, the variety of scene types burgeoned, as did the amount of detail within the individual desert hunt scenes themselves.[23] The number of 'scenic-elements' appears to have increased. In tombs containing a desert hunt scene, the number of different 'scene-elements' within each desert hunt scene was identified. The average quantity of components was plotted over time and presented in a scattergram (see Figure 10.2), to see the relative increase in scene elements within individual images. Each point represents the proportion of desert scenic elements for each king and the red line represents the average tendency. The data suggests a slight increase in the number of components that made up the individual desert hunt scenes (rising from 20% to 30% according to the trendline). As the land was slowly drying, the desert came closer to the river and the society interacted more with the desert environment.

7 Strandberg, *Gazelle*, 57–71; Fragments BM 20790 and BM 20792; and Louvre E 11254.
8 Quibell, *Hierakonpolis*, pl. 26; Adams, *Early Egypt*, 1.
9 Strandberg, *Gazelle*, 55, n. 43; Borchardt, *Sahure*, pl. 17.
10 Von Bissing, *Trois Saisons*, pl. 11, 21, 33.
11 Von Bissing, *Trois Saisons*, pl. XI a.; Smith, *Interconnections*, fig. 178; Edel & Wenig, *Königs Ne-user-Re*, pl. 14.
12 Strandberg, *Gazelle*, 53.
13 Petrie, *Medum*, 25, pl. xvii [left]; 23, 37, pl. ix [lower]; Harpur, *Maidum*, 77–91, 201–202, figs 80, 88, pls 11, 37; HESPOK, 151–152, fig. 60; Harpur, *Decoration*, 82.
14 Harpur, *Maidum*, 77, 79, fig. 80, pl. 11–12.
15 Harpur, *Maidum*, 89–90, 94, 201–202, fig. 88, pl. 37.
16 Harpur, *Maidum*, 61–62, 72–73, 183–184, fig. 74, pl. 5; also Klebs, *Reliefs*, 68.
17 Vachala, *Ptahshepses*, 120–121; Hayes, *Scepter of Egypt*, 98–99, fig. 56.
18 Harpur, *Decoration*, 82; Swinton, *Resources*, 27–50, tables 2–4; recently, Espinel, *Desert Hunt Iconography*.
19 Garnot, *Le lion*, 75–91; Evans, *Animals*, 9–10, 113–114; Evans & Woods, *Twins*, 64–66.
20 Newberry, *Beni Hassan* 1, 56–67; updated Kanawati & Evans, *Beni Hassan* 1.
21 See Kanawati & Woods, *Beni Hassan*, 34–35.
22 Kanawati, *Tomb & Beyond*, 9697.
23 Thanks to Miroslav Bárta for this idea.

Table 10.2: Desert hunt scene variants (OEE Scene-details Database): D4à6

Scene Details – desert related	Date																			
	IV.1	IV.2	IV.3	IV.4	IV.5	IV.6	V.1	V.2	V.3	V.4	V.5	V.6	V.7	V.8	V.9	VI.1	VI.2	VI.3	VI.4	VI.5+
2.1.1. Major figure holding leashes attached to dogs depicted in sub-registers	1																			
2.1.2. Hunter in the desert		1	1									4		1	4	3	1		7	6
2.1.3. Dogs hunting in the desert	4											4		2	6	3	1		7	6
2.1.4. Puppy accompanying a hunter															1				1	
2.1.5. Wild ox in the desert	1											2			4	1	1		3	3
2.1.6. Lion in the desert												1	1	1	5	2	1	1	1	1
2.1.7. Leopard in the desert	1											1			2	1			1	
2.1.8. Jungle cat in the desert												1			1					
2.1.9. Caracal in the desert															1					
2.1.10. Wild ass in the desert															1				1	
2.1.11. Deer in the desert	1														1	1	1			
2.1.12. Oryx in the desert	1											2			5	3	1	1	5	1
2.1.13. Roan antelope in the desert	1															1			1	
2.1.14. Ibex in the desert	2											4	1		3	3	1			2
2.1.15. Barbary goat in the desert												1			1					1
2.1.16. Addax in the desert												1		1	2	1				
2.1.17. Hartebeest in the desert	1											2		1	3	2	1	1	1	1
2.1.18. Gazelle in the desert (Dorcas gazelle)	3											4	1	1	6	3	1	1	3	2
2.1.19. Gazelle in the desert (Soemmering's gazelle)	1											1			1	1	1			
2.1.20. Gazelle in the desert (Isabella gazelle)												1								
2.1.21. Hyena in the desert												2			3		1			1
2.1.22. Jackal in the desert															1		1			
2.1.23. Fox in the desert	2											3			2				1	
2.1.24. Hare in the desert	3											3	1		1	1			1	
2.1.25. Ichneumon in the desert															1					
2.1.26. Jerboa in the desert	1											1			2					
2.1.27. Ratel in the desert	1														1					
2.1.28. Weasel in the desert	1											1			1					
2.1.29. Rat in the desert	1														1					
2.1.30. Porcupine in the desert												1								
2.1.31. Hedgehog in the desert	1											3		1	3	3	1	1		2
2.1.32. Ostrich in the desert															1					
2.1.33. Guinea fowl in the desert															1					
2.1.34. Young antelope, usually crouching in a sub-register, all species												4		1	4	3	1		1	
2.1.35. Animals with fatty appendages, or 'wattles', all species															1				1	

Table 10.2 continued

Scene Details – desert related	Date																			
	IV.1	IV.2	IV.3	IV.4	IV.5	IV.6	V.1	V.2	V.3	V.4	V.5	V.6	V.7	V.8	V.9	VI.1	VI.2	VI.3	VI.4	VI.5+
2.1.36. Mating animals, all species												1			3				1	
2.1.38. Animals suckling young, all species												2			2					
2.1.39. Animals defecating, all species													1		2	1				
2.1.40. Net barrier enclosing desert animals												1	1	1	2	2	1			
2.1.41. Wavy desert baseline	2											4		2	4	3	1		2	5
2.1.42. Sand grains or pebbles defined on the desert floor	1														1				1	2
2.1.43. Foliage, mainly shrubs and succulent plants, growing in the desert												3	1	2	4	3	1		1	1
2.2. Hunters returning from the desert with game	1											1	1		3	1	1	1	1	1
TOTAL	**31**	**1**	**1**									**59**	**8**	**14**	**91**	**42**	**18**	**6**	**59**	**35**

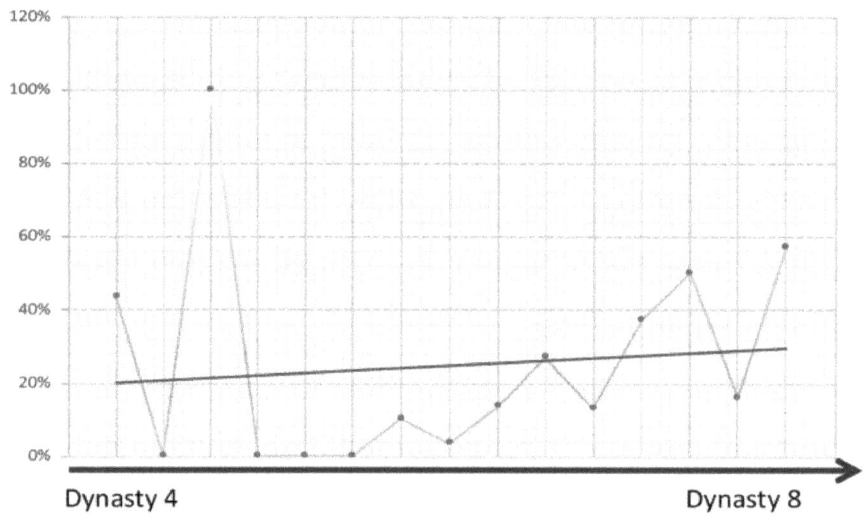

Figure 10.2: Component elements of desert hunt scene.

Table 10.3: Desert scenes: Depictions of hunters – pre-and post Niuserra

Desert hunters	Date		
	Pre-Niuserra	Niuserra → D6	D6 Onwards
Proportion of all scenes	2.6%	1.7%	2.4%
Theme = Desert Scene	2.2%	1.3%	1.8%
Theme = Hunters	0.4%	0.4 %	0.6%
Proportion = Hunters in Desert Scenes	15%	24%	25%

10.3.2. Desert Hunters

The desert hunt scene type chronicles a more significant change, however, with an increased proportion of the scenes' features focusing upon hunters returning with game. An earlier emphasis upon the hunt itself, inferring dominance, power and control[24] and what it represented had shifted to an increasing highlight on the rewards of the hunt (see Table 10.3). Later Old Kingdom hunting scenes depict more of the prey and convey a sense of a process more 'mass-produced' and 'industrialised' and less ritualistic. The changes in proportion of decorative elements depicting hunters is significant (see Figure 10.3).

10.3.3. The hunter as Archer

Bow-armed hunters appear for the first time in a non-royal context during the Sixth Dynasty.[25] In the tomb of

[24] Hassan, *Followers of Horus*, 316; Kemp, *Anatomy*, 81.
[25] Thompson, *El-Hawawish*, 63; Hoffmeier, *Hunting Desert Game*, 8–13; Espinel, *Desert Hunter Iconography*, 91; Borchardt, *Sahure* II, pl. 17.

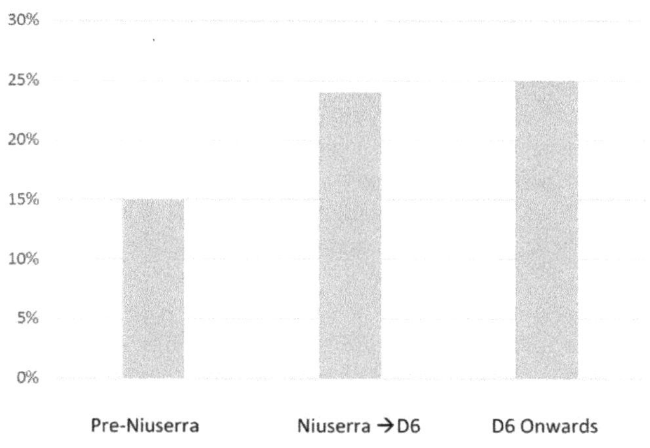

Figure 10.3: Desert scenes: proportion of hunter depictions change.

Khunes at Zawyet el-Maiyetin, craftsmen are depicted constructing a bow.[26] The context appears to be non-military, so might be related to hunting. The tombs of Ibi at Deir el-Gebrawi,[27] Idu/Seneni at el-Qasr wa-'l-Saiyad[28] and Meru at Naga ed-Deir[29] all depict hunters as archers. Among the earliest representations of non-royal archers in the Old Kingdom, however, who is the oldest may still be under debate.[30] The tomb of Khunes at Qubbet el-Hawa depicts an archer in the left background of the combined fishing and fowling scene.[31] Despite the growing ranks of archers in tomb scenes, no Old Kingdom tomb owner depicted himself in such an active role in the desert hunt.[32]

By the end of the Old Kingdom, bow and arrows had become status symbols,[33] becoming more frequently represented in funerary decorations and equipment during the First Intermediate Period and into the early Middle Kingdom.[34] The tomb of Ankhtify at Mo'alla is another that contains a desert hunt scene where a significant figure is depicted as an archer, however, Vandier does not identify the archer as Ankhtify.[35] If the person represented is indeed Ankhtify, then it is a unique representation of a non-royal individual representing himself as an archer in a desert hunt before the Middle Kingdom.[36] Perhaps we are seeing a developing self-perception of one who no longer sees himself as governing on behalf of the King but is rather himself acting in a king-like manner.[37]

As the importance of acquiring resources increased, so did a more pragmatic approach to the collection of food, as archers began to be depicted hunting. There appears to be an increasing tendency to represent hunters with markers suggesting Nubian ethnicity.[38] While Nubians have been identified as mercenaries,[39] perhaps a supplementary role as hunters can also be considered.[40] Espinel has plotted the change in abundance of hunting tools depicted in tomb scenes, producing results where the object in the scene, the 'tool' or 'weapon', is identified and recorded,[41] tracing the change in type of weapon and contrasts its aim to either the 'hunt' or the 'capture' of the animal.[42] The theme of hunter depicted in the desert is present in almost two-thirds of tombs dating to Dynasty 5 and more than three-quarters of tombs dating to Dynasty 6.[43]

10.3.4. Increasing Artistic Prioritization of Depicting the Desert Hunt?

In representing the desert floor in the form of a wavy line, skilled artists could identify singular characteristics in their surrounding environment.[44] Some tomb owners chose desert-related themes as their only form of tomb decoration within their tombs,[45] suggesting that, to at least those two named individuals, this particular type of tomb scene must have had an especial significance. Note, however, that only one Dynasty 6 tomb, that of Nehwet-desher-Meri at Akhmim, seems to have depicted no other scene than the desert hunt.[46] Boessneck identified the representations of various desert animals over dynastic history and, from this, calculated their apparent importance, identifying the relative importance of these animals at the end of the Old Kingdom.[47] The relative importance of the desert hunt may be measured in the comparative abundance of the animals within the scenes. Some desert animals appear to be more significant than others.[48] Otherwise, why would Nimaetre, for example, take the time to have an ostrich or a caracal so carefully rendered?[49]

10.4. Desert Resources and A River In 'Drought'

As a consequence of a weakened river diminishing the width of the valuable settled farming land between river and desert, and excess nutrient load increasing the riparian floral zone, desert animals would have ventures closer to the river's edges. This would have brought desert animals into increasing contact with the sedentary population bedside the river. Resources from the desert and desert trade increased during the time frame under

[26] Lepsius, *Denkmaler*, II, Band IV, Bl. 108.
[27] Kanawati, *Deir el-Gebrawi*, II, 44–45, pl. 52, 71b.
[28] Säve-Söderbergh, *Hamra Dom*, pl. 10.
[29] Peck, *Naga ed-Dêr*, 99, pl. 7, 10, 12.
[30] Kanawati & McFarlane, *Akhmim* I, n.108.
[31] de Morgan, *Catalogue*, 159.
[32] Decker & Herb, *Bildatlas zum Sport*, 313–314, pl. 142.
[33] Fischer, *Nubian Mercenaries*, 44–80; Fischer, *Egyptian Studies*, 84–85.
[34] Seidlmayer, *Gräberfelder aus dem Übergang*, 194; Grajetzki, *Burial Customs*, 37; Zitman, *Assiut*, 195.
[35] Vandier, *Mo'alla*, 93–95, figs. 29, 45–46.
[36] Vandier, *Mo'alla*, 271–274, fig. 81, pl. 42.
[37] Kanawati & Woods, *Beni Hassan*, 31–33.
[38] Pemler, *Looking for Nubians*, 443–449.
[39] Edel, *Qubbet el-Hawa* 3, 4, fig. 7, pl. 74; Morenó García, *Ancient Egyptian Administration*, 147–149.
[40] Espinel, *Desert Hunter Iconography*, 100.
[41] Espinel, *Desert Hunters Iconography*, tables 1–3.
[42] Espinel, *Desert Hunters Iconography*, 91–94.
[43] Depicted in ~60% of D5 tombs & ~80% of D6 tombs: Espinel, *Desert Hunters Iconography*, tables 1–3.
[44] Kanawati & Woods, *Artists*, 71–72.
[45] HESPOK, 151–152, fig. 60 and 167, 170, fig. 65.
[46] Kanawati, *Akhmim* VIII, 12, pl. 2a, 8a, fig. 3b.
[47] Boessneck, *Die Tierwelt*, 36.
[48] Herb & Förster, *Desert to Town*, 26–27.
[49] Roth, *Palace Attendants*, 132–3, pl. 95–97, pl. 189.

investigation.[50] The increasing representation of salted fish as a trade good with mobile communities based away from the river suggests that this communication was mutually beneficent. By the opening years of the Predynastic the land had already begun to dry out. The increasing hyper-aridity made the hunting of larger game easier as these animals became increasingly dependent on the water and plant resources of the river.[51]

10.4.1. Hunting or Domestication in a Time of a Changing Climate?

Significant also is the change from representations depicting sport to those using barriers or traps, serving to indicate a more 'industrialised' process in the 'harvesting' of desert produce. Swinton suggests a greater value was placed on desert animals in the funerary parade, identifying their attendants as of higher status, indicating these animals were becoming increasingly prized.[52] Whether the Egyptians domesticated the desert animals, or whether these animals had simply become more accustomed to human contact due to the forced interaction caused by the thinner ecotone and were therefore easier to capture, is an interesting investigation. The size of the stockades built to contain them suggests domestication was not very successful. Swinton considered an organised breeding program for desert ungulates, but suggests the scenes identified are more related to a process of 'taming' the animal.[53]

Herb and Förster and Linseele and van Neer believe that herds were kept on a more permanent basis,[54] as desert animals had become familiarised to humans and their cattle.[55] They have summarised the various archaeological and epigraphic evidence to develop a "*hierarchy of importance*" of large animals, identifying that cattle were the most significant large mammal, followed by the oryx, the ibex and then the gazelle.[56] Perhaps the term 'desert hunt' is inappropriate. It may seem more accurate to identify the desert sequences as the gathering and keeping of desert animals before their utilisation:[57] a form of desert husbandry, perhaps.

10.4.2. A Deepening Dependency on the Desert

During the Great Wet Phase of the Terminal Pleistocene, (~11,000 BP), it seems that humans had ready access to game,[58] with environmental conditions so favourable that there would have been little impetus to change their lifestyle. Aridification of the environment may have led to increasing attempts at domestication of desert animals, because desert animals are better adapted to irregular feed and variable-quality drinking water.[59] Perhaps the rapid increase in the representations of goats may be another indicator of the need to endure (stomach?) food grown in less desirable environments.[60] With an ability to follow the feed, animals that could forage as they moved may have become more viable in the increasing aridity at the end of the Old Kingdom.[61] Their increased exploitation may explain their increasing abundance in the archaeological record.[62] Mobile wealth may have been a strategy adopted at this time. It seems that cattle and desert mammals were kept in the same type of enclosure, suggesting that they were treated in a similar manner.[63]

Whatever the outcome of these attempts, desert animals became an increasingly important component of the resource mix. The epigraphic and archaeological evidence suggests that the exploitation of desert mammals increased over the Fifth Dynasty, through the Sixth Dynasty and into the Middle Kingdom.[64] The procession scenes depicted at Abusir[65] where cattle are accompanied by desert animals indicate an increasing habituation of the desert animals to cattle and their human handlers.

10.5. Summation: An Increasing Necessity to use Desert Resources?

The major points raised in this chapter include:

- Not as many scenes depicting desert animals are attested as would be expected for a time of 'drought'.
- All tombs dated to the Sixth Dynasty that depict desert scenes also depict marshland scenes.
- The variety and diversity of subjects within the desert hunt scene increased as did the richness and complexity of the compositions, suggesting a different attitude to the desert hunt.
- The emphasis appeared less focussed upon individual prowess and more on the success of the 'harvest'.
- Scenes depicting the intermingling of desert animals and 'regular' cattle, both big and small, suggest an overlapping of the marginal lands.
- Desert animals appear to have become habituated to humans and became increasingly portrayed in the procession scenes produced at the end of this period.

[50] Negus, *Great African Drought,* 320–321.
[51] Herb & Förster, *Desert to Town*, 19, 33–34; Moreno Garcia, *Paturages de vaches tachetes*, 241–242; Swinton, *Resources*, 244–245; Linseele & van Neer, *Archaeology of Exploitation*, 70–72; Pantalacci & Lesur-Gebremariam, *Wild Animals Downtown*, 253–256.
[52] Swinton, *Resources*, 248; see Altenmüller, *Mehu*, 156–162, 193–198, figs 12, 19, pl. 48–51, 72–74.
[53] Swinton, *Resources*, 246–247.
[54] Herb & Förster, *Desert to Town*, 22; Linseele & van Neer, *Archaeology of Exploitation*, 47.
[55] Pöllath, *Prehistoric Game Bag*, 79; Manlius, *Historical Biogeography*, 114–115, fig. 1; Linseele & Van Neer, *Archaeology of Exploitation*, 53, 70–71.
[56] Herb & Förster, *Desert to Town,* 27–33, fig. 14.
[57] Herb & Förster, *Desert to Town*, 17–19.
[58] Muzzolini, *Saharan Food Economy*, 228.
[59] Brass, *Cattle Cult Origins*, 104; Kay, *Responses to Drought*, 683–694.
[60] Shaker, *Salt Tolerant Goats*, 66–77.
[61] Wright, *Human Holocene Termination*, 5–8.
[62] Linseele & Van Neer, *Archaeology of Exploitation*, 47–78; Pollath, *Prehistoric Gamebag*, 79–108.
[63] Herb & Förster, *Desert to Town*, 22; Linseele & Van Neer, *Archaeology of Exploitation*, 47.
[64] Boessneck, *Die Tierwelt*, 35; Herb & Förster, *Desert to Town*, 32; Linselele & Van Neer, *Archaeology of Exploitation*, 47.
[65] Vymazalová & Arias Kytnarová, *Sheretnebty*, 41; Dulikova, *Ankhires,*11.

- As the river failed to deliver excess water, the width of the arable zone would have shrunk, bringing desert animals into increasing contact with the more sedentary population living bedside the river.

The increasing relative importance of the desert hunt to tomb owners is evidenced by an increasing complexity of desert hunt composition. Desert animals would have found themselves in more direct competition with the cattle, sheep and goats of the settled population. Their gradually more permanent presence, so close to the riparian zones, and their increasing familiarity with humans and their herds, would have made their capture easier.

Part D

Discussion and Conclusions

11

Art Imitating Life? Niuserra to the First Intermediate Period

During A.R.I.D. times, the river would have accumulated nutrients at a greater rate than normal. It would have developed ecological characteristics that were different to those from the normal situation, and as such would have exhibited environmental features different to those customarily demonstrated. It was expected that these changes should have impacted upon the visual culture produced at that time. A number of associations between the environmental changes occurring at the end of the Old Kingdom and variations in the decorative programmes produced in tombs at that time have been noted. While it is difficult, however, to progress to the next level of certainty and claim that this investigation has provided historical evidence of a significant correlation between the two factors: environmental and cultural – as with so very many coincidences being presented, then taken as a whole, some form of connexion appears highly likely. To further this study, it became important, therefore, to relate the environmental situation in which these tomb owners found themselves to the wider historical, social and economic context of this era. In this chapter, the inferences outlined by the A.R.I.D. hypothesis will be investigated in the broader context of ancient Egypt at this time.

11.1. The Ascent of the Marshland Economy?

The period under investigation witnessed a significant increase in the abundance of tomb scenes displaying representations of fishing and its associated techniques and technologies. This suggests that the river came into greater focus in the Egyptians' acquisition and exploitation of food resources, as has been suggested by the A.R.I.D. hypothesis. Many of these activities took place in areas that, with regard to agricultural purposes, Morenó García labels as "marginal".[1] These activities, based on living in the marshes and behaving as pastoralists herding goats instead of cattle, and perhaps swine, could more readily have supported alternative, more independent lifestyles, a likely response to an increasingly unreliable river. This appears to have had already occurred in the Levant, where some populations appeared to have decided to 'absent themselves' from the cycle of cultivation and its political responsibilities.[2] In some areas, such as the delta and along the desert margins, the collapse of central authority and its associated redistribution system, stimulated a return to more mobile lifestyles, based on the exploitation of a greater variety of resources that were more readily available in a local context.[3]

11.1.1. Resources Change in Response to a Changing Environment

The economy of the Old Kingdom was a delicate balance between those resources obtained from cultivation, cattle, the river and the desert.[4] This state of equilibrium was not static but existed in a constant state of flux, which passively adjusted itself according to circumstance. In the normal situation, resources were accessed from the river, its edges, and the desert, but primarily from the results of cultivation. If the river did not break its banks, then the overall yield from cultivation would have declined. Since the river had become more unreliable, then it may have become necessary to look for other food resources to supplement the regular diet. In times of an unreliable river, the ability for cultivation to remain the major source of resources was limited: consequently, the other sources of nutrition became relatively more important (see Figure 11.1).

11.1.2. Dietary changes in A.R.I.D. Times: Fish, Fowl and Cattle

As a regular agricultural surplus became less assured, it is possible that the flourishing marshland ecotone provided additional food resources.[5] In northern Sudan, evidence suggests that, in times of ecological stress, at about the same time Egypt was experiencing an increase in aridity, fishing strategies had been adapted to cope with the circumstances of a changing environment.[6] Fish bone concentrations have been shown to change in archaeological sites in parallel response to changing water levels.[7] The increasing proportion of catfish bones in some of the excavations from Abusir indicate a changing dietary balance, as fish more suited to brackish situations became more common.[8] A similar rapid increase in the abundance of scenes depicting the gathering and presentation of waterfowl has been identified in this study, and has been noted in some excavation reports.[9] The rapid increase in abundance of cattle attestations during this timeframe

1 Morenó García, *Terrtorial Administration*, 93.
2 Greenberg, *No Collapse: Transmutation*, 48–50.

3 Morenó García, *Terrtorial Administration*, 88–101.
4 Herb & Forster, *Desert to Town*, 19; Linseele & Zerboni, *Done with Fish*, 235.
5 Maltby & Acreman, *Ecosystem Services*, 1341–1359.
6 Linseele & Zerboni, *Done with Fish*, 237–238.
7 Willems etal., *Wadi Zabayda*, 318; Linseele & Zerboni, *Done with Fish*, 237–238.
8 Dulíková, *Ankhires*,27, 30; Willems etal., *Wadi Zabayda*, 318, 325.
9 Vymazalová & Arias Kytnarova, *Sheretnebty*, 441.

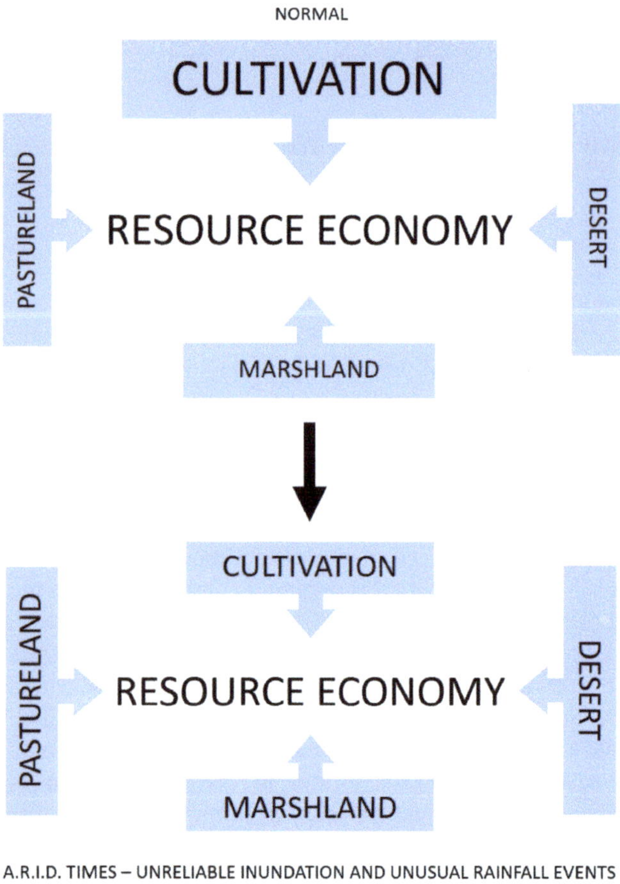

Figure 11.1: Changing Aspects of the Resource Economy.

likewise suggests that these animals had become relatively more important to the culture.

11.1.3. The Persistence of the Hippopotamus Hunt

The depiction of the hippopotamus hunt continued through the Old Kingdom from the time of Menkaura (IV.5) to beyond the time of Pepi II.[10] Originally symbolic of the power of the king over the forces of chaos,[11] depictions of the hunt continued, but with little suggestions of the royal ritual which earlier depictions evoke. In the great majority of tomb decorations depicting hippopotamus hunting scenes, the activities are performed by minor figures tasked with protecting the noble.[12] Of the major (non-royal) players depicting themselves as hippopotamus hunters, Khentika and Medou-nefer (Dynasty Six) at Balat in the Dakhlah Oasis.[13] Comparison with other tombs constructed during the time of Pepi II, the quality of the tombs prepared for the governors indicate the increased wealth flowing to the Oasis at this time,[14] and may also indicate a declining range of royal influence. They may have been far enough away not to have earned the ire of the King for assuming the royal prerogative as the killer of a hippopotamus.

It is possible to relate the persistence of the hippopotamus hunt to the A.R.I.D. hypothesis presented in this study. As a voracious, herbivorous, aquatic animal, the hippopotamus would have benefited from the thriving plant growth that resulted from the nutrient bloom in a shrinking Nile. Interestingly, hippopotamus grazing appears to enhance the variety of the riverine vegetation which make the area more attractive to other herbivores.[15] The ancient Egyptians prized the ivory teeth of the hippopotamus, which also provided meat, fat and skin,[16] so the continued hunting made nutritional sense.

Cattle would have taken advantage of the trails bulldozed through the marshes by the hippopotamus to reach the most succulent parts of the plant. This would have taken the hippopotamus into direct competition with the increased herds of cattle that were also thriving on a burgeoning plant growth.[17] In areas that are also exploited by hippopotamus cattle eat more of the available food before moving on: this leaves behind areas of stripped vegetation, typically less suitable for hippopotamus foraging.[18] With an increase in the number of secondary channels, incidences of contact between highly territorial hippopotamus[19] and cattle and their human handlers would have risen. The need for the Egyptians to protect cattle (and themselves)[20] from these competitors would have intensified, indicated by a number of tomb scenes depicting the hippopotamus hunt adjacent to cattle crossing scenes.[21]

Scenes also place the hunt adjacent to other swampland activities, such as spearfishing[22] or fowling,[23] or even a pleasure cruise.[24] All of these activities would have disturbed the hippopotamus in their natural habitat. The tomb of Shepsipumin/Kheni at el-Hawawish depicts the hunting scene on the same wall as other marsh activities, this time with dragnet haulers being protected by the

[10] Harpur, *Decoration*, 181, contra Hassan, *Gîza* IV, 175, fig. 121. See Hemre/Isi: Davies, *Deir el-Gebrâwi* II, pl. 20; Kanawati, *Deir el-Gebrawi* I, pl. 64. See Woods, *Marshes*, 2, table E: feature #66.

[11] Evans, *Animal Behaviour*, 15; Säve-Soderbergh, *Hippopotamus Hunting*, 15–19, 52–56; Hendrickx, *Hunting & Social Complexity*, 249–250.

[12] Woods, *Marshes* 2, 87–91, table A; Hamed, *Artistic Perspectives*, 11–12.

[13] Castel etal., *Khentika*, fig. 86 and Valloggia, *Medou-nefer*, pl. 41, respectively.

[14] Morris, *Art of Not Collapsing*, 78.

[15] Kanga, etal, *Landscape Ecological Engineers*, 2013, 9: 47; https://doi.org/10.1007/s11355-011-0175-y (03/05/2019).

[16] Stünkel, *Hippopotami*, http://www.metmuseum.org/toah/hd/hipi/hd_hipi.htm (10/06/2018).

[17] Hendrickx, *Hunting & Social Complexity*, 249; Muller, *Feindvernichtungsrituals*, 488.

[18] Perry, L. R., Observations of Hippopotamus H, https://digitalcommons.mtu.edu/etdr/41 (13/05/2019).

[19] Karstad & Hudson, *Riverine Hippopotami*, 154, 157–162; Dudley, etal., *Hippopotamus Carnivory*, 192, 196; Stommel etal., *Hippo Vulnerability*, 2.

[20] Evans, *Animal Behaviour*, 134.

[21] Hassan, *Giza* II, figs 140, 236; Schürmann, *Ii-nefret*, pl. 21; Kanawati, *Unis* II, pl. 54; Kanawati & Evans, *Meir* II, pl. 90; Davies, *Deir el Gebrâwi* II, pl. 20; Kanawati, *Deir el-Gebrawi* I, pl. 64.

[22] Kanawati & Hassan, *Ankhmahor*, pl. 37c; Duell, *Mereruka* I, pl. 9–13.

[23] Schürmann, *Ii-nefret*, pl. 6, 21; McFarlane, *Unis* I, pl. 41, 46; Kanawati & McFarlane, *Deshasha*, pl. 48.

[24] Wild, *Ti* II, pl. 119; Brovarski, *Senedjemib* II, fig. 42; Woods, *Marshes* I, 216–224; table 9.1; II, pl. 103–116.

hunters.[25] At Meir in the tomb of Pepyankh the Black, the hippopotamus hunt scene is adjacent to clapnetting, seine net hauling, cattle crossing and other swampland activities.[26] On the northern wall of room 6 in the tomb of Ty at Saqqara, the tomb owner is supervising a hippopotamus hunt.

The other large animal that shared the river with the humans and hippopotami was the crocodile. The increased movement of cattle across the small fords and secondary channels must have captured the interest of the crocodiles.[27] As the river shrank, the interactions between crocodiles and hippos must have become more noticeable and, coincidently or not, they entered the tomb decoration programme during the time frame under investigation.[28] Woods, updating Harpur, suggests that the scene of crocodile and hippopotamus fighting enters the repertoire at the same time as the river appears to become a more integral part of the resource cycle.[29] No hippopotamus hunt scene remains in the tomb of Pepyankh the Middle, but a crocodile is shown in its death throes in the jaws of a hippopotamus.[30]

11.1.4. Back to the Past: Return to a Semi-subsistence Existence?

With the new food arrangements, the economy appears to have become became more localised. If so, how did this more independent and semi-subsistence economy play out into larger Egyptian society at the time? New food resources would have been sought. The food sources that replaced grain would have been less storable for re-distribution, and subsequent re-distribution, so local redundancies may have developed.

While all this change may have been stressful for most people at the time, it may also have provided a bounty of socio-economic opportunities and greater health, at least for some localised populations. In a manner similar to earlier societies,[31] and in association with selective herding practices, some new hunting/gathering/foraging activities were adopted in order to utilise the extra produce arising from the newly bountiful 'drought'-ridden riverine habitats.[32] A partial return to a semi-sedentary way of life, or adoption of a "*periodic nomadism*"[33] may have occurred in response to a slight decline in cultivation success.[34] In a manner to the survival subsistence practices adopted by of the Mesopotamian marshland dwellers when mainstream agriculture collapsed,[35] low intensity cultivation with many and varied resources would have made the acquisition of adequate subsistence sustenance more readily achievable.

11.2. Tomb Decorations in pre-A.R.I.D. Cemeteries

Many tombs in these areas dating from before the time of Niuserra display a different decorative programme to that produced after this time.[36] A comparison of the different patterns of tomb decoration within these cemeteries provides a contrast between themes relevant to the tomb occupants at those earlier times and those whose later tomb owners experienced times of ecological hardship.

11.2.1. Pre-Niuserra Cemeteries: Tehna and El-Hammamiya

A number of cemetery sites rose to prominence before the end of the Fifth Dynasty and, therefore, theoretically before the impact of climate change was making its influence noticeable in the tomb scenes. However, the number of provincial cemeteries with tombs dating prior to the mid-Fifth Dynasty is limited, making comparison about the tomb scenes difficult. Both the cemeteries at Tehna and El-Hammamiya contain tombs dated to the time before Niuserra. This was period when, according to the artistic analysis presented earlier in this study, environmental factors had yet to have had an influence of the types of scenes that were depicted on tomb walls. If the changing riverine environment was, as yet, NOT an influence, then this may explain the lack of depictions of fishermen using small nets and these individuals displaying techniques more suited to a crowded riverbank before the tome of Niuserra.

The Old Kingdom cemetery at Tehna, in Middle Egypt, was initially described by George Willoughby Fraser,[37] with the cemetery currently being re-examined by an expedition based at Macquarie University, Sydney, Australia.[38]_This work is ongoing and has led to an increase in the total number of tombs recognised at this site.[39] The tombs in the cemetery are among the earliest burials of provincial officials in the Old Kingdom, dating from the end of the Fourth Dynasty and into the beginning of the Fifth. They include an exceptional administrative document outlining the distribution of temple income to family members.[40] While the tomb of Nikaiankh I at Tehna contains, perhaps uniquely an inscription of a royal endowment from Menkaure (IV.5) for land,[41] for a temple dedicated to Hathor, no other scenes depict any major acts of cultivation. The significance of this scene is that it was

[25] Kanawati, *El-Hawawish* II, fig. 22.
[26] Kanawati & Evans, *Meir* II, 48–52, pl. 90.
[27] Evans, *Animal Behaviour,* 127.
[28] Evans, *Animal Behaviour*, 14; Harpur, *Decoration*, 258.
[29] Woods, *Marshes* 2, table E, feature #177, updating Harpur, *Decoration*, 366–367; Evans, *Animal Behaviour*, 144–145.
[30] Kanawati, *Meir* I, pl. 79–80.
[31] Brass, *Cattle Cult Origins*, 102;Beierkuhnlein, *Occcurrence & Ecology*, 107–118.
[32] Bard, *Archaeology of Ancient Egypt*, 83–84.
[33] Brewer etal., *Domestic Plants & Animals*. 99.
[34] Daoud etal., *Adaptation & Resilience*, 209–250.
[35] http://www.clw.csiro.au/publications/consultancy/2004/Mesopotamian-marshlands-soil.pdf. (18/5/2019).
[36] Krejči, *Ptahshepses*, 258–259.
[37] Fraser, *Early Tombs*.
[38] Thompson, *Tehna* I.
[39] Lefebvre & Moret, '*Tehnèh'*, 30–38.
[40] Thompson, *Tehna Report*, 1–2.
[41] Thompson, *Tehna* I, 45–48, pl. 22–23, 56–57, updating Fraser, *Early Tombs*, 126–127 and Strudwick, *Texts*, 216–218.

prepared before the apparent environmental changes that are under investigation and suggests cultivation as the pre-eminent resource. They do not depict scenes associated with the acquisition and exploitation of marshland resources, and so follow the pattern proposed earlier. Due to the early date, none of the tombs at Tehna contains the fishing scenes that became more prevalent with the 'drought'.[42] This is even more noteworthy considering the site of the necropolis has ready access to the river.[43] The bulls portrayed in the tomb scenes display no signs of the aggression on show in the later tombs at Meir and El-Hawawish.[44]

Decorations prepared at the Fifth Dynasty necropolis of El-Hammamiya display features more similar to those at Tehna than to those prepared in the nearer cemetery of El-Hawawish,[45] which is dated later: to the second half of the Fifth and the Sixth Dynasties. Note, however, that some Memphite officials had begun to prepare their tombs in the provincial capital to which they had been assigned. Perhaps this an example of the transferral of artistic influence from the capital arriving with the entourage of the nobleman assigned.[46] Similar to those scenes at Tehna, the bulls depicted in the El-Hammamiya tombs of Kaikhent (A2) and Kaikhent (A3) appear quite docile and more passive than those depicted at El-Hawawish.[47] Interestingly, the tomb of Rehetep, dating closer to the time of Niuserra,[48] contains a number of scenes suggesting that environmental stimuli may have begun to have had an influence on the composition of its decorations. This appears to be the earliest example of a provincial tomb where scenes depicting the agricultural cycle have been added to the artistic sequence.[49] It also includes scenes of the cutting down of trees, though no goats are included.[50] While El-Khouli and Kanawati recognise the scene as incomplete,[51] they offer no suggestions as to what may be missing, although Harpur proposes that goats would have been present. Judging by the similarity of this scene to those at Zawyet el-Maiyetin,[52] El-Hawawish[53] and El-Hagarsa,[54] this seems the most likely interpretation. This addition of scenes depicting the agriculture cycle and goats in or near trees to the artistic repertoire at El-Hammamiya is consistent with the broader pattern identified earlier, whereby the new iconography may be part of a developing pictorial narrative influenced by environmental change.

11.3. Ra Setting... Osiris Rising in A.R.I.D. Times?

Concerns relating to an individual's religious beliefs may be identified upon tomb walls, as can concerns about the maintenance of one's good name: and other evidence of spiritual life and desires can be found in the tombs produced by the elite of the period.[55] The suggestion that the ancient Egyptians displayed a "*latent horror of death or extinction*"[56] may explain why so much effort was put into ensuring a successful afterlife. The linkages of art to society's beliefs, especially with regard to its ideas of a continued existence in the afterlife, allows for a simple understanding of their apprehensions.[57]

Due to Egypt's geographical features, it is no wonder that the religion of ancient Egypt reserved a special place for the role of the sun.[58] While the gods Ra, Ptah, Amun and Osiris appear to have been universal throughout the country,[59] the negative effects of the increasing aridity in the latter half of the Old Kingdom may have led to a decline in the relative importance of Ra, and an apparent increase in the veneration of Osiris, when compared to earlier times. The political machinations of the priests of Ra, for example, in influencing the enthronement of Userkaf as king[60] and providing legitimacy for his succession may have built up a resentment to their power.[61] The great wealth they accumulated[62] may have been another factor that led to a resistance to their active influence in the administration of the country.[63] Sun temples were no longer constructed after the time of Djedkara, and Menkauhor, for example, did not assume a 'Ra' name.[64]

It may be possible to ascribe an environmental influence behind a religious re-alignment from Ra to Osiris.[65] As the arid conditions continued, the accuracy of representations of particular birds associated with specific aspects of spirituality declined. By the end of the Old Kingdom, for example, the notion of the *ba* had entered the sphere of the private funerary ritual and mortuary cult,[66] but over this period, representations of the *ba* bird (and the *akh* bird) became less accurate, a consequence of the original birds becoming no longer regular visitors to Egypt.[67] Changing

---

[42] Thompson, *Tehna* I; neither Nikaiankh I, 30–48, Kaihep, 86–88, or Nikaiankh II, 66–76 depict river resources.
[43] Personal observation; also Thompson, *Tehna* I, pl. 2a.
[44] Thompson, *Tehna* I, pl. 14a.
[45] Thompson, *Tehna* I, 15–17.
[46] Swinton, *Dating Tombs*, 171.
[47] El-Khouli & Kanawati, *El-Hammamiya*, pl. 44–45 and 68, respectively.
[48] El-Khouli & Kanawati, *El-Hammamiya*, 16.
[49] El-Khouli & Kanawati, *El-Hammamiya*, pl. 73a (middle register).
[50] El-Khouli & Kanawati, *El-Hammamiya*, pl. 73a (bottom register).
[51] El-Khouli & Kanawati, *El-Hammamiya*, 74, pl. 73a.
[52] Lepsius, *Denkmäler*, B.II, Bl.II, Abt. 108, for Khunes; B.II, Bl.II, Abt. 111b for Niankhpepy.
[53] Kanawati, *El-Hawawish* I, fig. 15, pl. 10 for the tomb of Theti-iker; Kanawati, *El-Hawawish* VII, for Hesimin (fig. 3) and Gehesa/Nebi (fig. 30).
[54] Kanawati, *El-Hagarsa* III, pl. 22–23 for the tomb of Wahi and pl. 37, 40 for Mery-aa.

[55] Smith & Simpson, *Art & Architecture*, 104; Robins, *Egyptian Art*, 102; Kanawati, *Tomb & Beyond*, 112–122.
[56] Iverson & Shibata, *Canon & Proportions*, 6.
[57] Malek, *Egyptian Art*, 20–21; Shirai, *Ideal & Reality*, 325–333; Siebels, *Agricultural Series*, 55.
[58] Kahi, *Ra is my Lord*, 1.
[59] David, Religion & Magic, 57.
[60] Verner & Zemina, *Forgotten Pharaohs*, 102, 118; Goedicke, *Abusir–Saqqara–Giza*, 405–406; Malek, *Old Kingdom*, 98; David & David, *Biographical Dictionary*, 164; Verner, *Pyramids*, 263.
[61] David, *Religion & Magic*, 79; Gallardo, *Séquito de Horus*, 21–22.
[62] Dunard & Zivie-Cochie, *Gods & Men*, 85–86; Jacquet-Gordon, *Domains*, 106–108, fig. F; Janák etal., *Sun Temples*, 441–442.
[63] David, *Religion & Magic*, 111–112.
[64] Kanawati, *Conspiracies*, 3.
[65] Janák, *Extinction of Gods*.
[66] Altenmüller, *Sein Ba*, 1–15.
[67] Janák, *Extinction of Gods*, 125–129.

artistic representations of these birds has been identified, suggesting these birds appear to be no longer drawn from life,[68] linking this decline in artistic skill to the gradual disappearance of some bird species from the region; as due to a changing climate, presumably changing their migration routes.

11.3.1. Osiris in the Ascendant?

Osiris is the king under whom the blessed dead hope to spend eternity.[69] Osiris is an old god, found in early Egyptian religious thought, where he is more closely identified with the Delta than with locales further south and identified exclusively in private burials, not royal ones.[70] His first funerary representation appeared in a relatively insignificant manner in the reign of Djedkara, but he became more established in the later Sixth Dynasty.[71] Once his name began to appear in the tombs of high officials,[72] it became very common, appearing in almost every private tomb from late_Dynasty V onwards.[73] Eventually this deity assumed the role of the judge of the dead, indicating that he achieved an increasingly prominent role among the pantheon of the gods.[74] Osiris was originally identified as an exclusively Sixth Dynasty phenomenon[75] but there evidence indicating his worship beginning in the mid-Fifth Dynasty[76] – during the time frame of this study, or even earlier?[77] An increase in the importance of the cult of Osiris paralleled the decline in the construction of sun temples.[78] New, extravagant underground chambers that were added to tombs over this time frame may have linked the subterranean landscape to the domain of Osiris, as seen for example in the tomb of Kaiemankh at Giza.[79] It may be possible to attribute his 'ascendancy' in part to the potential environmental changes outlined in this investigation. The 'earthiness' of Osiris may be directly linked to the river, growth and fertility: a physical manifestation of a new, perhaps more pragmatic approach to worship, with people preferring the "*concrete humanity of Osiris over the abstract distance of Re.*"[80]

As well as the god of the dead, Osiris was the underworld identity who granted all life, including the sprouting of vegetation and the fertile flooding of the Nile River.[81] The linkage of Osiris to the mud of the inundation[82] his green skin colour may mimic the colour of green growth in a fecund river. 'Osiris burials,' where Nile silt in beer jars was placed alongside the body,[83] have been identified in a number of excavated tombs.[84] The layer of mud above the body is not the result of natural activity but has been shown to have been purposefully placed.[85] Since silt/mud is black and fertile, it contains within it the stuff of life,[86] these burials are possibly a response to the search for life and fertility; another potential link between the timing of Osirian influence and a changing river.

11.3.2. A Moral Re-alignment at the End of the Old Kingdom?

An apparent breakdown of order characterised the last years of the Old Kingdom, with the central government appearing to become more feeble and less proactive. Perhaps, due to this earthly disruption and to guarantee acceptance into the afterlife, there developed an increasing emphasis upon the need to live according to a strict moral code.[87] An increase in expressions of 'charity' became evident among the tomb wall scenes at the end of the Old Kingdom. Merer had recorded the following in his (Dynasty Eight) tomb at Dendera, "*...I gave bread to the hungry, clothing to the naked and protected the poor...*" [88] Perhaps this increase in personal charity developed in response to an increasing distrust in the government,[89] or perhaps a recognition of its inability to provide for its population. Perhaps the modest monuments produced by the kings during the latter stages of the Old Kingdom were not indicators of decreasing wealth but evidence of further efforts to minimise discontent[90] and to present a humbler attitude. Niuserra and Djedkara may or may not have had roles in disseminating this new religious understanding but, nevertheless, the struggles for authority[91] resulted in a shift in religious observances at about the same time as ecological change resulted in a shift in the variety of resources available for distribution among the population.

11.4. Trade, Warfare and Foreign Relations

Sowada suggests that the Egypto-Levantine trade had begun to diminish during the Fourth Dynasty and was virtually non-existent by the end of the Fifth Dynasty.[92] Since there appears to have been very little Egyptian

[68] Janák, *Early Ba Attestations,* 143–153; Janák, *Spotting the Akh,* 17–31.
[69] Andrews, *Book of the Dead,* 11–12.
[70] Griffiths, *Origins of Osiris,* 236–237; Shalomi-Hen, *Dawn of Osiris,* 460.
[71] Bárta, *Journey to the West,* 187; Mojsov, *Death & Afterlife,* 33.
[72] Hassan, *Gîza* II, fig. 228.
[73] Shalomi-Hen, *Writing of Gods,* 133–136; Mojsov, *Death & Afterlife,* 29.
[74] David, *Religion & Magic,* 95.
[75] Shalomi-Hen, *Dawn of Osiris,* 461; Dulíková, *Transformation,* 46–54.
[76] Griffiths, *Osiris Origins,* 44, 61–68; Bárta, *Architectural Innovation,* 120.
[77] Bolshakov, *Fourth Dynasty Osiris?* 65–80.
[78] Kanawati, *Government Reform,* 15.
[79] David, *Religion & Magic,* 135; Mojsov, *Death & Afterlife,* 30.
[80] Shalomi-Hen, *Dawn of Osiris,* 466.
[81] Wallis Budge, *Osiris & Resurrection,* 32.
[82] Mojsov, *Death & Afterlife,* 33.

[83] Bárta, *Qar,* 61.
[84] Dulikova, *Ankhires,*16, 30; Bárta etal., *Neferinpu,* 211; Bárta, *Shepseskafankh,* 25; Dulíková, *Transformation,* 56, fig. 7.
[85] Bárta, *Shepseskafankh,* 25.
[86] Bárta, *Journey to the West,* 211.
[87] Andrews, *Book of the Dead,* 12; David, *Religion & Magic,* 83; Kahl, *Ra is my Lord,* 51–60.
[88] Gee, *Old Kingdom Collapse?* 67–70; Petrie, etal., *Dendereh,* pl. 8; Franke, *Arme und Geringe im Alten Reich,* 104–120.
[89] Gee, *Old Kingdom Collapse?* 69–70.
[90] Morris, *Art of Not Collapsing,* 75–76.
[91] Gallardo, *Séquito de Horus,* 45–46; Bárta & Dulikovà, *Divine & Terrestrial,* 41–43; Schneider, *Existence Now Silenced,* 315.
[92] Sowada, *OK Eastern Mediterranean,* 11–12; Andrassey, *Ägyptens zu Vorderasien,* 130–132; Mazar, *Land of the Bible,* 141–142.

trade with the Levant by the end of the Fifth Dynasty,[93] it suggests that the Levantine territory had already begun to experience a decline in regional cohesion and an increasing fragmentation.[94] As such, luxury or exotic items from Egypt were no longer needed or could no longer be afforded,[95] or were no longer accessible. More inward-looking societies may have been less willing to receive trade envoys, they may have been less able to receive them with the same finesse and efficiency as previously, they may have been unable or reluctant to exchange luxury goods and may have had less surplus available for trade. Was it possible that the environmental challenges faced in the Levant were similar to those experienced in Egypt, and a non-sedentary mode of living was becoming more attractive with increasing power in the hands of the pastoralists?[96] If this was the case, then it is unlikely that an excess of supplies or stores would have been available for trade with Egypt.[97]

While ocean-going boats are attested from the reign of Sahura, fewer are attested in later periods of the Old Kingdom,[98] perhaps suggesting less sea-borne expeditions, thereby becoming less conspicuous as subjects for tomb scenes. As anticipated by the hypothesis, a changing river would have resulted in fewer navigable sections of the river, especially within the delta and outwards to the sea, decreasing the ability of deep-hulled vessels to negotiate an increasingly shallow and more plant-cluttered river, making these objects less common and therefore, less prominent.

11.4.1. Fall of the Cities in the Levant

The destruction layers excavated in many Levantine cities suggest regional unrest.[99] Höflmayer suggests that the fall of the first city-states of the Levant happened about 2300 BCE, which may be earlier than the dates suggested for the climatic events that may have brought about the decline of the Old Kingdom.[100] It is difficult to accept the proposition that Egyptian military activity, prompted literally by hunger, was involved directly in the fall of the first cities in the Levant, although Egyptian forces' increased activity in the latter half of the Old Kingdom, from the reign of Niuserra, including a most likely campaign by Djedkara,[101] may well have helped speed up the decline. It is important to consider the timing of the emergence of siege scenes within elite tombs, and what they represent.[102] Perhaps the siege scenes depicted in the tombs of Inti and Kaiemheset are of a forcible collection of 'tribute' that was no longer a regular part of the relationship between the two regions.

This Levantine disorder is unlikely to have arisen as a consequence of Egyptian foreign policy, for it followed a time when many cities in the region had completed a period of fortification.[103] The changing military architecture may suggest that fortifications became more necessary as urban populations built to protect themselves from worsening socio-political conditions.[104] The walls of these urban centres experiencing attack appear to be represented differently, with that depicted in the tomb of Kaiemheset, dating to the time of Niuserra, appearing a more primitive design than that depicted in the tomb of Inti, dated to the reign of Djedkara.[105] The latter scenes depict more substantial walls, more suitable to withstand a determined enemy: these are artistic representations, and their military effectiveness have been commented upon by Mourad and de Miroschedji as more substantial than those of the earlier depictions.[106] This may indicate an increasing effectiveness in defensive capabilities of small urban areas. Perhaps the difference in the military architecture represents one group of Asiatics had occupied an Egyptian fortified settlement and was in the process of being expelled. In order to distinguish between the context of the representations, these differences deserve a more detailed investigation by experts in the military architecture of the time.

11.4.2. Trade or Conquest?

This is a widely discussed issue in relation to Egypt and Levantine interactions.[107] The way out of Egypt, known in the New kingdom as the 'Way of Horus' followed pathways known from much earlier ways.[108] Porous borders, arising out of a diminished royal power, would have enabled mobile populations and traders to act in a more enterprising manner.[109] It may be that Egypt was no longer the dominant power in the Nile Valley with Nubian Chiefs and Caravan leaders becoming more locally pre-eminent.[110] More and more trade was being taken out of government control and usurped by adventurous individuals, and these actions may have encouraged the kings as attempts to minimise the increasing acquisition by individual Egyptians of more direct control over these areas of trade.[111] Perhaps the reason for the diminishing importance of the Abu Ballas Trail at the end of the Sixth

[93] de Miroschedji, *Egyptian-Canaanite Interactions*, 43–48; Mourad, *OK Siege Scenes*, 149; Sowada, *OK Eastern Mediterranean*, 251–253.
[94] Greenberg, *No Collapse, Transmutation*, 50.
[95] de Miroschedji, *Egyptian-Canaanite Interactions*, 47–48.
[96] Greenberg, *No Collapse, Transmutation*, 48–49.
[97] Kohl, *Archaeology of Trade*, 47.
[98] Bader, *Egypt & the Mediterranean*, 8; for Byblos; Espinel, *OK Egypt & Byblos*, 103–119.
[99] Mourad, *OK Siege Scenes*, 149.
[100] Höflmayer, *Dating Catastrophes*, 118–125, 132.
[101] Mourad, *OK Siege Scenes*, 149; Anđelković, *Hegemony for Beginners*, 789–808.
[102] Mourad, *OK Siege Scenes*, 148–149.
[103] Mourad, *OK Siege Scenes*, 149; Callaway, *Et-Tell*, 292–293. See also di Miroschedji, *Unis' Asiatic Campaign*, 273.
[104] See Burke, *Evolution of Warfare*, 64–65; Sowada, *OK Eastern Mediterranean*, 12; de Miroschedji, *Yarmuth*, 6–7.
[105] Kanawati & McFarlane, *Deshasha*, 17–19.
[106] Mourad, *OK Siege Scenes* and de Miroschedji, *Unis' Asiatic Campaigns*.
[107] Czarnowicz, etal., *Trade or Conquest?* 113; 122; Jirásková, *Egypt & Syria-Palestine*, 539–568.
[108] Förster, *Abu Ballas Trail*, 3, fig. 7; Hoffmeier & Moshier, *Highway out of Egypt*, 485–486; Stewart etal, *Tracing Ancient Tracks*, 197–221.
[109] Moreno García, *Territorial Administration*, 93.
[110] Morris, *Art of Not Collapsing*, 78–79; Snape, *Of Life and Death*, 111.
[111] Anđelković, *Hegemony for Beginners*, 789–808.

Dynasty suggests that this area had become increasingly difficult to police.[112]

The title 'Overseer of Troops', is noted as occurring with increased frequency at this time.[113] As Egyptian government control became more direct, tomb decorations began to include military themes,[114] starting from the mid-Fifth Dynasty, with tomb scenes depicting cities being besieged with the opponents are identified as Asiatics.[115] It appears these individuals were involved in and were commemorating the same campaign,[116] which Sowada has suggested was one directed by Niuserra against populations to the northeast of the delta.[117]

Those individuals from Elephantine who dealt with the Nubians also appear to have dealt with Punt.[118] Harkhuf, for example, mentions several caravans led by land and river to Nubia, the inscription of Khnumhotep in the tomb of Khui of Elephantine also suggests visits to Byblos and Punt; while another official from Elephantine, Ankhti, was killed by Asiatics, which indicates a location somewhere in the Eastern Delta, with Morenó García suggesting Ayn Sukhna.[119] The destruction at Mendes dated to the end of the Old Kingdom,[120] may be an indication of attacks from foreign powers ("*...the east abounds with bowmen...*")[121] or may be indications of the need for a violent re-affirmation of direct control by the central government.[122] A number of the injuries observed could have been caused by the attempted warding off of blows,[123] suggesting a violent end for the individual.

It is important to remember that armour was quite rudimentary at the time, with the technological advancements of Middle Kingdom military equipment a little way off, so simple arrows would have made quite effective weapons,[124] as the primary tactics for most of these societies was archery.[125] Even by the Middle Kingdom, armoured individuals were very scarce, and the weapons relied on were mostly effective against un-armoured opponents.[126] While the decorations in the tomb of Khunes at Zawyet el-Maiyetin display the making of bows and arrows,[127] the most significant weapon for the Egyptian military at this time,[128] those decorations adjacent are related to the acquisition of food and resources, so could presumably be representing their use in hunting.

11.4.3. Foreign Relationships in A.R.I.D. Times

While the central government may have been losing control over the peripheries,[129] beyond the eastern Delta, the oases and into the Sudan, the Delta itself was controlled differently to the rest of the country, being managed directly from the capital.[130] As a result of declining trade, by sponsoring their own expeditions to acquire resources directly, the kings may have been attempting to balance the decline in Egypt's own tax revenues brought about by diminishing yields from the centres of agriculture.[131] Was this a drawn-out campaign or one that developed spontaneously to take advantage of weakening control within the city states of the southern Levant? A lack of evidence of this occurring in the latter stages may suggest that their attempts at minimising the influence that powerful individuals may have been building were unsuccessful.

With the reduction in centralised authority in the region, it is more likely that the resources that were available would have been less well protected and, consequently, more vulnerable.[132] As one society deteriorated, others seem to take advantage of that situation. In the "dawn of internationalism,"[133] perhaps as a part of the attempted consolidation of power, a more aggressive policy was adopted towards neighbouring states.[134] Similar to the Early Dynastic government's aggressive stance towards Nubia[135] and the Delta regions during that period,[136] it is possible that an offensive strategy may have been adopted by the rulers at the latter stages of the Old Kingdom. It may be that Egyptian forces moved into the southern areas of the Levant to take advantage of the climate-induced disorder that was developing at this time, perhaps mimicking strategies that were applied further south.[137]

Perhaps this behaviour indicates a dynamic similar to that played out by those earlier rulers with the military escalades against the northeast undertaken by Weni from Abydos[138], in the time of Pepy I, designed to take advantage of the ongoing disruption. Whether the opponents were

[112] Förster, *Abu Ballas Trail*, 9.
[113] Moreno Garcia, *Sociopolitical Transformation*, 5.
[114] Mourad, *OK Siege Scenes*, 135–144.
[115] Strudwick, *Administration*, 150–151, Kanawati etal., *Saqqara* I, 7–8; Kanawati & McFarlane, *Deshasha*, 19; McFarlane, *Saqqara Mastabas*, 15; El-Khadragy, *Asyut's Soldiers Tomb*, 147–149; Fischer, *Army Scribe*, 233–272.
[116] Mourad, *OK Siege Scenes*, 145.
[117] Sowada, *Synchronisation*, 55–56.
[118] Müller-Wollermann, *End of the Old Kingdom*, 4; Moreno García, *Territorial Administration*, 96–97.
[119] Moreno García, *Territorial Administration*, 93.
[120] Adams, *OK Mendes*, 90–94.
[121] Lichtheim, *Literature* I, 103–104; Jansen-Winkeln, *Alten Reiches Untergangun*, 302–303; see also commentary from Förster, *Abu Ballas Trail*, 6–7.
[122] Snape, *Of Life and Death*, 113–114; Adams, *OK Mendes*, 91–92.
[123] Mant, Mendes Palaeopathology, 11–14.
[124] Faulkner, *Military Organization*, 32–47; Miller etal., *Ancient Near Eastern Archery*, 179, fig. 1.
[125] Miller etal., *Ancient Near Eastern Archery*, 180–187; Moreno Garcia, *OK Warfare*, 6–7.
[126] Dean, *Dynastic Warfare*, 19–23, 63–72; Fields etal., *Pharaoh's Soldiers*, 13–21; Moreno Garcia, *OK Warfare*, 31–35.
[127] Lepsius, *Denkmäler*, B.IV, Bl.II, Abl. 108.
[128] El-Khadragy, *Asyut Soldier's Tomb*, 147–164; Partridge, *Fighting Pharaohs*, 21–28; Miller etal., *Ancient Archery*, 181.
[129] Morris, *Art of Not Collapsing*, 78–79; Snape, *Of Life and Death*, 111.
[130] Müller-Wollermann, *End of the Old Kingdom*, 4; Moreno Garçía, *Territorial Administration* 96–97.
[131] Redford, *OK Egypt & Western Asia*, table 1; Sowada, *OK Eastern Mediterranean*, 251–253.
[132] Greenberg, *No Collapse, Transmutation*, 36; Mumford, *Ras Budran*, 43–46; Hoffman & Cohen, *Chronological Conundrums*, 3.
[133] Hoofman & Cohen, *Chronological Conundrums*, 4.
[134] Seidlmayer, *OK Elephantine*, 112–114.
[135] See Bietak, *Nubia in Review*, 385–391.
[136] Mortensen, *Settlement Pattern Change*, 29.
[137] Smith, *Egyptian Imperialism*, 77–102.
[138] Strudwick, *Texts*, 354–357; Richards, *Weni*, 75–102.

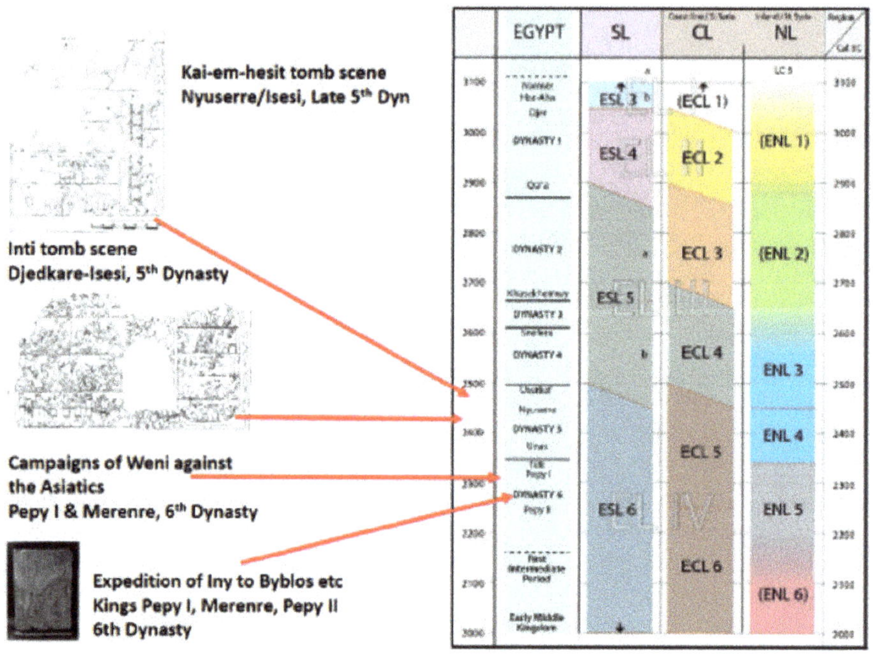

Figure 11.2: Early Bronze Age Egypt and the southern Levant: latest dating correlation.

city-states or smaller entities recovering from previous disorders is unclear. The many campaigns of Harkhuf and Pepynacht-Heqaib, from Aswan,[139] for example, may be interpreted as protracted campaigns against a few opponents or examples of numerous minor campaigns against a number of smaller disparate entities?

11.5. Chronological Re-Appraisals?

Refinements of the dating of contiguous civilisations has led to modifications of their chronologies. The work published by ARCANE (Associated Regional Chronologies for the Ancient Near East and the Eastern Mediterranean)[140] may indicate that a slightly different chronology for ancient Egypt is more likely[141] (see Figure 11.2).[142] The table represents a recent effort to consolidate current chronological understanding into a more reliable correlation in order to better understand the international inter-relationships within the region at this time.

Currently, it is believed that the urban areas of the southern Levant had already begun to decline BEFORE experiencing those Egyptian campaigns that were directed against them. These areas were hit by the same 'drought' as Egypt but did not have the luxury of extra resources from a river fecund with nutrients and blooming with resources. This means that the very many military interventions of Weni and Iny in the Sixth Dynasty against the southern Levant may not have been directed against a few powerful city-states requiring extended campaigns, but against a multiplicity of numerous smaller states, states already weakened, perhaps by those campaigns memorialised by Kaiemheset and Inti, or perhaps undermined by the environmental deterioration described earlier.

Earlier in this study, the reign of Pepy I was identified as the most likely era for the impact of the '4200BP event'. Judging by the numerous variations in the tomb decoration programme, the reign of Niuserra appears more likely to be the one where a potential threat to the stability of the society was identified. Perhaps these different environmental conditions had begun to have a minor impact upon the society, and this concern began to be represented within the visual culture.

11.6. Reign of Niuserra: A Re-Alignment?

Scholarly commentary suggests that Niuserra was a vigorous and pro-active ruler whose many administrative reforms[143] may be seen in the new light of responding to the changing environmental situation that was developing. The reforms undertaken have been associated with the goal of strengthening royal power and re-establishing the authority of the king in the face of growing dominance of officials and their families.[144] In order to minimise the excessive accumulation of power and wealth, different families were given particular activities in which to become 'specialists', theoretically making it difficult to broaden their powerbases within the administration.[145] These changes heralded the further introduction of

[139] Strudwick, *Texts*, 328–335.
[140] http://www.arcane.uni-tuebingen.de/index.html.
[141] Sowada, *Synchronisation*, 55–56, fig. 1.
[142] Sowada, K, Perspectives on Egypt in the Southern Levant in Light of the High Early Bronze Age Chronology.' (with permission).

[143] Duliková, *Transformation*, 56; Bárta, *Punctuated Equilibrium*, 8–11.
[144] Bárta & Duliková, *Divine & Terrestrial*, 46–47.
[145] Krečji, *Ptahshepses*, 358–360; Moreno García, *Territorial Administration*, 97.

social[146] and administrative change,[147] with the reforms (in theory, at least) remaining in place until the king could no longer project his power, that is, into the First Intermediate Period.[148] Some of these administrative reforms may have been designed to wean the society off relying on a regular inundation. The rapid growth in importance of the Dakhlah Oasis may indicate one such strategy.[149] Since the Oasis' water supply was less reliant on an annual flood, the increasing colonisation and cultivation of the oases be seen as another strategy to compensate for an unreliable riverine water supply and ensure a more dependable food supply.

As well as administrative reforms, social change occurred with some evidence suggesting that individuals seemed to have had a greater say in the design and organisation of their own tombs.[150] The furnishing of decorated burial chambers from the late Fifth Dynasty and into the Sixth Dynasty may be an indication of a greater desire for a more secure resting place, one that would be better protected underground.[151] These changes may be indicators of elite individuals becoming less confident of their funerary homes lasting for eternity.[152] This may have been an indicator of changes in the religious and administrative arrangements of the time.[153]

11.7. Summation: Re-adjusting for a New 'Normal'?

The major points raised in this chapter include:

- A wider range or resources appear to have been exploited at the end of the Old Kingdom and into the First Intermediate Period.
- Necropolises whose use had ceased before the Sixth Dynasty display 'old-fashioned' tomb decoration programmes with few of the 'new' decoration themes.
- Osiris, and his association with the river and the marshlands, appears to have become a more significant figure in the pantheon of the gods.
- Egypt's foreign inter-relationships display an inconsistent approach, perhaps as a result of less reliable environmental circumstances, indicating shifting power vacuums.
- Administrative reforms suggest a pro-active response to a perceived environmental concern.

The relative consistency of the themes displayed in the tomb decoration programme from the time of Niuserra to the end of the Old Kingdom suggests, perhaps, that a new paradigm of cultural depiction had been reached. While cultivation was still important, other activities were necessary to provide the shortfall that developed as a result of an unreliable water supply. The presence of the river, despite the land being in drought, meant that an adaptiveness within of the population enabled a societal response more flexible than previously expected. Despite the apparent development of discontinuity within the country, the intrinsic resilience of the society meant that the culture continued, though perhaps in a form less recognisable.[154]

[146] Schneider, *Existence Now Silenced*, 315–316.

[147] Bárta & Duliková, *Divine & Terrestrial*, 32–37; Krečji, *Niuserra Revisited*, 513; Morris, *Art of Not Collapsing*, 75–78.

[148] Morenó García, *Territorial Administration*, 100.

[149] Morris, *Art of Not Collapsing*, 81–82.

[150] Bárta, *Architectural Innovation*, 106, 114; Jánosi, *Houses of Eternity*, 34–36.

[151] Kanawati, *Decorated Burial Chambers*, 57–63: Dawood, K., 2005, in Publications de la Maison de l'Orient et de la Méditerranée, 40, 1: 107–127.

[152] Bárta, *Filling the Chambers*, 10.

[153] Schneider, *Existence Now Silenced*, 315.

[154] Butzer, *Organisms or Systems*? 8.

12

The A.R.I.D. Project: Extending the Investigation

As the river became less reliable, certain flow-on consequences to be expected. More accurately, the term "A.R.I.D" Should relate more to A River In **Deviance**, since it was not behaving as expected. Beyond a simple proposed hypothesis, and on-going A.R.I.D. (A River In Deviance) Project should look beyond the theoretical and challenge more widely. In this chapter, additional areas for consideration will be identified. These tangents vary across many disciplines requiring a wide variety of expertise.

12.1. Middle Kingdom Tomb Decoration Programmes

Succession studies can be traced from a source of disruption, in both time AND distance, and this approach has been applied to tomb scenes produced over the time frame of this investigation. The potential exists for this sort of succession to be further studied beyond the timeframe of this investigation, which ends with the onset of the First Intermediate Period. The further from the time of environmental dislocation, the more likely it would be that the ecological conditions had returned to equilibrium, that is, the inundation levels should have become more regularised, land would have once again been fertile with grain surpluses occurring more regularly. As the river returned to a less erratic flow, and the unusual rainfall events became more intermittent, the annual cycle would be expected to return to one more recognisable. Perhaps artistic expressions of the culture returned to styles and forms more conventional, traditional decorations more in line with those 'traditional' scenes produced during the Fourth and early Fifth dynasties. If the suppositions made by this study are valid, then we would expect to identify a 'reverse-image' of the pattern identified in this current study. If this was the case, then the further away from the time of disruption, the more likely it would be to recognise a more 'conventional' pattern of tomb decoration, with the decorations produced after this time returning to a more orthodox repertoire (could this be a potential origin of 'archaising'?).

12.1.1. Art Succession and A River In 'Drought'

Good candidates for this sort of examination are those sites with burials dated from the second half of the Old Kingdom into the early Middle Kingdom. While some of these sites have been comprehensively excavated over the last one hundred years, ongoing and more sophisticated re-investigations are being undertaken, as though a new 'lens' was being applied. The unusual artwork produced at places like Beni Hassan has been interpreted in a socio-cultural context with regard to the increasing contact with Asiatics and their introduction into the tomb decoration repertoire.[1] Re-assessment of these unique images at the necropolis of Beni Hassan was presented in 2010 as an art-based investigation into society and culture in a provincial centre.[2] This new project, currently underway by Macquarie University, Australia, has the long-term goal of publishing all of the decorated tombs at the site.[3] With the benefit of improved technology, enhanced photography techniques and more accurate dating criteria, the production of a more reliable timeline is imminent.[4] Eventually, this updated artistic analysis will enable a clearer light to be shed on the social history of this period.[5]

The necropolis of Meir was extensively surveyed by Blackman and Apted[6] and has now undergone a re-evaluation under the auspices of Macquarie University.[7] Tombs date from the latter half of the Old Kingdom onwards and follow an extended family over generations into the 12th Dynasty, from Ni'ankh-Pepy-kem (Pepi I), to Pepy-ankh Heni-kem and Pepy-ankh Heri-ib (Pepi II) to Senbi (Amenemhat I) to the later tombs of Ukh-hotp and . Kha'kheperre-sonb, dating to the reign of Senusret II.[8] New artistic evidence has already arisen as a result of this re-examination; for example, some grid systems displayed in Old Kingdom tombs are actually the result of Middle Kingdom artists setting up a grid *over* the artwork in order to copy the Old Kingdom decoration.[9]

Deir el-Barshā was the chief cemetery of the nomarch of the 15th Upper Egyptian (Hare) nome during the Middle Kingdom.[10] The origin of Deir el Barshā has now been located in the Old Kingdom, however, with traces of activity dated to the Fifth and Sixth Dynasties.[11] Recent archaeology suggests Old Kingdom formations also for the sites of Edfu and Dendera.[12] The necropolis of Gebelein, in Middle Egypt, has recently been identified as a site with evidence suggesting it has experienced continuous

[1] Cohen, *Uses & Abuses*, 19–20.
[2] Newberry, *Beni Hasan* I and II; Griffith, *Beni Hasan* III and Carter etal., *Beni Hasan* IV.Kanawati & Woods, *Beni Hassan.*
[3] Kanawati & Evans, *Beni Hassan* I and III; Lashien, *Beni Hassan* II.
[4] Evans & Mourad, *D-Stretch & Tomb Paintings*, 81–83.
[5] Evans & Mourad, *D-Stretch & Tomb Paintings*, 82–83.
[6] Blackman & Apted, *Meir* I; Blackman, *Meir* II, III and IV; Blackman & Apted, *Meir* V and VI.
[7] Kanawati & Evans, *Meir* I and *Meir* II.
[8] Blackman, *Meir* III and V.
[9] Kanawati, *Art & Gridlines*, 483–496.
[10] *The KU Leuven Dayr el-Barsha Project*, https://www.universiteitleiden.nl/en/nvic/research/ archaeology--egyptology/ongoing-projects/the-ku-leuven-dayr-al-barsha-project (22/05/2019).
[11] Vereecken etal., *Dayr al-Barsha Pottery*, 204; de Meyer, *Two Cemeteries, One Capital*, 42–43, fig. 1; de Meyer, *Restoring the Tombs*, 125–135; de Meyer, *Shadows of the Nomarchs*, 421–427.
[12] Moeller & Marouard, *Edfu & Dendara Development*, 33–56; Snape, *Of Life and Death*, 107, 115.

occupation over Pharaonic times,[13] and may be added to the list of possible study sites. The current survey may uncover decorations produced across a time frame similar to the one of this investigation. Due to the small sample size, however, it may be difficult to achieve significant results from individual locations, but when added to other sites developed across a similar time frame, it may be possible to discern glimpses of a new pattern unfolding.

Once the investigations outlined above are completed, with newly refined dates and a more accurate chronology, an investigation into the evolution of tomb scene types in these areas can be undertaken allowing the progression of the decorative programme to be plotted; in a manner similar to that in Chapter Five. The plotting of the distribution and abundance of tomb scenes from a number of contemporary sites, the transitional sites, from the late Old Kingdom to the Middle Kingdom, may help the understanding of the evolution the decorative programmes in this area.

12.1.2. Art Databases Plotting the Evolution of Artistic Themes

Unfortunately, the OEE database is no longer being updated.[14] The prospect exists, still though, to link this repository with the database "MastaBase" that was produced by Leiden University and is in the process of being resurrected.[15] In order to investigate if the evolution of the decorative programmes within tombs can be linked to external environmental factors, the current statistical analysis of the distribution and abundance of wall scenes using the database of the Oxford Expedition to Egypt (OEE) could be extended beyond the time frame of that particular database. Data contained in the database 'MEKETREpository' for tomb decorations, for example, whose goals include the collation of First Intermediate Period and Middle Kingdom decorations,[16] could widen the range of this information. If these databases were to be merged a decoration database over many dynasties could be developed and the understanding of the art history of the time enhanced.

12.2. Geoarchaeology: Messages in the Mud?

Changes to the distribution of Nilotic sediments and flora as a consequence of changing riverine flood patterns should leave corroborating evidence in the soil structures,[17] for example, how early industry had affected local water quality.[18] If the environment changed in the ways proposed, then the consequences would have had societal implications[19] by impacting how the landscape was modified as the population re-organised their use of space in response to the new circumstances that were unfolding.[20]

The Arid Climate Adaptation & Cultural Innovation in Africa [ACACIA] project, for example, with its goal to link the historical process of cultural development and the emergent social dynamics to the ecological background[21] is one such project. Of particular relevance to this study is Project A5: Environmental Situation and Change in Northeastern Africa: The Special Example of Ancient Egypt,[22] whose final results are highly anticipated. By applying a 'landscape-integrated' approach, it can integrate various geological, geographical and ecological aspects to the investigation of societies in the region.

12.2.1. Geoarchaeology

Geoarchaeology applies earth sciences to inform archaeological knowledge and thought.[23] It is a multidisciplinary approach investigating the natural physical processes affecting archaeological sites,[24] synthesising the information and then presenting it in a holistic manner.[25] Over the last six thousand years, the Nile floodplain has shifted significantly,[26] with its geomorphology and local environments displaying a variety of patterns and characteristics[27] unique to each stage in its geological history.[28] The formation of sites and how geological processes changed these landforms are studied, as well as how these geological processes may have affected cemeteries, and the artefacts deposited within, many years after they were constructed.[29] Geological evidence can be applied to the development of a number of sites over the time frame of this investigation. Finding sites with precise chronological dates, while difficult, is becoming more reliable and should allow a more accurate picture of any population shifts (forced or otherwise) during the later Old Kingdom. The value of the work at Edfu is considerable, because it represents a site occupied before and after the time frame under investigation.[30] Edfu, for example was founded in a 'flood-safe' area of the Nile, whereas nearby

[13] Esmond, etal., 2017. *Report on the archaeological survey at Gebelein in the 2014, 2015 and 2016 seasons,* Polish Archaeology in the Mediterranean, 26: 1, see 239–240, 265.
[14] http://www.oxfordexpeditiontoegypt.com/forthcoming-books.html.
[15] https://digitalegyptology.org/mastabase/.
[16] http://meketre.org/repository/.
[17] Ghilardi & Desruelles, *Geoarchaeology*, 2–3; Fuller & Lucas, *Archaeobotany*, 305.
[18] Bardají etal., Dra Abu el-Naga Geomorphology, 242–243.
[19] Goldberg & Macphail, *Geoarchaeology*, 225–246; Ghilardi etal., *Geoarchaeology*, 227; Bunbury etal., *Holocene Nile @ Karnak*, 358–359; Wright, *Human Holocene Termination*, 1–3; Qin, *Landscape Change*.
[20] Riemer, *Risks & Resources*, 124.
[21] http://www.uni-koeln.de/sfb389/a/a5/index.htm (13/05/2018).
[22] http://www.uni-koeln.de/sfb389/a/a9/index.htm (13/05/2018).
[23] Butzer, *Cross-Disciplinary Geoarchaeology*, 403–404.
[24] See, for example, Ghilardi etal., *Nile Evolution,* Love, *Abusir Drill Core*, and Bardají etal., *Dra Abu el-Naga Geomorphology*.
[25] Verstraeten etal., *Nile Floodplain Transition*, 239–241; Bunbury, *Memphite Floodplain Landscape*, 61.
[26] Bunbury, *Memphite Floodplain Landscape*, 71.
[27] Butzer, *Nil*, 480–483; Zahran & Willis, *Vegetation*, 251–254; Bunbury, *Memphite Floodplain Landscape*, 75–78, figs. 5, 6.
[28] Bunbury etal., *Holocene Nile @ Karnak*, 358–359, 364–366; Butzer, *Urban Geoarchaeology*, 3345–3346, 3354–3355, figs 3, 11; Verstraeten etal., *Nile Floodplain Transition*, 247, identifying some places where the Nile has moved in the opposite direction to that previously understood, and 252–253, summarising the different behaviours displayed by the two different banks of the river.
[29] Bardají etal., *Dra Abu el-Naga Geomorphology*, 248–249.
[30] Schneider, *Existence Now Silenced*, 313.

Dendera was built in an area further from the river.[31] Both of these sites developed in the late Fifth Dynasty and then grew in the Sixth Dynasty, but Dendera declined during the Middle Kingdom as a result of a greater threat from the strengthening inundations.[32]

Some geoarchaeological assessments made in earlier times, relying on more rudimentary equipment, may have under-estimated the rate of movement of the river across the floodplain.[33] Improved techniques and technologies can be utilised to re-examine previous commentary to further refine our understanding of this area of investigation. Bunbury's water-based perspective of the First Intermediate Period suggests that the climate had begun changing much earlier, more likely from earlier in the Old Kingdom.[34] Significantly, the amount of marshland was much less than most scholars anticipate,[35] a proportion that according to the A.R.I.D. hypothesis, should have increased over the time frame of this investigation. Since water, or a lack of it, is a significant underlying theme of this investigation, a water-based geoarchaeological emphasis may be of value as the space that could be exploited changed in response to water availability.[36]

12.2.2. Archaeobiology

Archaeobiology studies, in an archaeological context, the remains of plants and animals exploited by humans. It is a composite discipline, combining botanical and zoological knowledge with archaeological materials, as it attempts the reconstruction and interpretation of past human-animal-plant inter-relationships[37] by studying the behavioural and ecological interactions between past peoples and the living things around them. This includes the role of pastoralism and animal husbandry in the survival of populations and sometimes includes the representation of animals in rock art and on portable materials. This allows for the reconstruction of the diet, the subsistence activities practiced, those agricultural strategies employed, the social and cultural role of food, the exploitation of wild resources, the procurement of fodder, the aspects of seasonality, and the environment in which people and their animals dwelt,[38] including all the organic remains left in the soil after the death and decay of plants and animals.[39] Giuseppe Passalacqua, the leader of an excavation of tombs at Deir el-Bahri, is believed to have directed the first archaeobotanical research in Egypt by having dried plant remains analysed by a botanist.[40] Plant macrofossils are preserved through four main modes. Firstly, plant remains can be examined; usually cereal grains, chaff, seeds and charcoal that may be 'charred' as a result of cooking.[41] Secondly, plant remains deposited in permanently waterlogged conditions are preserved because microbial decay is limited due to a lack of oxygen.[42] Thirdly, mineralisation of plant remains occurs, usually in latrine pits and middens, with the plant remains being replaced by the mineral calcium phosphate.[43] Finally, plant remains are preserved by desiccation in arid environments, where the absence of water limits decomposition.[44] Since water is a significant underlying theme of this investigation, the identity of the plants growing at particular locations and dates becomes particularly salient.[45] Analyses of core drilling samples of the Nile sediment should be able to distinguish between the relative proportions of floral organisms present at the time the sediments were laid down. With regard to the A.R.I.D. hypothesis, the recovery and identification of plant and animal remains can be used to make inferences about the ecology and culture of the site, as well as proposing plant succession.[46]

Using the above approaches, alterations to the habitat of a region or changes in the diet of its occupants can therefore be traced by observing changes in the archaeobotanical record and comparing them to those remains from earlier times.[47] The movements of peoples as a consequence of changing circumstances, either environmental or political, has been traced using chemical means. By investigating the faunal remains, for example, it has been possible to identify population shifts during the New Kingdom.[48]

The residue in the bones of the animals used for dairy and meat can indicate where the animal lived in the early stages of its life. Significant advances in genetics now allow for the identification of organic remains previously unidentifiable.[49] The prospects of this technique are thought to be vast,[50] with the ability to re-examine previous evidence using these new techniques, new levels of understanding and interpretation can be elicited. Importantly, the potential negative impact of humans on their surroundings may be clarified.[51]

---

[31] Moeller & Marouard, *Edfu & Dendera Development*, 33–53.
[32] Moeller & Marouard, *Edfu & Dendera Development*, 54–55.
[33] Butzer, *Hydraulic Egypt*, 134; updated by Verstraeten etal., *Nile Floodplain Transition*, 244–245.
[34] Bunbury, *River Nile Development*, 52–53, fig. 1; Bunbury, *Memphite Floodplain Landscape*, 73.
[35] *ACACIA project A5:* http://www.uni-koeln.de/sfb389/a/a5/a5_picture6.htm (13/05/2018).
[36] Riemer, *Risks & Resources*, 123–124.
[37] Fuller & Lucas, *Archaeobotany*, 305.
[38] http://www.oxfordhandbooks.com/view/10.1093/oxfordhb/9780199573493.001.0001/oxfordhb-9780199573493-e-30 (28/05/2018).
[39] http://www.oxfordhandbooks.com/view/10.1093/oxfordhb/9780199927142.001.0001/oxfordhb-9780199927142-e-001 (06/02/2019).
[40] Stuart, *Paleoethnobotany*, 5756. See also Abdel-Magid, *Archaeo-ethnobotanical Plant Domestication.*
[41] Märkle & Rösch, *Carbonization Effects*, 257–263; Van der Veen, *Carbonized Plant Remains*, 977–979.
[42] Kenward & Hall, *Urban Organic Archaeology*, 584–596; Jacomet, *Archaeobotany*, 497–514.
[43] Smith, *Identifying Cess-pits*, 526–543.
[44] Van der Veen, *Carbonized Plant Remains*. 970–977.
[45] See Ritchie, *Dakhla Pollen Spectra*, 1–6 and Ritchie, *Dakhleh Palaeobotany*, 74–80, as examples.
[46] Riehl, *Archaeobotanical Evidence*, 93–114; Cappers, *Reconstructing Agriculture*, 429–446.
[47] Fahmy, *Predynastic Plant Exploitation*, 14: 4; 287–294.
[48] Buzon & Simonetti, *Mobility Studies*, 7.
[49] Palmer etal., *Plant Archaeogenetics*, 146–56.
[50] Fuller & Lucas, *Archaeobotany*, 310.
[51] Fernandez etal., *Degradation & Recovery*, 297–323.

12.3. Human Archaeology: Beacons in the Bones?

Duhig's work, the goal of which was to address the question… "do the skeletal remains of those living during the First Intermediate Period support the documentary references to societal breakdown at this time?"[52] found that, *"…no physical evidence for the famine assumed by Bell…"*[53] This data, a study of more than eight hundred individuals implies that famine was less common than anticipated, suggesting that the diet would have been suitable, under ordinary situations, for the maintenance of health, with most basic food types available to the great majority of the population due to the environmental uniformity of the country.[54]

While this is very important to the context of this investigation, very little indicators of changes to the diet were identified. According to the A.R.I.D. hypothesis, it would be expected that the diets should have changed. As the Nile weakened and papyrus and reeds flourished, the sluggish river would have changed the food web profile. Fish numbers would have increased, and bird populations altered, while cattle would have flocked to the riverbanks to feast upon the newly growing plants. Desert animals would have infringed upon this shared space. This would have impacted upon the eating habits of human populations and the consequences of these adaptations should be identifiable in the remains of their bodies and the residues of their meals. As the society adapted to changing socio-ecological conditions, corroborating osteo-archaeological evidence of changes in diet and lifestyle behaviours should be identifiable in the bodily remains of these populations.[55] Bio-archaeological studies can now identify particular nutrients as well as more accurately defining instances of nutritional stress and high mortality.[56] Applying these analyses to different populations over different stages of the Old Kingdom and into the Middle Kingdom should provide evidence of changes in diet as well as suggest movements of populations,[57] movements that may be linked to relocations in response to changing environmental circumstances.

12.3.1. Dental Health

From the very earliest times, humans have been beset by dental problems and for those living along the Nile valley in ancient times, this was no different.[58] Fortunately the environmental conditions meant that much has been preserved, with teeth a good source of information because the tissue maintains its quality for longer than other hard remains. The diagnosis and interpretation of dental disease from ancient remains is able to provide indicators of the general health of a population as well as furnishing information about diet, oral hygiene, the environment and prevailing economic conditions. It is able to provide a record of childhood stress, useful in evaluating the general health of a population, linking poor levels of juvenile nourishment to the development of hypoplasia at later ages.[59] As well as serving as an indicator of poor diet, poor dental health indicators can be used as 'stress markers' indicating difficult living conditions.[60]

In Egyptian history, there is palaeo-pathological evidence of diseases such as anaemia, diabetes and tuberculosis, all of which leave scarring on the teeth.[61] Drought and famine, and their resultant nutritional deficiencies, would also impact upon dental hygiene.[62] The ability to identify different ethnic groups by the relative vulnerability to dental disease can enable the migration of such groups to be traced and could be used to identify migration patterns that changed in response to changing climatic conditions.

Those samples preserved are usually from those people with the best means to survive, so the remains should be understood as 'atypical'. It seems, however, that all Egyptians, from the peasantry to the elite, suffered from extremely worn teeth, periodontal problems and numerous dental abscesses,[63] with women appearing to suffer worse than men.[64] Early Nubians and those of the Western Desert had better dental health than their Early Dynastic and Predynastic Egyptian counterparts, due most likely to a more varied diet.[65] However, The dental health of the non-Egyptian populations gradually declined so that, by early Dynastic times, all those populations displayed similar dental health problems,[66] presumably due to an increasing reliance on the products of cultivation.

Because of its association with biological ageing, the degree of dental wear has been recognised as a useful method of estimating the age at death of an individual.[67] As well as wear, other dental issues that serve as helpful evidence include diseases of the enamel, tooth loss, tooth fractures and abscesses.[68] The principal cause of

[52] Duhig, *Eating People Here*, 1.
[53] Duhig, *Eating People Here*, 115.
[54] Duhig, *Eating People Here*, 4.
[55] Cable etal., *Global Change & Disease Control*, 1–3. Veiga, *Prevalent Pathogens*, 65; Zakrzewski, *Life Expectancy*, 1–4.
[56] Zakrzewski. *Continuity or Change*, 501–509; Zakrzewski, *Variation in Ancient Egyptians*, 219–229; Raxter, *Egyptian Body Size*; Kumar, *Health at Hierakonpolis*; Cable etal., *Global Change & Disease Control*, fig. 1; Richards, *Mortuary Landscapes*, 169.
[57] Buzon & Simonetti, *Strontium Isotope Variability*, 6–7.
[58] Forshaw, *Dental Health*, 421.
[59] King et.al., *Enamel Hypoplasia*, 29–39; Lovell & Whyte, *Dental Enamel Defects*, 76.?
[60] Lovell & Whyte, *Mendes Dental Defects*, 69–80; Goodman etal, *Hypoplasias Development*, 7–19; Kumar, *Health at Hierakonpolis*, 206.
[61] Miller, *Skulls & Dentition*, 15–21.
[62] Miller, *Skulls & Dentition*, 23–27.
[63] Triambelas, *Caries Prevalence*,192–194; Forshaw, *Dental Health*, 421.
[64] Fields et.al., *Agricultural Transition*, 43; Cramer et.al, *Paleopathologies in Mummies*, 7; Antoine & Ambers, *Scientific Analysis*, 20–30.
[65] Buzon & Bombak, *Nile Valley Dental*, 15–16 and Nikita et.al., *Saharan Dentition*, 392.
[66] Thompson et.al., *Kerma Diet*, 380–385; Lukacs and Joshi, *Ethnic Hypoplasias*, 359–377; Triambelas, *Caries Prevalence*, 182–184, fig. 8; Kanchan etal., *Enamel Hypoplasia*, 101; Gerloni et.al., *Dental Status*, 58–64; Lovell & Whyte, *Dental Enamel Defects*, 76.
[67] Forshaw, *Dental Health*, 424.
[68] Schwarz, *Dentaire dans l'Égypte Pharaonique*, 37–42; Veiga, *Health & Medicine*, 55–56; Veiga, *Prevalent Pathologies*, 74.

periodontal disease is a bacterial irritation brought about by the accretion of plaque. The condition was known in antiquity and is traceable back to prehistory.[69] It appears less frequently in ancient Egypt than it does currently in modern times, perhaps because the diet did not include simple carbohydrates.[70] Enamel hypoplasia, the formation of a hard but thin enamel leading to less protected teeth,[71] appears to have been a fairly common occurrence in many ancient Egyptians.[72] It was thought to be an indicator of malnutrition but can also indicate major infections, diseases, or vitamin deficiencies.[73]

The new technique of dental microwear texture analysis (DMTA) is enabling different lifestyles to be distinguished between different types of societies.[74] Groups that exploited their surrounds as foragers, farmers or pastoralists can now be identified. Unfortunately, the study investigated different groups from different geographical locations from around the globe.[75] The Egyptian sites selected were used for the representation of farmers. But there were no sites selected to form comparison groups; no Egyptian site composed of foragers or pastoralists. Despite that, enough differentiation was found to exist between tooth wear patterns of the different lifestyle groups, and [76] importantly, this technique appears to be able to distinguish between those societies that had a diverse resource base compared to those with simpler practices.[77] This technique, therefore, could be applied to populations whose remains date to the time frame under investigation and valid inferences about potential lifestyle changes could be made. The barley/wheat problem should also consider culinary changes; one can think in the cultural and social implications of the preference for sorghum instead of barley or wheat among Nubian peoples in Antiquity.

12.3.2. Skeletal Health

Bone samples allow for the identification of factors that may have interfered with the quality of life of the individual.[78] Overall trends in changes to the population of the Nile valley from Predynastic to well into Classical times have been identified,[79] with the general results echoing the dental evidence: that Nubians were generally more robust than Egyptians and Egyptian women, when adapting to a diet based on the products of cultivation, displayed the most rapid decline in stature size of all.[80] Studies at sites of long-running occupation can trace the health of the population across its history; the changing health outcomes of the workers at Deir el-Medina, for example,[81] can provide an overview of almost five hundred years. However, this site was occupied at a time of relative stability: ideally, sites with long-term occupation during times of environmental or economic or political stress would be most suited to investigating the validity of the hypothesis made in previous chapters of this book.

As yet only a few detailed studies have been conducted on a site with large numbers of remains from the late Old Kingdom AND from the First Intermediate Period.[82] The site at Mendes does not seem to be demographically representative,[83] since a broad cross-section of the population is not present, suggesting that the sample may make precise inferences invalid. The lack of large numbers of ancient Egyptian skeletons available for detailed investigation[84] makes it difficult to clearly identify and articulate the changes that may have occurred to localised populations during the time frame under investigation. While inferences may be made from extrapolated or estimated data, actual evidence is less substantial.[85] CT scans, for example, give great insights but still cannot allow the researcher to detect infectious diseases.[86] As more precise dating technologies come online, a more accurate health picture of the end of the Old Kingdom and into the First Intermediate Period will emerge.

12.3.3. Parasites

As environmental conditions change, the diversity of parasites well-suited to that environment changes, leading to increases in varieties better adapted to the new ecology.[87] As the Nile weakened and the marshlands flourished, the lethargic river would have become more pestilent, developing conditions favourable to an increase in malaria and bilharzia. As such it would be expected that in A.R.I.D. times, an increase in water-borne and water-related pathogens and their vectors should have occurred.[88] Malaria was, and still is, a disease endemic to the region,[89] and 'marshland conditions' would have favoured its spread, evidence of which could be identified in skeletal remains, with most remains suggesting encounters with this parasite.[90] Identifying an increase in diseases brought about by sluggish waterflow or stagnant water during the

[69] Clement, *Caries*, 115–123. Adler, *Ancient DNA*, 205.
[70] Rateutschak-Pluss & Guggenheim, *Dental Plaque Accumulation*, 239–244; contra Adler, *Ancient DNA*, 203.
[71] Kanchan etal., *Enamel Hypoplasia*, 99–100.
[72] Hillson, *Diet & Dental Disease*, 147–162.
[73] Langsjoen, *Diseases of Dentition*, 393–412.
[74] Schmidt etal., *Dental Microwear*, 208.
[75] Schmidt etal., *Dental Microwear*, Table 1.
[76] Schmidt etal., *Dental Microwear*, 217–220, figs. 3, 4.
[77] Schmidt etal., *Dental Microwear*, 222.
[78] Fritsch et.al., *Orthopedic Diseases*, 1036–1037.
[79] Zakrzewski, *Variation in Ancient Egyptians*; Raxter, *Egyptian Body Size*.
[80] Raxter, *Egyptian Body Size*, 124–126, fig. 7; Zakrzewski, *Variation in Ancient Egyptians*, 223–224, figs 3–6.
[81] See Austin, *Deir el-Medina Health Care*.
[82] Summer, *Excavating Mendes*, 1–2; Mant, *Mendes Palaeopathology*.
[83] Mant, *Mendes Palaeopathology*, 5; Adams, *Mendes Stratification*, 90–94.
[84] Schultz, *Paleohistopathology*, 106–109; Raxter, *Egyptian Body Size*, 174.
[85] Raxter *Egyptian Body Size*, 174; Zakrzewski, *Variation in Ancient Egyptians*, 225–227; Zakrzewski, *Continuity or Change*, table 2.
[86] Cramer etal., *Paleopathologies in Mummies*, 6.
[87] Cable etal., *Global Change & Disease Control*, 3–6.
[88] Veiga, *Health & Medicine*, 46–48; Veiga, *Prevalent Pathogens*, 65–66.
[89] Kenawy, *Malaria in Ancient Egypt*, 7; Smith-Guzmán, *Nile Valley Malaria*, 11.
[90] Brier, *Infectious Diseases*, 17–27; Nerlich etal., *Plasmodium falciparum*, 1317–1319; Bianucci etal., *Malaria in Paleopathology*, 176–80.

time frame under investigation would provide supportive evidence to the ideas developed in this study.

Unfortunately, precise dating may make this identification problematic as does the ability to actually determine WHEN the individual was infected. Predynastic evidence of parasites in human remains suggests that schistosomiasis, or bilharzia, was an infection endemic to the region,[91] as mummies are found infected with it.[92] Schistosomiasis, also known as snail fever and bilharzia, is a disease caused by a number of species of parasitic flatworms.[93] Symptoms of schistosomiasis include blood in the urine, changes to urinary frequency and chronic cystitis.[94] In some cases, it can cause genital disease in both men and women, resulting in numerous symptoms including infertility.[95] Since many of these symptoms are associated with the genitalia, one may be tempted to ask if the circumcision scenes depicted on a number of tomb walls[96] are 'commemorating' a treatment for this condition?

As well as the increasing likelihood of infection due to the development of agriculture and the growth of population densities,[97] schistosomiasis, a significant environmental and occupational hazard in ancient Egypt,[98] would be expected to flourish in the developing conditions suggested by the A.R.I.D. hypothesis. When the Aswan High Dam was first constructed, instances of schistosomiasis increased rapidly in the sluggish waters.[99] With a greater focus on marshland activities, it would be expected that this exposure should have increased over the time frame of this investigation. Schistosomes leave evidence on the skeleton and an increased proportion of evidence in remains dating to the time frame under investigation would suggest an increased level of exposure.[100] A similar problem exist to that identified above, the difficulty of determining when the first infection occurred, how many times was the individual infected and for how long did they experience the symptoms?

The regular cycle of an increase of disease as a consequence of the inundation, the 'annual pestilence,'[101] may not have occurred as regularly in the delta with 'marsh-like' conditions existing all year around. Mendes, for example, is a site located within the Delta, which suggests that the basic parameters of 'normal' health would be different to that expected of someone living within the Nile Valley.[102] Bilharzia is less likely in areas of flowing water, so this may explain the lack of infestation in remains from this area. In some areas, for example, Hierakonpolis, burials did not display as great a preponderance of these parasites.[103] Perhaps, in times of low flood, there were not enough puddles in which the population might immerse themselves to become (potentially) infected.[104]

One unique species of parasitic flatworm that produces infections like bilharzia, *Schistosoma mansoni,* however, is endemic to the delta and rarely found in remains from the Nile Valley.[105] This is another potential direction for further investigation, though the precision of its identification is problematic as it is also difficult to determine when in an individual's life did the infection occur, as well as how many times the individual was infected. Hierakonpolis form the end of the Old Kingdom burials may display evidence of *Schistosoma mansoni,* due to the increasing sluggishness of the river at this time. With more people coming into increasing contact with cattle, it is expected that this would lead to an escalating incidence of tuberculosis,[106] which, while not always fatal, would have led to difficulties in everyday activities.[107] Tuberculosis is an infectious disease usually caused by a Mycobacterium that is not unique to humans.[108] Bovine tuberculosis can jump the species barrier and can affect humans, especially those with a long exposure.[109] Tuberculosis is identifiable in skeletal remains if it is specifically looked for.[110] As the environment dried up, the tuberculosis bacterium would have become more susceptible to desiccation,[111] perhaps limiting the spread expected as a consequence of increased cattle numbers. Anthrax, another disease carried by cattle,[112] originated in ancient Egypt,[113] so should be also identifiable in ancient remains, providing the correct tests are performed. Since the A.R.I.D. hypothesis suggests an increase in the cattle population at this time, then it would be expected to see an increase in the proportion of remains from this time displaying this pathology.

12.3.4. Anaemia

Anaemia occurs due to a dietary iron deficiency which reduces the capacity of the blood to carry oxygen, causing fatigue, pale features, shortness of breath and, in

[91] Barakat, *Schistosomiasis in Egypt*, 425–426; Ruffer, *Bilharzia in Mummies*, 16; Farley, *Bilharzia*, 5–8.
[92] Baralat, *Schistosomiasis in Egypt*, 425–432.
[93] Colley, etal., *Human Schistosomiasis*, 2253–2264.
[94] Spiegelman, *Emergency Surgery*? 91–100.
[95] Merck Manual, '*Schistosomiasis*', https://www.merckmanuals.com/professional/infectious-diseases/trematodes-flukes/schistosomiasis (04/12/2017).
[96] Kanawati & Hassan, *Ankhmahor*, 49–50, pl. 19, 55b; See also Herodotus, *Histories* 109.
[97] Veiga, *Prevalent Pathogens*, 65–66.
[98] Deelder etal., *Schistome in Mummies*, 724–725; Miller etal., *Predynastic Schistosomiasis*, 54–60.
[99] Abd-El Monsef etal., *Aswan Dam Impact*, 1877–1878, 1884.
[100] Silverman, *Anthropology & Circumcision*, 419–445.
[101] Contardi, *Rhythm of the Nile*, 13, 20–23, Duhig, *Eating People Here*, 16.
[102] Mant, *Mendes Palaeopathology*, 1, 19.
[103] Kumar, *Health at Hierakonpolis*, 199–200.
[104] Keita & Boyce, *Dental Hypoplasias*, 733–743.
[105] Duhig, *Eating People Here*, 16.
[106] Veiga, *Health & Medicine*, 49; Veiga, *Prevalent Pathologies*, 70–73.
[107] Cramer etal., *Paleopathologies in Mummies*, 7.
[108] https://www.who.int/en/news-room/fact-sheets/detail/tuberculosis (06/05/2018).
[109] Olea-Popelka, etal., *Human Tuberculosis*, 16.
[110] Zink, etal., *Tuberculosis from Mummies*, 380–391.
[111] Kumar, *Health at Hierakonpolis*, 232–233.
[112] https://www.who.int/csr/resources/publications/anthrax_web.pdf, fig. 1 (08/05/2018).
[113] https://www.cdc.gov/anthrax/resources/history/index.html (08/05/2018).

some people, heart palpitations.[114] Children with chronic anaemia are often malnourished, increasingly prone to infections and more vulnerable to learning problems.[115] While learning problems are not readily identifiable within the archaeological evidence, malnutrition is.[116] Evidence of anaemia in remains is a good indicator of environmental stress due to increased rates of infection or poor food supply or periodic malnourishment.[117] Anaemia can develop in populations where increasing population density leads to increasing rates of parasitic infection,[118] so could also be an indicator of more people coming in to contact with one another, suggesting increasing mobility in the population.

However, it has recently become recognised that low blood iron levels can be interpreted as an adaptive response to pathogen attack, as people with anaemia are somewhat protected from developing malaria.[119] The incidence of anaemia may not an illness per se, it may be an evolutionary survival strategy to minimise the debilitating effect of malaria. This seems relevant to this investigation as this condition is evidenced in populations living in or around marshlands,[120] and could be a significant factor in furthering this investigation. A measurable increase in this condition around the time frame of our investigation may aid in the identification of an increase in mosquito-borne illnesses.[121] Currently the accuracy of the above diagnosis is in dispute,[122] so it would be necessary to undertake this investigation in conjunction with a number of different markers of a satisfactory diet.[123] This data could be used in conjunction with investigations whose goal would be to identify bilharzia in order to identify an increase in marshland-related parasitic infections.

12.4. Identifying A Changing Pattern of Health?

Dental indicators from remains at Mendes suggest poor nutritional stress at the end of the Old Kingdom,[124] while other data from the same site suggests that there was less nutritional stress early in the First Intermediate Period than at the end of the Old Kingdom.[125] Studies have failed to identify significant changes in remains suggesting that the late Old Kingdom was a time of nutritional stress.[126] If this was a time of drought and famine, then indicators should exist identifying significant developmental harm. A change in diet, as a consequence of adaptation to new environmental factors, may be identified if well-designed investigations are performed. Duhig noticed changes in the physical conditions of remains dating to the Archaic Period when compared to those dated to the late Old Kingdom, but these differences weren't explicitly stated, and it was recommended that these changes be further investigated.[127]

12.4.1. Dietary Considerations

The diet of the ancient Egyptian was thought to consist of primarily of bread and beer,[128] accompanied by onions and other vegetables, supplemented with some fish and fowl, and some meat.[129] Overall it would display more vegetarian-like characteristics than not.[130] In the Neolithic period (c. 10,000 BCE–c. 4,000 BCE) Egyptian society adapted from a hunter-gatherer to a more sedentary agricultural lifestyle dwelling in settled communities along the Nile valley. One consequence was that their diet would have changed from one rich in protein to one where carbohydrates played a predominant role.[131] This may have had a negative impact upon the general health of the society,[132] especially with regard to the female members of the population.[133]

Cereals do not contain all of the necessary amino acids for a healthy diet,[134] so, if the assumption that an increase in protein uptake may have occurred, then the result should have been a population who's biological remains displayed higher concentrations of iron. Evidence of diabetes has been found in the remains of Egyptian mummies,[135] with papyrus Ebers describing its symptoms.[136] This has been associated with a diet high in carbohydrates. A diet with a consistently high carbohydrate loading would be healthier if it became increasingly supplemented by food foraged from the edges of the burgeoning marshlands.[137] If the consequences of a weakened river led to an increasingly varied diet, it would be expected that cases of diabetes would have declined. As yet, no investigations tracing the change in incidence of diabetes over the time frame under investigation have occurred. In a similar manner, cultural preferences for particular types of foods, cereals for example, need to be considered when forensically analysing remains.

[114] Shiel, *'Anemia'*, https://www.medicinenet.com/script/main/art.asp?articlekey=15491 (02/02/2018).
[115] Stuart-Macadam, *Porotic Hyperostosis*, 39–40.
[116] Fujita & Adachi, *Qar Palaeohealth*, 40–41.
[117] Stuart-Macadam, *Porotic Hyperostosis*, 43–45.
[118] Duhig, *Eating People Here*, 102.
[119] Walker etal., *Anemia Hypothesis Reappraisal*, 109–125; Wapler etal., *Sudanese Cranial Pathology*, 333–339.
[120] Mant, *Mendes Palaeopathology*, 16; Angel, *Anemias, Malarias, & Marshes*, 760–763; Angel, *Porotic Hyperostosis*, 10–16.
[121] Bianucci etal., *Malaria in Paleopathology*, 176–80.
[122] Oxenham & Cavill, *Porotic Hyperostosis Response*, 199–200, contra Rothschild, *Porotic Hyperostosis*, 417–420. McIlvaine, *Implications*, 997–1000.
[123] Gowland & Western, *Morbidity in the Marshes*, 302–305.
[124] Lovell & Whyte, *Mendes Dental Defects*, 69–80.
[125] Mant, *Mendes Palaeopathology*, 3–4.
[126] Lawler, *Die - or Fade Away?* 7.
[127] Duhig, *Eating People Here*, 116–117.
[128] Samuel, *Ancient Egyptian Nutrition*, 579–580.
[129] Saffiro, *Dietary Habits*, 297–305.
[130] Touzeau etal., *Ancient Egyptian Diet,* 121.
[131] Wells, *Changes in Nutritional Diseases*, 1–12.
[132] Rothschild, *Shift to Maize*, 157–160.
[133] Fields etal., *Sex & Agricultural Transition*, 42–51; Williams & Hill, *Meat & Evolution*, 14–16; Watson etal., *Agriculture & Women's Health*, 92–102.
[134] Williams & Hill, *Meat & Evolution*, 12; Duhig, *Eating People Here*, 14.
[135] Forshaw, *Dental Health*, 422; Kamal, *Pharaonic Medicine*, 134.
[136] Ebbel, *p.Ebers*, 115.
[137] Williams & Hill, *Meat & Evolution*, 11.

12.4.2. Meat, Dairy, Diet and Health Indicators

If cultivation experienced a downward trend, then the proportion of the diet that came from cereal-based energy would have declined, being replaced, as suggested by a slightly more meat-based diet. With the proposed increase in availability of fowl and fish and, to a lesser extent, meat, the skeletal remains and the teeth, and associated coprolites, should indicate if this dietary adaptation took place.[138] Perhaps the dietary evidence from this time will suggest a reversion to a herder-forager lifestyle, not as nomadic as before but with an increased mobility.[139]

Fish was a consistent component of the diet of the ancient Egyptian, and with an apparent increase in 'incidental aquaculture' identified in the artistic record, the proportion of fish protein in the diet should have increased. An increase in protein generally increases the overall health of an individual,[140] although it may have been that an increase of fish protein rather than red meat protein meant that anaemia remained a problem within the population as there is less iron in aquatic meat. With an increase of meat in the diet, humans develop a more robust body shape,[141] so we could expect to see some similar changes in populations along the river during the time frame in question. Interestingly, some dietary analysis suggests that less fish was consumed than expected,[142] a result most likely related to the very few samples taken. More likely, it is due to the difficulty of identifying fish bones and the possibility that collection of this data was less prioritised in earlier studies. This conclusion seems to disagree with the apparent rise in fishing as suggested by the artistic record, and more importantly, as perceived in the empirical evidence from community spoil heaps being currently investigated at Edfu and Deir el-Bershā.[143] With a specific investigation in mind, a greater focus on this aspect of data collection; and applying improved excavation techniques may allow for the trend to be investigated more reliably. A similar investigation could be undertaken on the waterfowl bird remains uncovered at these sites.

If cattle, both large and small, and their products were becoming an increasingly significant proportion of the diet, then long-term health consequences would be expected.[144] High cholesterol levels have been identified in the remains of high-status individuals in ancient Egypt.[145] If the proportion of red meat in the diet increased for the majority of the population, then archaeological evidence should show a mounting incidence of cholesterol in individuals of lesser status. In this geographical area, cattle herders later developed an increasing lactose tolerance.[146] Perhaps this characteristic entered the Egyptian gene pool at this time; and with that particular goal in mind, it should be possible to identify its presence in remains from the time frame under investigation.

The scientific examination of the Dynasty VI mummy of Djau conducted by Schulz and Walker[147] described the remains of an individual living in an era that was supposedly contending with the trauma of famine. There were no signs of major parasitic infestation. While Djau experienced two or three childhood traumas, none were able to be precisely identified as being related to malnutrition; however, the impact of diseases is exacerbated by irregular nourishment. Despite the traumas of his youth, his diet was ample enough to facilitate the obesity that resulted in osteoarthritis. The orthoses that developed were not as a consequence of poor diet. It was suggested that Djau's diet seemed to have included less bread than expected with a larger number of calories derived from protein, though the type of protein has not been identified.[148] Similarly, the teeth seemed 'younger' in appearance than the rest of the skeletal structure, indicating limited abrasion from bread. As tempting as it may be to insinuate such results strengthen the 'low-carb' high-protein diet suggested by the A.R.I.D. hypothesis, it is important to recognise that Djau was among the upper echelons of society, where a more varied diet and a more refined and less coarse bread may have been more readily available. A future study may investigate Djau's teeth to identify the exact site of his youth, and his diet at various stages in his development.

A corresponding examination of the skeleton of the '*wife of Kahai*' has suggested that she died after having reached her fifth decade and that she belonged to the "lower level of the elite."[149] However, the internal structure of her teeth was not investigated, not through oversight but due to lack of technological ability. A more modern study would have been able to provide more precise details about these factors. Improved technology will improve the understanding of the lives of these individuals.

12.4.3. À la Famine?

Famines were experienced in Egypt throughout its history and are not unique to the end of the Old Kingdom. Some administrators claimed to have managed the situation, promoting themselves as successful despite the dark times of drought and famine.[150] Many scholars remain unconvinced as to the veracity of many of these claims: these mentions of 'famine' have less to do with hunger

[138] Smith, *Identifying Cess-pits*, 526–543.
[139] Copley, etal., *Foraging & Foddering*, 1273–1286; Brewer etal., *Domestic Plants & Animals*, 99.
[140] Houston etal., *Protein Intake*, 150–155; Shimazu etal., *Dietary Patterns & Mortality*, 600–609; Donaldson etal., *Proteins & Health*, 32–36; Lonnie etal., *Protein for Life*, 1–18; Williams & Hill, *Meat & Evolution*, 1–3, figs. 1, 2, 3. See *WHO*, *Dietary References*, 240–241.
[141] Pfeiffer & Sealy, *Foragers Body Size*, 7–8.
[142] Touzeau etal., *Ancient Egyptian Diet*, 123.
[143] M G. Marouard, N. Moeller and M. de Meyer, (pers. comm. 12/05/2017). Note Duhig, *Eating People Here*, 9 identifies these sites as ones suggesting that a good lifestyle was available.
[144] Smith, *African Herders*, 1–30, 85–90; Chan etal., *Diverged Maternal Lineages*, 12–13; Makarewicz, *Pastoralist Manifesto*, 159–174.
[145] Veiga, *Prevalent Pathologies*, 76–77.
[146] Smith, *African Herders*, 144–145.
[147] Schultz & Walker, *Mummy of Djau*, 67–78.
[148] Forensic tests can now identify the type of protein consumed (R. Power, Macquarie, pers. comm.).
[149] Schultz, *Biological Reconstruction*, 165–169.
[150] Vandier, *Mo'alla, inscriptions* 1–12.

and environmental crises and more to do with self-aggrandisement.[151] In a similar manner, those 'famine' victims depicted in royal complexes, such as those belonging to Djoser, Sahura and Unas, can be seen as 'clichés' rather than true representations of the actuality of the event.[152] It is difficult to accept the notion that a herder who is leading hundreds of kilograms of good food by a strap can end up 'starved'. Not all the cattle are female, so it is invalid to argue that the breeding stock is being preserved. Even if it was forbidden to eat the cattle they were guarding, it seems obvious that 'accidents' must have occurred, and, for a pragmatic people, the meat of injured animals would not have been wasted.

Skeletal analysis should be able to identify if famine was indeed a problem. Some studies have suggested that the diet changed for the worse during the First Intermediate Period,[153] yet studies that have investigated mummy health suggest a relatively homogeneous diet across the entirety of ancient Egyptian civilisation.[154] Duhig, in her study ' *"They are eating people here!" skeletal indicators of stress in the Egyptian First Intermediate Period*' found very little evidence of famine in the remains that she studied from this time frame.[155] Evidence from Aswan identifies changes to the diet that are attributed to the First Intermediate Period,[156] with the differences dissimilar to those identified during the Second Intermediate Period. Remains from both Intermediate periods, First and Second, contain evidence suggesting dietary modifications, but the data suggest that the sources of disruption were different: aridification in the earliest example and foreign invasion in the Second.[157]

Chronological discrepancies make using samples from the period of this investigation problematic: improved dating techniques may allow samples from this time frame to be incorporated into future similar studies. However, it is important to be cautious about such statements: thousands of years of history and a few dozen mummies do not make a highly detailed analysis. While it is important to use samples from all periods of time, especially if statements are to be made about changes in diet, it is equally important to avoid leaving out huge swathes of data from important historical eras. Touzeau etal., for example, in their study of the Ancient Egyptian diet, used data from remains ranging from the Pre-Dynastic era to Byzantine Egypt[158] (see Figure 12.1).[159] When the data is investigated more thoroughly, however, it can be noticed that it does not include samples from the time period with which we are interested. Similarly, none of the samples in the 'Horus Study'[160] are dated to the time frame under investigation.

12.4.4. Future Explorations

Population mortality rates could be used to compare the impact of the 'drought' upon society during the time frame under investigation. The most overwhelming data with regard to palaeo-pathologies is the very high incidence of dental disorders and degeneration. Whether these disorders contributed to a weakening in overall health or merely made life less comfortable[161] is difficult to ascertain. While disorder and diseases can be identified, it is difficult to identify the overall cause of mortalities.[162] A disparity can develop between suggested cause of death and actual cause of death, with the actual cause of death as ascertained by CT scans sometimes disputing the cause of death that had been suggested by earlier forensic examinations.[163] Older studies may have been less careful with regard to the recording of the context of the finds, perhaps leading to inaccurate conclusions.[164] While most records are related to the temple and its offerings to the king or the gods, to better understand the consequences of climate change on a population, more detailed investigation with regard to settlement archaeology and their food remains are needed.[165] Some changes may not have been an environmental response but perhaps a social one, such as the changes to the population at Gebelein that occurred when a large body of Nubian mercenaries, with their families and different cultural needs, arrived to fight for Ankhtify.[166]

12.5. Network Analyses

In his study of late Old Kingdom inheritance practices, Kanawati noted: "Whether the practice of nepotism was wide-spread throughout the Old Kingdom or was particularly common in the troubled first half of Dynasty 6, needs further documentation and research."[167] Possibly as part of a response to a changing climate and/or in reaction to a weakening central government, family ties became more durable with many administrative positions becoming 'inherited'.[168] At the same time the kings were shoring up their power by marrying their offspring to powerful nobles.[169] The tracing of interrelationships and familial links (the social 'network') may shed some light on the response processes to changing environmental circumstances. While the term 'networks' has been used

[151] Coulon, *Véracité et Rhétorique* 128–132; Coulon, *Famine*, 3.
[152] Franke, *Arme im Alten Reich*, 104–120.
[153] Kumar, *Health at Hierakonpolis*, 231–232; Zakrzewski, *Variation in Ancient Egyptians*, 219–229.
[154] Touzeau etal., *Ancient Egyptian Diet*; Adel etal., *The Horus Study*.
[155] Duhig, *Eating People Here*, 109–111, 115.
[156] Al-Khafif & El-Banna, *Ancient Egyptian Diet*, 1–4.
[157] Al-Khafif & El-Banna, *Ancient Egyptian Diet*, 5.
[158] Touzeau etal., *Ancient Egyptian Diet*, 118, 123.
[159] Touzeau etal., *Ancient Egyptian Diet*, fig. 2.
[160] Adel etal., *The Horus Study*, table 1.
[161] Cramer etal., *Paleopathologies in Mummies*, 6.
[162] Forshaw, *Dental Health*, 421.
[163] Cramer etal., *Paleopathologies in Mummies*, 6.
[164] Stevenson, *Museum Archaeology*, 7–8; Cramer etal., *Paleopathologies in Mummies*, 7.
[165] Morenó Garciá, *Microhistory*, 12.
[166] Ejsmond, *Nubians of Gebelein*, 11; Schneider, *Existence Now Silenced*, 317.
[167] Kanawati, *Nepotism*, 57.
[168] Duliková & Mařík, *Complex Network Analysis*, 63, 74.
[169] Duliková, *Old Kingdom Viziers*, 327–336; Bárta & Duliková, *Divine & Terrestrial*, 31–48.

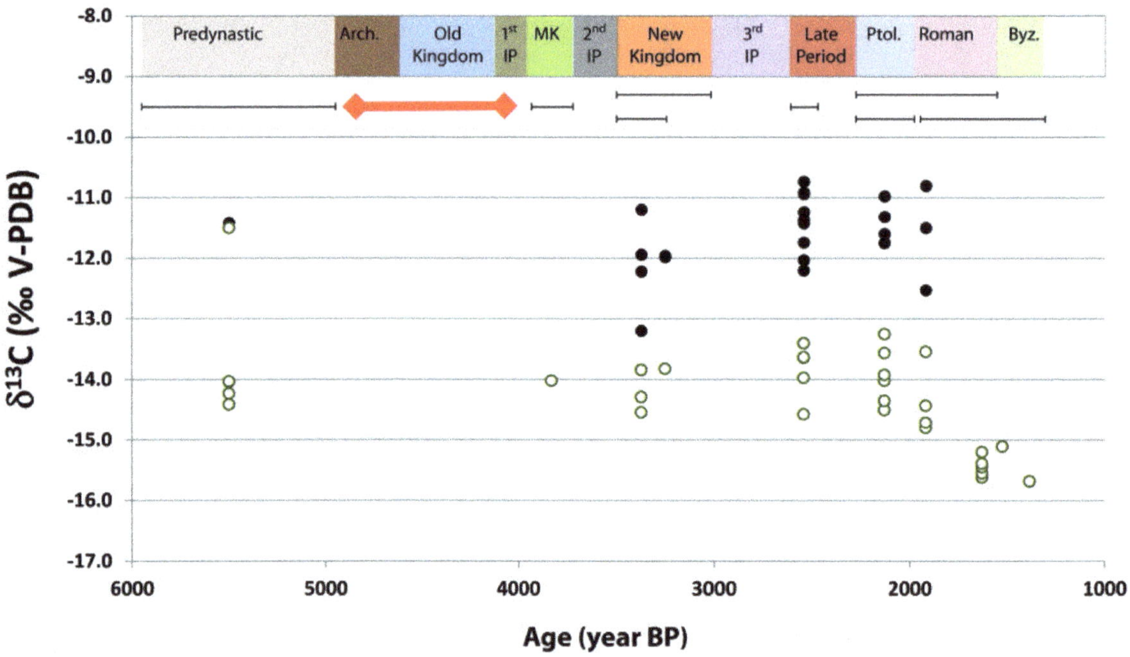

Figure 12.1: Modified copy of figure 2, Touzeau et al., *Ancient Egyptian Diet* (used with permission). Note the lack of samples from late Old Kingdom to early First Intermediate Period. (Red interval bar is the current author's addition.)

in archaeology for many years,[170] a more formal approach to the idea of networks has arisen recently, especially with regard to the application of many computer programs specifically written to identify the relationships and 'connectivity' within social systems.[171] Some projects investigate societal response to external stressors,[172] using some form of 'social index' to measure a society's inherent toughness.[173] The FRAGSUS project, for example, brings together many disciplines in order to explore the multifarious factors that underpinned the sustainability of a specific European island society, prehistoric Malta.[174]

12.5.1. Tracing Change

The interlocking influences and interactions of what van der Leeuw has conceptualised as 'Settlement, Sustainability, Subsistence and Society' can impact upon the long-term survival of a society.[175] Some cultures manage to sustain their civilisation for centuries or millennia, while others collapse in response to the impact of changing conditions in the wider environment.[176] How some societies successfully adapted to change, while others failed, is becoming an important issue in understanding the persistence of some ancient cultures, with the interaction between the society and its environment recognised as a dynamic consisting of a continually changing balance between influences sometimes leading in opposite directions.[177] Change and instability imposed stress on these societies, causing either extinction or adaptation and evolution.[178]

Despite a wide variety of network types existing in ancient Egypt,[179] there seems, however, when compared to other aspects of archaeology, a hesitation to apply these methods in the field of Egyptology.[180] This will slowly change as more cross-discipline analyses are undertaken.[181] Recently, a detailed investigation of nepotism in Old Kingdom society has been undertaken by Duliková & Mařík,[182] and this, along with newly published studies using multi-disciplinary approaches,[183] has led to a number of conclusions that may be relevant to the current study. Duliková's & Mařík's analysis traced titles

[170] Brughmans, *Connecting the Dots*, 289–291; Knappett, *Why Networks?*, 1–2; Terrell, *SNA & History*, 17–42; van der Leeuw, *Archaeology, Networks & Beyond*, 335; Mills, *SNA in Archaeology*, 380; Collar etal., *Networks in Archaeology*, 1; For Egypt, Cline & Cline, *Amarna Social Networks*, 17–44.

[171] Östborn & Gerding, *Archaeological Data Network Analysis*, 87; Mills, *SNA in Archaeology*, 381. Brughmans, *Connecting the Dots*, 288; Collar etal., *Networks in Archaeology*, 1–2, 17–25.

[172] See Mills, *SNA in Archaeology*, 383; Pearce, *Hunter-Gatherer Networks*, 410–411; Brughmans etal., *Inter-settlement Visibility*, 89–94.

[173] Borck etal., *Social Networks as Survival Networks*, 51.

[174] https://www.arch.cam.ac.uk/research/projects/fragsus, (04/04/2017).

[175] van der Leeuw, *Archaeology, Networks & Beyond*, 336; Middleton, *Nothing Lasts Forever*, 257–307; Borck etal., *Social Networks as Survival Networks*, 33–57.

[176] Trøjelesgaard & Olesen, *Networks in Motion*, 1926.

[177] van der Leeuw, *Archaeology, Networks & Beyond*, 338–339.

[178] Bárta, *Punctuated Equilibrium*, 1–17.

[179] Brughmans, *Connecting the Dots*, 288–292; Collar etal., *Networks in Archaeology*, fig. 2; Evans, *Which Network Model?*, 149–74; Peeples & Roberts, *To Binarize or Not*, 3001–3008; Trøjelsgaard & Olesen, *Networks in Motion*, 1931–1934.

[180] Brughmans, *Thinking through Networks*, 632–635; Brughmans, *Connecting the Dots*, 293; Östborn & Gerding, *Archaeological Data Network Analysis*, 77–79; Terrell, *SNA & History*, 17–42; Trøjelsgaard & Olesen, *Networks in Motion*, 1926–1928; Mills, *SNA in Archaeology*, 380–384; Collar etal, *Networks in Archaeology*, 6–8.

[181] Brughmans, *Thinking Through Networks*, 630; Östborn & Gerding, *Archaeological Data Network Analysis*, 75, 87; Brughmans, *Connecting the Dots*, 299; Collar etal., *Networks in Archaeology*, 15–16; Mills, *SNA in Archaeology*, 380.

[182] Duliková & Mařík, *Complex Network Analysis*, 63–83.

[183] Bárta, Kings, *Viziers & Courtiers*, 153–175; Bárta, *Abusir Paradigm*, 51–73; Cline & Cline, *Amarna Social Networks*, 17–44.

and relationships, especially marriage alliances.[184] The network analysis map that was produced[185] identified the familial inter-relationships and the societal linkages that Ptahhotep enjoyed and clearly show the intrinsic power base behind him that provided the impetus for him to inherit power and become the vizier.

12.5.2. Continuity, Change or Collapse during A River In Deviance?

Using network analyses to test the A.R.I.D. hypothesis would entail having to incorporate a number of interdisciplinary approaches: including the disciplines of ecology, environmental studies, chronology, sociology, archaeology and anthropology, among others.[186] Investigating linkages between artistic themes may provide another insight into the interrelationships and responses to external factors that took place:[187] As well as cultural change due to ecological responses,[188] evidence of conflict as a cause of collapse should also be looked for.[189]

Considering the length of the river, different responses may be identified along the length of the river, as latitude has been shown to have some variable impact upon the development of social networks.[190] Bearing in mind the overwhelming influence of the river upon the Egyptian civilization, it seems most fitting that the 'ecological' framework of network analysis may be best suited to explore how environmental changes impacted upon local population networks. Changes in settlement sites due to changing local situations can be plotted.[191] Populations with significant reliance on a local resource would require a specifically designed series of questions to develop a network analysis.[192]

As well as geographical position being a factor, social networks vary over time, and this variance can also be plotted.[193] Also, the interaction with non-riverine human populations may also affect the social networks that develop in times of water stress. Societal inter-relationships would have changed as they responded to this new situation, with the significant discrepancies identified able to lead to new pathways of investigation.[194] All of this data, when added together could be used to eventually measure the resilience of the population during the time frame under investigation.[195]

12.6. Summation: An Ongoing (or never-ending?) Investigation?

The major points raised in the chapter include:

- The unlikelihood of scientific 'proof.'
- Despite this, many tangents exist to enable the testing of the A.R.I.D. hypothesis across a number of fields of study.
- With improved technological expertise, geoarchaeological and bioarchaeological testing will enable greater precision to the identification and dating of archaeological findings, and this will enable a more accurate rendering of the diet at specific times.
- Greater use of dental evidence will enable the more precise localisation of individuals and provide a more precise interpretation of the dietary changes they experienced.
- Greater exploration of the role that DNA investigations can play in this field will allow for a more explicit refinement of human evidence from archaeological findings.

A number of future projects have been identified, pointing out that with increasing sophistication of the techniques available, the likelihood traditional archaeological methods can now be supplemented by scientific analyses of soil, seeds and skeletons to identify diet and disease. It is expected that within the near future, scientific means will be more readily available to test the hypothesis that, at the end of the Old Kingdom and into the First Intermediate Period, the ancient Egyptian had a more diverse diet than that in the Early dynastic and early Old Kingdom. With the continually increasing intricacy of technologies, new data analyses will result in many of the current understandings and existing interpretations of the 'decline' at the end of the Old Kingdom becoming superseded.

---

[184] Dulikovà & Mařík, *Complex Network Analysis*, 76–78, fig. 17.

[185] Dulikovà & Mařík, *Complex Network Analysis*, 70, fig. 9.

[186] As those outlined, for example, in Borck etal., *Social Networks as Survival Networks*, 51–52.

[187] As outlined in Brughmans, *Connecting the Dots*, 293–294, fig. 10; Collar etal., *Networks in Archaeology*, 9, and Mills, *SNA in Archaeology*, 386.

[188] Trøjelsgaard & Olesen, *Networks in Motion*, 1931–1934, figs 1, 3 & 4.

[189] Borck etal., *Social Networks as Survival Networks*, 34.

[190] Pearce, *Hunter-Gatherer Networks*, 412–413.

[191] Brughmans etal., *Inter-settlement Visibility*, 70–71, fig. 13 and Mills, *SNA in Archaeology*, 389–391.

[192] Pearce, *Hunter-Gatherer Networks*, 404–405.

[193] Borck etal., *Social Networks as Survival Networks*, 54; Trøjelsgaard & Olesen, *Networks in Motion*, 1928–1930; Brughmans, *Connecting the Dots*, 277; Brughmans, *Thinking Through Networks*, 655–658; Östborn & Gerding, *Archaeological Data Network Analysis*, 76–77; Collar etal, *Networks in Archaeology*, 15–16.

[194] Östborn & Gerding, *Archaeological Data Network Analysis*, 82–86.

[195] Borck etal., *Social Networks as Survival Networks*, 33; Daoud etal., *Adaptation & Resilience*, 209–250.

13

Conclusion: Resilience, <u>not</u> Regression

While it is imprudent to say that the goal of 'proving' a record of Old Kingdom environmental change as being depicted in the art has been completely achieved, a number of highly suggestive circumstantial relationships and connexions between the ecology and tomb wall scenes have been identified. Despite the inability to precisely link these correlations to a definitive cause, the associations identified may help improve our understanding of the cultural circumstances in Egypt at the end of the Old Kingdom.

The main points are summarised below.

13.1. Riverine Ecosystems and A River In 'Drought' (A.R.I.D.)

With the focus shifted away from the land and the analyses directed towards 'A River In Drought,' a re-evaluation of the significance of the 'drought' and the subsequent repercussions this may have had upon the society as a whole, has been advocated. Drought was recognised as not constituting the primary cause for the decline of the Old Kingdom. The term 'drought' was identified as inappropriate because of its land-based bias. It was realised that in times of weak or no inundation, the nutrients normally lost from the river to the surrounding land were retained in the river, thereby altering the biological balance. An 'excess' of nutrients in the river would have changed its biological profile. This led to the expectation of a different floral diversity developing along the edge of the river, with modern research identifying papyrus and phragmites as species that respond well to those hypothesised conditions, while typha and lotus appear to respond with less vigour. An increasing density of plant growth along the riverbanks may have required changes to some traditional fishing and boating practices. The findings may lead to certain aspects of the culture of late Old Kingdom/First Intermediate Period Egypt requiring reconsideration.

13.2. Tomb Decoration Analyses Suggest a Changing Ecology

A distribution and abundance analysis that was performed on tomb decorations showed changes within artistic themes over the time frame under investigation. Importantly, instead of focussing upon the quality or execution of the art, this study focussed on the significance of the *subject* of the scene, especially with regard to the representation of resources. A comparative decline in the proportion of representations relating to the agricultural cycle and other aspects of cultivation was identified, with an increasing emphasis on decorations relating to the river and its associated products recognised. The distribution of the data suggested that the 'watershed' reign of Niuserra was apparently a turning point where artwork whose subject was related to resources predominate. Similarly, but less dramatically, from the time of Teti to the time of the so-called 'drought', other depictions became relatively more important.

13.3. A Developing Pictorial Narrative within Tomb Decorations

With regard to the choices made for the composition of wall scenes, subconsciously or not, the tomb owners or artists were, with regard to the choices made for the composition of wall scenes, indicating a changing environmental awareness. Papyrus thickets began to be depicted in a larger, more chaotic manner, perhaps representing uncontrolled growth due to an excess of nutrients. No representations of gardens and orchards occurred during the Sixth Dynasty and into the early First Intermediate Period. A greater variety of fishing apparatus was depicted, and an assortment of varied fishing activities became apparent. Catfish swimming upside down were first depicted at this time. Goats browsing upon trees felled specifically for their use began to be depicted more readily. Representations of cattle became more significant at this time. Depictions of the poultry yard entered the lexicon at this time. Waterfowl became a predominant factor in offering processions. This developing pictorial narrative could be related to societal responses to an environment that was slowly changing around them. The changes observed allow the inferences below to be made.

13.4. New Depictions of Fishing Indicate its Increasing Significance

Fishing scenes increased in overall proportion over the time frame of the investigation. Some fishing scenes enter the lexicon from this time, while others declined in apparent importance: for example, dragnet scenes. Increasing numbers and varieties of fish bones found in the waste heaps of excavated settlement sites dating to the end of the Old Kingdom and into the First Intermediate Period have been noted. This suggested that fish had become an increasingly important protein source during the time frame under investigation. The decrease in depictions of larger nets was explained by the suggestion that unrestrained plant growth would have resulted in a more cluttered river, making it less likely for large nets to remain untangled. The increasing proportion of depictions of other types of fishing behaviours, including the use of smaller, more manoeuvrable nets, can be explained by the same reasoning. The number of individuals depicted working these nets decreased, suggesting more family-sized groupings rather than the 'industrialised' efforts needed to handle the larger drag and seine nets that were depicted at the start of the era.

13.5. Increased Depictions of Waterfowl suggest Thriving Birdlife

As the amount of plant cover increased, the diversity of habitats available for organisms living below the water would have increased. Animals living above the water would have experienced a similar increase in the amount and variety of habitats of available for them. Birds specifically associated with the riverbank, the waterfowl, were noted as becoming increasingly present in the tomb scenes produced at this time, perhaps explaining why fowling scenes came to outnumber spearfishing scenes on tomb walls. The significant increase in the relative prominence of birds in wall scenes was almost certainly related to an increase in habitats, a phenomenon known in the modern era in central and southern Africa, where greater avian biodiversity during times of decreasing water availability is experienced. Poultry yard depictions and processions, in the form of waterfowl parades, also increased in relative proportion when compared to other types of decorations depicting birds. Parallel to the increasing frequency of waterfowl depictions, attestations of their mammalian predators were added to the decorative repertoire at this time.

13.6. Cattle in Marshlands taking Advantage of New Feed Sources

As the riverbank ecotone became more crowded, the floral balance changed. Some plants benefit from the mild pruning that grazing animals provide, as this further encourages their growth, while others did not. A weaker, shallower river would have become criss-crossed with more secondary channels as the erosive force of water diminished. The rapid increase in cattle herd size may be a consequence of an increase in feed becoming available. Cattle were suggested as being better able to take advantage of these conditions than crops, due to the fact that they are a more mobile form of food. The ability to move cattle meant that the likelihood of localised damage would be better controlled. Cattle travel with the 'flood' allowing the grazed land behind a chance to re-invigorate itself. As well as the Delta making a feeding ground for cattle, from this time it was recorded that cattle moved to marshland both up and down the river, grazing as they went. The rise of the 'cattle barons' may be traced on the walls as these scenes became more varied and widespread.

13.7. Desert Hunt Scenes Imply a Rising Reliance on the 'Wild'

While the desert hunt had been an iconic tomb scene early in the Old Kingdom, its composition began to change over the time frame represented by this study. When this scene type re-entered the tomb decorative repertoire, the scenes produced appeared to suggest a more pragmatic approach to the hunting process, with the outcome of the desert hunt related more to the production and supply of food resource than purely for symbolic representation. Traps and hunting dogs give the impression of an 'industrialised' acquisition process glorifying the volume of the harvest and not the skill. The depiction of the mingling of desert animals and other pastureland animals in processions and field representations may suggest that, as climatic conditions changed, the desert and riverbank ecotones became less distinct from one another and the zones occupied by the animals exploiting these areas overlapped. Eventually, desert animals appear to have become habituated to the presence of humans and their domesticated stock.

13.8. A Partial Reprise of the Herder-Forager Lifestyle?

As the ability of the river to provide enough water for an agricultural surplus declined, its irregularity may have resulted in the general population needing to become more flexible with regard to the resources that were now available, how they were acquired and how these were exploited. The menu may have needed to broaden to compensate for times when an agricultural surplus was undependable, expanding the food base to more transient and varied resources. Earlier, Badarian hunter-foragers had utilised the produce arising from riverine exploitation, crop cultivation and selective herding practices, which resulted in a semi-sedentary way of life. Since the evidence suggests that, during this era the society utilised a similar resources base to this earlier time, so it may have witnessed a partial return to these practices on a wider scale, where groups within society practiced a periodic nomadism.

13.9. Final Statements: Innovation and Invention, NOT Inertia

The idea of a drought changing the history of the latter part of the Old Kingdom should be seen as a simplification of the overall situation. The end of the Old Kingdom and into the First Intermediate Period has been thought by many to represent a time of regression to a culture and administration less sophisticated than that celebrated in the Old Kingdom itself. This may not be the case. While it is true that the amount of water available was less than at the beginning of dynastic times, most current scientific evidence suggests that ancient Egypt had been experiencing a gradual aridification since the Third Dynasty. Despite some nearby cultures failing to respond adequately, generally, Egyptian society seems to have adapted to this change. The changing artistic evidence, as attested at the time under investigation, suggests that the ancient Egyptians were able to respond to changing environmental circumstances by adapting the way they exploited local food resources. While there may have been a developing political vacuum, rather than reverting to a more primitive stage of existence, the latter stages of the Old Kingdom and early First Intermediate Period may, in fact, display evidence of the great resilience of the civilisation of ancient Egypt.

Bibliography

Abd-El Monsef etal., *Aswan Impact.* Abd-El Monsef, H., Smith, S. and K. Darwish, 2015. Impacts of the Aswan high dam after 50 years. Water Resources Management, 29: 6, 1873–1885.

Abdel-Magid, *Archaeo-ethnobotanical Plant Domestication.* Abdel-Magid, A., 1989. Plant domestication in the Middle Nile Basin. An archaeo-ethnobotanical case study. British Archaeological Reports IS 532.

Abdo, *Water Quality.* Abdo, M. 2013., Physico-chemical studies on the pollutants effect in the aquatic environment of Rosetta branch river Nile, Egypt. Life Science Journal, 104, 493–501.

Abtew & Melesse, *Nile River Basin.* Abtew, W., and A. Melesse. 2014. The Nile River Basin. In Melesse, A., Abtew, W. and S. Setegn, (eds). Nile River basin: ecohydrological challenges, climate change and hydropolitics, 7–22. Dordrecht: Springer.

Abtew, *Land & Water.* Abtew, W. 2014. Land and Water in the Nile Basin. In Melesse, A., Abtew, W. and S. Setegn, (eds). Nile River basin: ecohydrological challenges, climate change and hydropolitics, 119–129. Dordrecht: Springer.

Abu El-Kasim, *Delta Problems.* Abu El-Kasim, R. 1988. The Northwestern Nile Delta: Problems and Priorities. The Archaeology of the Nile Delta, Egypt: Problems and Priorities. In Brink, E. van den (ed.) The archaeology of the Nile delta, Egypt: problems and priorities: proceedings of the seminar held in Cairo, 19-22 October 1986, on the occasion of the fifteenth anniversary of the Netherlands Institute of Archaeology and Arabic Studies in Cairo, 279–282. Amsterdam: Netherlands Foundation for Archaeological Research in Egypt.

Adams, *OK Mendes.* Adams, M. 2007. The early dynastic through Old Kingdom stratification at Tell er-Rub'a, Mendes. Thesis. The Pennsylvania State University.

Admasu etal, *Genetta in Farmlands.* Admasu, E., Thirgood, S., Bekele, A. and M. Laurenson. 2004. A note on the spatial ecology of African civet Civettictis civetta and common genet Genetta genetta in farmland in the Ethiopian Highlands. African Journal of Ecology 42: 160–162.

Ailstock, *Phragmites Adaptions.* Ailstock, S. 2000 Adaptive strategies of Common Reed Phragmites australis. In Arnold, M. (ed.) The Role of Phragmites in the Mid-Atlantic Region, Princess Anne, MD, Environmental Center, Anne Arundel College.

Al-Hassan & Ofori-Danson, *Plankton Abundance Factors.* Alhassan, E. and P. Ofori-Danson. 2016. Plankton abundance in relation to physicochemical factors in the Bui reservoir of Ghana's Black Volta River. African Journal of Ecology: 1–9.

Al-Khafif & El-Banna, *Ancient Egyptian Diet.* Al-Khafif, G. and R. El-Banna, 2015. Reconstructing ancient Egyptian diet through bone elemental analysis using LIBS Qubbet el Hawa Cemetery. BioMed Research International, 1–7.

Aldred, *Days of the Pharaohs.* Aldred, C. 1996. Egypt art in the days of the Pharaohs. London: Thames & Hudson.

Aldred, *Old Kingdom.* Aldred, C., 1965. Egypt to the End of the Old Kingdom. London: Thames & Hudson.

Allan & Castillo, *Stream Ecology.* Allan, J. and M. Castillo, 2007. Stream ecology: structure and function of running waters. Dordrecht: Springer.

Allan, *Stream Ecology.* Allan, J., 1995. Stream Ecology: structure and function of running waters. London: Chapman & Hall.

Allen & Pye, *Saltmarsh Morphodynamics.* Allen, J. and K. Pye, 1992. Saltmarshes: Morphodynamic, Conservation and Engineering Significance. Cambridge University Press.

Allen, *Pyramid Texts.* Allen, J., 2005.The Ancient Egyptian Pyramid Texts. Atlanta: Society of biblical literature.

Allen, *Reading a Pyramid.* Allen, J., 1991. Reading a Pyramid. In Berger, C., Clerc, G. and N. Grimal (éds), Hommages à Jean Leclánt, 1: 5–28. Cairo: IFAO.

Alpert etal., *Paradoxical Rainfall.* Alpert P., Ben-Gai, T., Baharad, A., Benjamini Y., Yekutieli, D., Colacino, M., Diodato, L., Ramis, C., Homar, V., Romero, R., Michaelides, S. and A. Manes, 2002. The paradoxical increase of Mediterranean extreme daily rainfall in spite of decrease in total values. Geophysical Research Letters 29: 135–154.

Altenmüller, *Mehu.* Altenmüller, H., 1998. Die Wanddarstellungen im Grab des Mehu in Saqqara, Archäologische Veröffentlichungen. Mainz: von Zabern.

Amills etal., *Domestic Pigs in Africa.* Amills, M., Ramírez, O., Galman-Omitogun, O. and A. Clop. 2013. Domestic pigs in Africa. African Archaeological Review, 30: 1, 73–82.

Anđelković, *Hegemony for Beginners*. Anđelković, B. 2012. Hegemony for Beginners: Egyptian Activity in the Southern Levant during the Second Half of the Fourth Millennium BC. Issues in Ethnology and Anthropology, 73: 789–808.

Andrassey, *Ägyptens zu Vorderasien*. Andrassy, P. 1991. Die Beziehungen Ägyptens zu Vorderasien bis zum Ende des Alten Reiches. In Probleme der frühen Gesellschaftsentwicklung im Alten Ägypten, 103–154. Berlin: Institution für Sudanarchäologie u. Ägyptologie.

Angel, *Anemias, Malarias, & Marshes*. Angel, J., 1966. Porotic hyperostosis, anemias, malarias, and marshes in the prehistoric eastern Mediterranean. Science, 153: 3737, 760–763.

Angel, *Porotic hyperostosis*. Angel, J., 1978. Porotic hyperostosis in the eastern Mediterranean. MCV/Q, Medical College of Virginia Quarterly, 141: 10–16.

Angelier, *Streams & Rivers*. Angelier, E., 2003. Ecology of Streams and Rivers. Boca Raton: CRC Press.

Archer, *PlantZAfrica*. Archer, C., 2004. PlantZAfrica. http://pza.sanbi.org.

Armelagos etal., *Historical Perspectives*. Armelagos, G., Brown, P. and B. Turner, 2005. Evolutionary, historical and political economic perspectives on health and disease. Social Science & Medicine, 61: 4, 755–765.

Arnold, *Early Animal Trade*. Arnold, E., Hartman, G., Greenfield, H., Shai, I., Babcock, L. and A. Maeir, 2016. Isotopic evidence for early trade in animals between Old Kingdom Egypt and Canaan. PloS one, 11(6), p.e0157650.

Art, *Dictionary of Ecology*. Art, H., 1993 A dictionary of ecology and environmental science. New York: Holt.

Arthington etal., *River Ecosystems*. Arthington, A., Bunn, S., Poff, N. and R. Naiman, 2006. The challenge of providing environmental flow rules to sustain river ecosystems. Ecological Applications, 164: 1311–1318.

Arz etal., *Red Sea Dry Event*. Arz, H., Lamy, F. and J. Pätzold, 2006. A pronounced dry event recorded around 4.2ka in brine sediments from the northern Red Sea. Quaternary Research 66: 3, 432–441.

Ash, *North Australia Food Supply*. Ash, A., 2014. Factors driving the viability of major cropping investments in northern Australia – a historical analysis. Canberra: CSIRO.

Ashley, *Tracks, Trails & Trampling*. Ashley, G. and C. Liutkus, 2003. Tracks, trails and trampling by large vertebrates in a rift valley paleo-wetland, lowermost Bed II, Olduvai Gorge, Tanzania. Ichnos, 91: 2, 23–32.

Austin, *Deir el-Medina Health Care*. Austin, A., 2014. Contending with illness in ancient Egypt: A textual and osteological study of health care at Deir el-Medina. PhD, University of California, Los Angeles.

Avnaim-Katav, etal., *Shallow Marine Record*. Avnaim-Katav, S., Almogi-Labin, A., Sandler, A., Sivan, D., Porat, N. and A. Matmon, 2012. The chronostratigraphy of a Quaternary sequence at the distal part of the Nile littoral cell, Haifa Bay, Israel. Journal of Quaternary Science, 27: 7; 675–686.

Ayyad etal., *Mudbrick Pollen Information*. Ayyad, S., Krzywinski, K. and R. Pierce, 1992. Mudbrick as a bearer of agricultural information: an archaeopalynological study. Norwegian Archaeological Review, 24: 2; 77–96.

Ayyad, *Pollen Grain Ecology*. Ayyad, S., 1988. Pollen Grain Ecology of the Mediterranean Sea Coast, Egypt. PhD, University of Mansoura, Egypt.

Ayyad, *Vegetation & Environment*. Ayyad, M., 1973, Vegetation and Environment of the Western Mediterranean Coastal Land of Egypt: I. The Habitat of Sand Dunes. Journal of Ecology. 61: 2; 509–523.

Azza etal., *Papyrus Permeability*. Azza, N., Kansiime, F., Nalubega, M., and P. Denny, 2000. Differential permeability of papyrus and Miscanthidium root mats in Nakivubo swamp, Uganda. Aquatic Botany, 673: 169–178.

Badawy, *Nyhetep-Ptah & Ankhmahor*. Badawy, A., 1978. The Tomb of Nyhetep-Ptah at Giza and the Tomb of Ankhmahor at Saqqara. Berkeley: University of California Press.

Bailey, *Circumcision*. Bailey, E., 1996. Circumcision in ancient Egypt. Bulletin of the Australian Centre for Egyptology 7: 15–28. Sydney: ACE.

Bailleul-Lesuer, *Avian Resources*. Bailleul-Lesuer, R., 2016. The exploitation of live avian resources in Pharaonic Egypt: A socio-economic study. Thesis; The University of Chicago.

Baines & Malek, *Cultural Atlas*. Baines, J. and J. Malek ,1980. The Cultural Atlas of Ancient Egypt. Oxford University Press.

Baines, *Communication & Display*. Baines, J., 1989. Communication and Display: The Integration of Early Egyptian Art and Writing. Antiquity. 63: 240, 471–482.

Baines, *Religious Practice*. Baines, J., 1991. Society, Morality and Religious Practice. In Schafer, B. E., (ed.) Religion in Ancient Egypt. London: Routledge.

Baines, *Schäfer & Egyptian Art*. Baines, J., 1988. Theories and Universals of Representation: Heinrich Schäfer and Egyptian Art. Art History. 8: 1, 1–25.

Baines, *Status & Purpose*. Baines, J., 1994. On the Status and Purposes of Ancient Egyptian Art. Cambridge Archaeological Journal, 4: 67–94.

Balagurunathan & Shanmugasundaram, *Microbial Biodiversity*. Balagurunathan, R. and T. Shanmugasundaram., 2015. Microbial biodiversity of selected major river basins of India. In Ramkumar,

M., Kumaraswamy, K. and R. Mohanraj, (eds.) Environmental Management of River Basin Ecosystems, 575–591. Dordrecht: Springer.

Baralat, *Schistosomiasis in Egypt*. Baralat, R., 2013. Epidemiology of Schistosomiasis in Egypt: travel through time: review. J Adv Res 45: 425–432.

Baran, etal., *Reed as Roughage*. Baran, M., Váradyová, Z., Kráčmár, S. and J. Hedvábný., 2002. The common reed Phragmites australis as a source of roughage in ruminant nutrition. Acta Veterinaria 71: 4, 445–449.

Bard, *Archaeology*. Bard, K., 2008. An introduction to the archaeology of ancient Egypt. New York: Wiley-Blackwell.

Bard, *Encyclopaedia*. Bard, K., 1999. Encyclopaedia of the Archaeology of Ancient Egypt. New York: Routledge.

Bardaji etal., *Dra Abu el-Naga Geomorphology*. Bardají, T., Martínez-Graña, A., Sánchez-Moral, S., Pethen, H., García-González, D., Cuezva, S., Cañaveras, J.C. and A. Jiménez-Higueras, 2017. Geomorphology of Dra Abu el-Naga Egypt: The basis of the funerary sacred landscape. Journal of African Earth Sciences, 131, 233–250.

Bares, *Introduction Abusir & Saqqara 2000*. Bares, L., 2000 Introduction. n Bárta, M. and J. Krejci, (eds), Abusir and Saqqara in the Year 2000, 1–12. Prague: Charles Univeristy Press.

Barich, *People, Water & Grain*. Barich, B., 1998. People, Water and Grain: The Beginning of Domestication in the Sahara and the Nile Valley Vol. 98. Roma: Studia Archaeologica.

Baridon, *Les Jardins*. Baridon, M., 1998. Les jardins: paysagistes, jardiniers, poètes. Paris: Laffont.

Barkworth etal., *Flora*. Barkworth, M., Capels, K., Long, S. and M. Piep, (eds), 2003. Flora of North America north of Mexico. Volume 25: Magnoliophyta: Commelinidae in part: Poaceae, part 2. Oxford University Press.

Barros, etal, *Egyptian Mongoose Expansion*. Barros, T., Carvalho, J., Pereira, M., Ferreira, J. and C. Fonseca, 2015. Following the trail: factors underlying the sudden expansion of the Egyptian mongoose Herpestes ichneumon in Portugal. PloS one, 10: 8, 1–18.

Bárta & Bezdek, *Beetles & Decline*. Bárta, M. and A. Bezdek, 2008. Beetles and the decline of the Old Kingdom: climate change in ancient Egypt. In Vymazalová, H. and M. Bárta, (eds), Chronology and Archaeology in Ancient Egypt The Third Millennium B.C. Proceedings of the Conference Held in Prague June 11–14, 2007, 214–222. Prague: Charles Univeristy Press.

Bárta etal., *Abusir South 2012*. Bárta, M., Vymazalová, H., Dulíková, V., Arias, K., Megahed, M., and L. Varadzin, 2014. Exploration of the Necropolis at Abusir South in the Season of 2012. Preliminary Report. Ägypten und Levante, 24: 15–38.

Bárta, *Abusir Paradigm*. Bárta, M., 2016. 'Abusir Paradigm' and the Beginning of the Fifth Dynasty. In Hein, I., Billing, N. and E. Meyer-Dietrich (eds), The Pyramids: Between Life and Death. Proceedings of the Workshop Held at Uppsala University, Uppsala, May 31st – June 1st, 2012, 51–74. Uppsala University Press.

Bárta, *Abusir* V. Bárta, M., 2001. Abusir V; The Cemeteries at Abusir South I. Prague: Charles Univeristy Press.

Bárta, *Architectural Innovations*. Bárta, M., 2006. Architectural Innovations in the development of the non-royal tomb during the reign of Nyuserra. In Janosi, P. (ed.), Structure and Significance; Vol. 25, Thoughts on Ancient Egyptian Architecture, 105–130. Vienna: Verlag der Österreichischen Akademie der Wissenschaften.

Bárta, *Collapse Hidden in Success*. Bárta, M., 2014. Collapse Hidden in Success: Rise and Fall of the Old Kingdom. KMT 251: 18–28.

Bárta, *In Mud Forgotten*. Bárta, M., 2013. In mud forgotten: Old Kingdom palaeoecological evidence from Abusir. Studia Quaternaria, 302: 75–82.

Bárta, *Journey to the West*. Bárta, M., 2011. Journey to the West: The World of the Old Kingdom Tombs in Ancient Egypt, Prague: Charles Univeristy Press.

Bárta, *Kings, Viziers & Courtiers*. Barta, M., 2013. Kings, Viziers and Courtiers. In Morenó García, J. C. (ed.), Ancient Egyptian Administration, 153–176. Leiden/Boston: Brill.

Bárta, *Long term Short term*. Bárta, M., 2015. Long Term or Short Term? Climate Change and the Demise of the Old Kingdom. In Kerner, S., Dann, R. and P. Bangsgaard, (eds), Climate and ancient societies, 177–196. Copenhagen: .

Bárta, *Old Kingdom Kingship*. Bárta, M., 2013. Egyptian Kingship during the Old Kingdom. In Hill, A., Jones, P and J. Morales (eds), Experiencing power, generating authority. Cosmos, Politics, and the Ideology of Kingship in Ancient Egypt and Mesopotamia, 257–283. Pennsylvania University Press.

Bárta, *Ptahshepses Junior II*. Bárta, M., 2000. The Mastaba of Ptahshepses Junior II at Abusir. In Bietak, M. (ed.), Ägypten und Levante/Egypt and the Levant, 10: 45–66.

Bárta, *Punctuated Equilibrium*. Bárta, M., 2015. Ancient Egyptian History as an Example of Punctuated Equilibrium: An Outline. In Der Manuelian, P. and T. Schneider, (eds), Towards a New History for the Egyptian Old Kingdom: Perspectives on the Pyramid Age, 1–17. Leiden/Boston: Brill.

Bates, *Ancient Egyptian Fishing*. Bates, O., 1917. Ancient Egyptian Fishing. Cambridge: Peabody Museum.

Beare & Zedler, *Cattail Invasion & Resistance*. Beare, P. and J. Zedler, 1987. Cattail invasion and persistence in a coastal salt marsh: the role of salinity reduction. Estuaries, 102: 165–170.

Beierkuhnlein, *Occurrence & Ecology*. Beierkuhnlein, C., 2015. Bos primigenius in Ancient Egyptian art–historical evidence for the continuity of occurrence and ecology of an extinct key species. Frontiers of Biogeography, 7: 3, 107–118.

Beinart & Hughes, *Environment & Empire*. Beinart, W. and L. Hughes, 2007. Environment and Empire. Oxford University Press.

Bell, *Dark Age Egypt*. Bell, B., 1971. The Dark Ages in Ancient History, I. The First Dark Age in Egypt. American Journal of Archaeology, 75: 1–26

Bell, *Oldest Nile Floods*. Bell, B., 1970. The Oldest Records of the Nile Floods. The Geographical Journal, 136: 4, 569–573.

Ben-Yaacov & Yom-Tov, *Egyptian Mongoose*. Ben-Yaacov, R. and Y. Yom-Tov, 1983. On the biology of the Egyptian mongoose, Herpestes ichneumon, in Israel. Zeitschrift für Säugetierkunde, 481, 34–45.

Bernhardt etal., *Rapid Pollen Response*. Bernhardt, C., Stanley, J. and B. Horton, 2011. Wetland vegetation in Manzala lagoon, Nile delta coast, Egypt: rapid responses of pollen to altered Nile hydrology and land use. Journal of Coastal Research, 27: 4, 731–737.

Bernhardt, *Delta Vegetation*. Bernhardt, C., Horton, B., and J. Stanley, 2012. Nile Delta vegetation response to Holocene climate variability. Geology, 407, 615–618.

Binder, *The Physical World*. Binder, S., 2000. representing the Physical World. In Donovan, L. and K. McCorquodale (eds), Egyptian Art, Principles and Themes in Wall Scenes Prism Archaeological Series 6, 29–36. Guizeh: Foreign Cultural Information Department.

Bintliff, *Time, Process & Catastrophism*. Bintliff. J., 2002. Time, process and catastrophism in the study of Mediterranean alluvial history: A review, World Archaeology, 33: 3, 417–435.

Blackman & Apted, *Meir* I. Blackman, A. and M. Apted, 1914. The Rock Tombs of Meir. Part I; The Tomb-Chapel of Ukh-hotep's Son Senbi. London: EES.

Blackman, *Meir* III. Blackman, A., 1915. The Rock Tombs of Meir. Part III; The tomb-chapel of Ukh-hotp son of Ukh-hotp and Mersi (B, No. 4). London: EES.

Blackman, *Meir* IV. Blackman, A., 1924. The Rock Tombs of Meir. Part IV; The Tomb-Chapel of Pepi'onkh the Middle Son of Sebkhotpe and Pekhernefert. London: EES.

Blackman, *Meir* V. Blackman, A., and M. Apted, 1953. The Rock Tombs of Meir. Part V; The Tomb-Chapels A, No. 1 that of Ni-'ankh-pepi the Black A, No. 2 that of Pepi'onkh with the "Good Name" of Heny the Black, A, No. 4 that of Hepi the Black, D, No. 1 that of Pepi, and E, Nos. 1-4 those of Meniu, Nenki, Pepi'onkh and Tjetu. London EES.

Blench & MacDonald, *Chickens*. Blench, R. and K. MacDonald, 2008. Chickens. Cambridge World History of Food, 496–499. Cambridge University Press.

Blench, *Pigs in Africa*. Blench, R., 2000. A history of pigs in Africa. In Blench, R. M. and K. MacDonald (eds), Origins and development of African livestock: Archaeology, genetics, linguistics and ethnography; 355–367. Oxford University Press.

Boar etal., *Biomass Allocation*. Boar, R., Harper, D., and C. Adams, 1999. Biomass Allocation in Cyperus papyrus in a Tropical Wetland, Lake Naivasha, Kenya1. Biotropica 313: 411–421.

Boar, *Papyrus Water Level Response*. Boar, R., 2006. Responses of a fringing Cyperus papyrus L. swamp to changes in water level. Aquatic Botany 84: 2, 85–92.

Boessneck, *Riesige Hausgänse*. Boessneck, J., 1991. Riesige Hausgänse aus der Spätzeit des alten Ägypten. Archiv für Geflügelkunde 55: 105–10.

Boessneck, *Die Haustiere*. Boessneck, J., 1953. Die Haustiere in Altägypten. Munich: Pfeiffer.

Bolshakov, *Boatmen Jousting*. Bolshakov, A., 1993. The Scene of the Boatmen Jousting in Old Kingdom Tomb Representations, Bulletin de la Société d'Égyptologie de Genève 17: 29–39.

Bolshakov, *Fourth Dynasty Osiris?* Bolshakov, A., 2001. Osiris in the Fourth Dynasty Again? The false door of Jntj, MFA 31.781. In Győry, H. (ed.), Mélanges offerts à Edith Varga le lotus qui sort de terre", Budapest: [Bulletin du Musée Hongrois des Beaux-Arts. Supplément-2001], 65–80.

Borchardt, *Sahure II*. Borchardt, L., 1913 Das Grabdenkmal des Königs Sahu-re II. Leipzig: Hinrichs.

Borck, etal, *Social networks as survival networks*. Borck, L., Mills, B., Peeples, M. and J. Clark, 2015. Are social networks survival networks? An example from the Late Prehispanic U.S. Southwest. Journal of Archaeological Method and Theory 22: 1, 33–57.

Boulos & Fahmy, *Ancient Egyptian Grasses*. Boulos, L. and A. Fahmy, 2007. Grasses in Ancient Egypt. Kew Bulletin 62: 3, 507–511.

Bowe, *Early Garden Making*. Bowe, P., 2017. The early development of garden making c. 3000–c. 2000 BCE. Studies in the History of Gardens and Designed Landscapes 37:3, 231–241.

Boyd, *Water Quality*. Boyd, C., 2015. Water quality: an introduction. Dordrecht: Springer.

Brass, *Cattle Cult Origins*. Brass, M., 2003. Tracing the origins of the Ancient Egyptian cattle cult. In A. Eyma & C. Bennett (eds), A delta-man in Yebu: occasional volume of the Egyptologists' Electronic Forum 1: 101–110. Parkland: Universal Publishers.

Brass, *Cattle Domestication Reassessment*. Brass, M., 2018. Early North African Cattle Domestication and Its

Ecological Setting: A Reassessment. Journal of World Prehistory 31: 81–115.

Bravard & Petit, *Geomorphology*. Bravard, J. and F. Petit, 2015. Geomorphology of Streams and Rivers. In Likens, G. (ed.), The Encyclopedia of Inland Waters, 387–395. Amsterdam: Elsevier.

Brewer & Friedman, *Fishing*. Brewer, D., and R. Friedman, 1989. Fish and fishing in ancient Egypt. London: Warminster: Aris & Phillips.

Brewer etal., *Domestic Egypt*. Brewer, D., Redford, D., and S. Redford, 1994. Domestic plants and animals: the Egyptian origins. Warminster: Aris & Phillips.

Brewer, *Predynastic Temperatures*. Brewer, D., 1991. Temperature in Predynastic Egypt inferred from the Remains of Nile Perch. World Archaeology 223: Archaeology and Arid Environments, 288–303.

Brier, *Infectious Diseases*. Brier, B., 2004. Infectious diseases in ancient Egypt. Infectious disease clinics of North America 181: 17–27.

Brierly etal., *Green Sahara Pastoralism*. Brierley, C., Manning, K. and M. Maslin, 2018. Pastoralism may have delayed the end of the green Sahara. Nature communications 91: 1–9.

Britton, *Papyrus Swamps*. Britton, P., 1978. Seasonality, Density and Diversity of Birds of a Papyrus Swamp in Western Kenya. Ibis 120: 4, 450–466.

Bronk Ramsay etal., *Radiocarbon Chronology*. Bronk. F., Ramsay W., Dee, M., Rowland, J., Higham, T., Harris, S., Quiles, A., Wild, E., Marcus, E. and A. Shortland, 2010. Radiocarbon-based Chronology for Dynastic Egypt. Science 328: 1554–1557.

Brovarski, *Senedjemib* I. Brovarski, E., 2000. The Senedjemib Complex Part I. Boston: Museum of Fine Arts.

Brown etal., *DNA & Archaeobotany,* Brown, T., Cappellini, E., Kistler, L., Lister, D., Oliveira, R., Wales, N. and A. Schlumbaum, 2015. Recent advances in ancient DNA research and their implications for archaeobotany. Vegetation History and Archaeobotany, 24(1), 207–214.

Brughmans, *Connecting the Dots*. Brughmans, T., 2010. Connecting the dots: towards archaeological network analysis. Oxford Journal of Archaeology 293: 277–303.

Brughmans, etal., *Inter-settlement Visibility*. Brughmans, T., Keay, S. and G. Earl, 2015. Understanding Inter-settlement Visibility in Iron Age and Roman Southern Spain with Exponential Random Graph Models for Visibility Networks. Journal of Archaeological Method and Theory 221: 58-143.

Brughmans, *Thinking Through Networks*. Brughmans, T., 2013. Thinking through networks: A review of formal network methods in archaeology. Journal of Archaeological Method and Theory,204: 623–662.

Bruins, *Dating Pharaonic Egypt*. Bruins, H., 2010. Dating Pharaonic Egypt. Science 328; 1489–1490.

Brundage, *Grazing as Management*. Brundage, J., 2010. Grazing as a management tool for controlling Phragmites australis and restoring native plant biodiversity in wetlands. PhD, University of Maryland.

Brunner-Traut, *Epilogue*. Brunner-Traut, E., 1986, Epilogue. In Schäfer, H., Brunner-Traut, E. and J. Baines (trans.), Principles of Egyptian Art, 421–426. Oxford University Press.

Bryant el al., *Nutritive Content*. Bryant, F., Kothmann, M., and L. Merril, 1980. Nutritive content of sheep, goat and white-tailed deer diets on excellent condition rangeland in Texas. J. Range Manage. 33: 410–414.

Bukaveckas, *Rivers*. Bukaveckas, P., 2015. Rivers. In Likens, G. (ed.), Encyclopedia of inland waters Vol. 1, 721–732. Amsterdam: Elsevier.

Bunbury & Jeffries. *Ancient Landscapes*. Bunbury, J. and D. Jeffreys, 2011. Real and literary landscapes in ancient Egypt. Cambridge Archaeological Journal, 21(1), pp.65-76.

Bunbury & Lutley, *Nile on the Move*. Bunbury, J. and K. Lutley, 2008. The Nile on the move. Egyptian Archaeology 32: 3–5.

Bunbury etal., *Holocene Nile @ Karnak*. Bunbury, J., Graham, A. and M. Hunter, 2008. Stratigraphic landscape analysis: charting the Holocene movements of the Nile at Karnak through ancient Egyptian time. Geoarchaeology 233: 351–373.

Bunbury, *Development in Nile Flood Zone*. Bunbury, J., 2010. The Development of the Capital Zone within the Nile Floodplain. In Subías, E. Azara, P., Carruesco, J., Fiz, I. and R. Cuesta (eds), The Space of the City in Graeco-Roman Egypt Image and Reality, 211–218. Tarragona: Institut Català d'Arqueologia Clàssica.

Bunbury, *Memphite Floodplain Landscape*. Bunbury, J., 2013. Geomorphological development of the Memphite floodplain over the past 6,000 years. Studia Quaternaria, 30: 2, 61–67.

Bunbury, *Nile River Development*. Bunbury, J., 2010. The Development of the River Nile and the Egyptian Civilization: A Water Historical Perspective with Focus on the First Intermediate Period. A History of Water, Series II, 2: 52–71.

Bunn & Arthington, *Altered Flow Regimes*. Bunn, S. and A. Arthington, 2002. Basic principles and ecological consequences of altered flow regimes for aquatic biodiversity. Environmental Management 30: 4, 492–507.

Bunson, *Encyclopaedia*. Bunson, M., 1991. The Encyclopaedia of Ancient Egypt. New York: Facts on File.

Büntgen etal., *Climate & Human Susceptibility*. Büntgen, U., Tegel, W., Nicolussi, K., McCormick, M., Frank, D., Trouet, V., Kaplan, K., Herzig, F., Heussner, K., Wanner, H., Luterbacher, J. and J. Esper, 2011. 2500 years of European climate variability and human susceptibility. Science 3316017: 578–582.

Burchard, *Hunting Climatic Impact*. Burchard, I., 1998. Anthropogenic impact on the climate since man began to hunt. Palaeogeography, Palaeoclimatology, Palaeoecology, 1391: 1-14.

Burn, *Environment*. Burn, J., 2012. Ancient Egypt and its Environment. In McFarlane, A. and A. Mourad eds, Behind the scenes: Daily life in Old Kingdom Egypt, 1–12. Oxford University Press.

Burn, *Fishing & OK Climate Change*. Burn, J., 2014. Climate Change, Fishing and the Nile: Changes in Fishing Techniques and Technologies at the End of the Old Kingdom. In Bartá, M. and J. Janák (eds), Profane Landscapes, Sacred Spaces. Publication of the Proceedings of the Conference, Charles University, Prague, 26–27th June 2014. Equinox eBooks Publishing, United Kingdom. May 2019. ISBN 981781794098. https://www.equinoxpub.com/home/view-chapter/?id=29015.

Burn, *Marshlands & OK Famine*. Burn, J., 2014. Marshlands, drought and the great famine: on the significance of marshlands to the end of the Old Kingdom. In Accetta, K., Fellinger, R., Lourenco Goncalves, P., Musselwhite, S. and W. P. Van Pelt (eds), Publications of the Proceedings of CRE XIV – Crossing Boundaries, 34–48. Cambridge University Press.

Burn, *Mehu*. Burn, J., 2011. The Pyramid texts and tomb decoration in Dynasty Six: the tomb of Mehu at Saqqara. In Binder, S. (ed.), Bulletin of the Australian Centre for Egyptology, 22: 17–34.

Burn, *Pyramid Texts*. Burn, J., 2012. An Ecological Approach to Determine the Potential Influence that The Pyramid Texts had upon Dynasty Six Tomb Decorations. In Bárta, M., Coppens, F., and J. Krejčí, (eds), Abusir and Saqqara in the Year 2010, 233–245. Prague: Charles Univeristy Press.

Butzer & Enfield, *Critical Perspectives*. Butzer, K. and G. Engield, 2012. Critical Perspectives on Historical Collapse. Proceedings of the National Academy of Sciences, March: 3628–3631.

Butzer & Hansen, *Desert & River*. Butzer, K. and C. Hansen, 1968. Desert and river in Nubia: geomorphology and prehistoric environments at the Aswan Reservoir. Madison: Plenum Press.

Butzer etal, *Urban Geoarchaeology*. Butzer, K., Butzer, E. and S. Love, 2013. Urban geoarchaeology and environmental history at the Lost City of the Pyramids, Giza: synthesis and review. Journal of Archaeological Science 408: 3340–3366.

Butzer etal., *East African Lake Levels*. Butzer, K., Isaac, G., Richardson, J. and C Washbourn-Kamau, 1972. Radiocarbon dating of East African lake levels. Science 175: 4026, 1069–1076.

Butzer, *Before Agriculture*. Butzer, K., 1970 Physical Conditions in Eastern Europe, Western Asia and Egypt Before the Period of Agricultural and Urban Settlement. In Edwards, I., Gadd, C. and N. Hammond (eds), The Cambridge Ancient History, 35–69. Cambridge University Press.

Butzer, *Archaeology & Geology*. Butzer, K., 1960. Archaeology and geology in ancient Egypt. Science 1323440: 1617–1624.

Butzer, *Collapse*. Butzer, K., 2012. Collapse, Environment, and Society. Proceedings of the National Academy of Sciences. March: 3632–3639.

Butzer, *Context in Archaeology*. Butzer, K., 1980. Context in archaeology: an alternative perspective. Journal of Field Archaeology 7: 417–422.

Butzer, *Desert in Flood*. Butzer, K., 2001. When the desert was in flood. Environmental history of the Giza Plateau. Aeragram: Newsletter of the Ancient Egypt Research Associates 5: 1, 3–5. Boston: AERA.

Butzer, *Discontinuity*. Butzer, K., 1997. Sociopolitical discontinuity in the Near East c. 2200 BCE: Scenarios from Palestine and Egypt. In Dalfes, N., Kukla, G. and H. Weiss (eds), Third Millennium BC Climate Change and Old World Social Collapse, 245–295. Berlin: Springer.

Butzer, *Ecological Archaeology*. Butzer, K., 1975. The Ecological Approach to Archaeology: Are We Really Trying? American Antiquity: 106–111.

Butzer, *Environment & Human Ecology*. Butzer, K., 1959. Environment and human ecology in Egypt during predynastic and early dynastic times. Reprinted from Bulletin de la Societé de Géographie d'Egypte, 32, 43–87.

Butzer, *Geological Deposits*. Butzer, K., 1959. Some recent geological deposits in the Egyptian Nile Valley. Geographical Journal: 75–79.

Butzer, *Human Ecology*. Butzer, K., 1982. Archaeology as human ecology: method and theory for a contextual approach. Cambridge University Press.

Butzer, *Hydraulic Egypt*. Butzer, K., 1976. Early Hydraulic Civilisation in Egypt: A Study in Cultural Ecology. Chicago University Press.

Butzer, *Near East Environmental Change*. Butzer, K., 1995. Environmental change in the Near East and human impact on the land. In Sasson, J. M. (ed.), Civilizations of the Ancient Near East 1: 123-51. New York: Scribner.

Butzer, *Nile Valley Pleistocene*. Butzer, K., 1980. Pleistocene history of the Nile Valley in Egypt and

Lower Nubia. In Williams, M. and H. Faure (eds), The Sahara and the Nile: Quaternary Environments and Prehistoric Occupation in Northern Africa, 253–280. Rotterdam: Balkema.

Butzer, *Organisms or Systems?* Butzer, K., 1980. Civilizations: Organisms or Systems? American Scientist 68: 44–58.

Butzer, *Variation & Discontinuities*. Butzer, K., 1984. Long-term Nile flood variation and political discontinuities in pharaonic Egypt. In Clark, J. and S. Brandt (eds), From Hunters to Farmers: The Causes and Consequences of Food Production in Africa, 102–112. Berkeley: University of California Press.

Buzon & Bombak, *Nile Valley Dental*. Buzon, M. and A. Bombak, 2010. Dental disease in the Nile Valley during the New Kingdom. International Journal of Osteoarchaeology 204: 371–387.

Buzon & Simonetti, *Nile Valley Strontium Isotope Variability*. Buzon, M. and A. Simonetti, 2013. Strontium isotope 87Sr/86Sr variability in the Nile Valley: identifying residential mobility during ancient Egyptian and Nubian sociopolitical changes in the New Kingdom and Napatan periods. American Journal of Physical Anthropology 151: 1, 1–9.

Cable etal., *Global Change & Disease Control*. Cable, J., Barber, I., Boag, B., Ellison, A. R., Morgan, E. R., Murray, K., Pascoe, E. L., Sait, S. M., Wilson, A. J. and M. Booth, 2017. Global change, parasite transmission and disease control: lessons from ecology. Philosophical Transactions of the Royal Society B: Biological Sciences 372: 1719, 1–17.

Callaway, *Positive Interactions*. Callaway, R., 1995 Positive interactions among plants. Botanical Review 61: 306–349.

Campbell etal., *Concepts & Connections*. Campbell, N., 2003 Biology: concepts and connections. San Francisco: Pearson.

Cant & Morris, *Geographies of Art*. Cant, S. and N. Morris, 2006. Geographies of art and the environment. Social and cultural geography 7: 6, 857–861.

Cappers, *Reconstructing Agriculture*. Cappers, R., 2006. The reconstruction of agricultural practices in Ancient Egypt: an ethnoarchaeobotanical approach. Palaeohistoria, 47: 48, 429–446.

Caracuta etal., *Dakhleh Archaeobotancal*. Caracuta, V., Fiorentino, G., Davoli, P. and R. Bagnall, 2015. The archaeobotanical analysis at Amheida Dakhleh Oasis–Egypt, preliminary results. Supplemento Vol. CXL: 45–48.

Cary & Weerts, *Growth & Water Temperature*. Cary, P. and P. Weerts, 1983. Growth of Salvinia molesta as affected by water temperature and nutrition I. Effects of nitrogen level and nitrogen compounds. Aquatic Botany 162: 163–172.

Chambers etal., *Phragmites Ecology*. Chambers, R., Meyerson, L. and K. Dibble, 2012. Ecology of Phragmites australis and responses to tidal restoration. In Roman, C., and D. Burdick (eds), Tidal Marsh Restoration, 81–96. Washington: Island Press.

Chambers etal., *Phragmites Expansion*. Chambers, R., Osgood, D., Bart, D. and F. Montalto, 2003. Phragmites australis invasion and expansion in tidal wetlands: interactions among salinity, sulfide, and hydrology. Estuaries 262B: 398–406.

Chapman etal, *Swimming Upside Down*. Chapman, L., Kaufman, L. and C. Chapman.,1994. Why swim upside down? A comparative study of two Mochokid catfishes, Copeia 1: 130–135.

Chapman etal., *Papyrus Swamp Variations*. Chapman, L, Chapman, C., Brazeau, D, McLaughlin, B. and M. Jordan, 1999. Papyrus Swamps, hypoxia and faunal diversification: variation among population of Barbus Neumayer. Journal of Fish Biology 54: 310–327.

Chapman, etal., *Limnological Observations*. Chapman, L., Chapman, C. and T. Crisman, 1998. Limnological observations of a papyrus swamp in Uganda: Implications for fish faunal structure and diversity, Verh. Internat. Verein. Limnol. 26: 1821–1826.

Chase-Dunn etal., *Power & Size*. Chase-Dunn, C., Alvarez, A. and D. Pasciuti, 2005. Power and Size: Urbanization and Empire Formation in World-Systems. In Chase-Dunn, C. and E. Anderson (eds), Historical Evolution of World Systems, 92–112. Basingstoke: Palgrave Macmillan.

Chew, *Harappa to Mesopotamia*. Chew, S., 2005. From Harappa to Mesopotamia and Egypt to Mycenae: Dark Ages, Hegemonial Shifts, and Environmental/Climactic Changes, 2200 B.C.-700 B.C. In Chase-Dunn, C. and E. N. Anderson, (eds), Historical Evolution of World Systems, 52–74. Basingstoke: Palgrave Macmillan.

Church & Bell, *Settlement Patterns*. Church, R. and T. Bell., 1988. An Analysis of Ancient Egyptian Settlement Patterns Using Location-Allocation Covering Models. Annals of the Association of American Geographers 784: 701-714.

Cílek etal., *Lake Abusir*. Cílek, V., Bárta, M., Lisá, L., Pokorná, A., Juříčková, L., Brůna, V., Moniem, A., Mahmoud, A., Bajer, A., Novak, J. and J. Beneš, 2012. Diachronic development of the Lake of Abusir during the third millennium BC, Cairo, Egypt. Quaternary International 266: 14-24.

Clarke etal., *Climate Change & Social Transformation*. Clarke, J., Brooks, N., Banning, E., Bar-Matthews, M., Campbell, S., Clare, L., Cremaschi, M., di Lernia, S., Drake, N., Gallinaro, M., Manning, S., Nicoll, K., Philip, G., Rosen, R., Schoop, U., Tafuri, M., Weninger, B. and A. Zerboni, 2015. Climatic changes and social transformations in the Near East and North Africa during the 'long'4th millennium BC: A comparative study of environmental and archaeological evidence. Quaternary Science Reviews: 1–26.

Clayton, *Chronicle*. Clayton, P., 1994. Chronicle of the Pharaohs. London: Thames & Hudson.

Clement, *Caries*. Clement A., 1958. The antiquity of caries. Br Dent Journal104: 115–123.

Cleuziou, *Archaeology & Climate*. Cleuziou, S., 2009. Extracting wealth from a land of starvation by creating social complexity: A dialogue between archaeology and climate? Comptes Rendus Geoscience 3418: 726–738.

Cline & Cline, *Amarna Social Networks*. Cline, D. and E. Cline, 2015. Text Messages, Tablets, and Social Networks: The "Small World" of the Amarna Letters. In: Mynářová, J., Onderka, P. and P. Pavúk (eds), There and Back Again–Crossroads II. Proceedings of an International Conference held in Prague, September 15–18, 2014, 17–44. Prague: Charles Univeristy Press.

Clutton-Brock, *Domestic Animals*. Clutton-Brock, J., 2014. The spread of domestic animals in Africa. In Andah, B., Okpoko, A., Shaw, T. and P. Sinclair (eds), The Archaeology of Africa: food, metals and towns, 61–70. London: Thames & Hudson.

Clutton-Brock, *The Walking Larder.* Clutton-Brock, J., 1989 The Walking Larder. Patterns of Domestication, Pastoralism, and Predation. London: Routledge.

Collar etal., *Networks in Archaeology*. Collar, A., Coward, F., Brughmans, T. and B. Mills, 2015. Networks in Archaeology: Phenomena, Abstraction, Representation. Journal of Archaeological Method and Theory 221: 1–32.

Collon, *Ancient Near Eastern Art*. Collon, D., 1995. Ancient Near Eastern Art. Berkeley: University of California Press.

Coltherd, *Domestic Fowl*. Coltherd, J., 1966. The domestic fowl in ancient Egypt. Ibis 1082: 217–223.

Conard, *Waterlilies*. Conard, H., 1905. The waterlilies: a monograph of the genus Nymphaea. Washington: Carnegie Institute.

Connor & Marks, *High Water Adaptations*. Connor, D. and A. Marks, 1986. The terminal pleistocene on the Nile. In Straus, L. G. (ed.), The End of the paleolithic in the Old World. British Archaeological Reports: 171–199.

Contardini, *Rhythm of the Nile*. Contardini, F., 2015. Disasters Connected to the Rhythm of the Nile in the Textual Sources. In Vittozz, G. C. (ed.), A Research on Ancient Catastrophes, 11–26. Roma: Istituto di Studi sul Mediterraneo Antico.

Copley, etal., *Foraging & Foddering*. Copley, M., Jim, S., Jones, V., Rose, P., Clapham, A., Edwards, D., Horton, M., Rowley-Conwy, P. and R. Evershed, 2004. Short- and long-term foraging and foddering strategies of domesticated animals from Qasr Ibrim, Egypt. Journal of Archaeological Science, 319: 1273–1286.

Corby etal., *Archaeology & Art*. Corbey, R., Layton, R. and J. Tanner, 2004. Archaeology and Art. In Bintliff, J. L., Earle, T. K., and C. S. Peebles (eds), A companion to archaeology, 357–379. Oxford University Press.

Cornevin, *Paléoclimatologie*. Cornevin, M., 1996. Paléoclimatologie et peuplement de l'Égypte ancienne. Revue d'égyptologie, 47: 183–203.

Coulon, *Famine*. Coulon, L., 2008. Famine. In Frood, E. and W. Wendrich (eds), UCLA Encyclopedia of Egyptology, Los Angeles. http://escholarship.org/uc/item/2nv473z9.

Coulon, *Véracité et Rhétorique*. Coulon, L., 1997. Véracité et rhétorique dans les autobiographies égyptiennes de la Première Période Intermédiaire. Bulletin de l'Institut français d'archéologie orientale 97: 109–138.

Cramer etal., *Paleopathologies in Mummies*. Cramer, L., Brix, A., Matin, E., Rühli, F. and K. Hussein, 2018. Computed Tomography–Detected Paleopathologies in Ancient Egyptian Mummies. Current problems in diagnostic radiology 474: 225–232.

Crumley, *Historic Ecotonal Shifts*. Crumley, C., 1993. Analyzing historic ecotonal shifts. Ecological Applications 3: 3, 377–384.

Currie etal., *Agricultural Productivity*. Currie, T., Bogaard, A., Cesaretti, R., Edwards, N.R., Francois, P., Holden, P., Hoyer, D., Korotayev, A., Manning, J., Morenó García, J. C. Petrie, C., Turchin, P. Whitehouse, H., Williams, A. and O. K. Oyebamiji, 2015. Agricultural productivity in past societies: Toward an empirically informed model for testing cultural evolutionary hypotheses. Cliodynamics: The Journal of Quantitative History and Cultural Evolution: 6: 24–56.

Cushing & Allan, *Streams*. Cushing, C. and J. Allan, 2001. Streams: their ecology and life. San Diego: Elsevier.

Czarnowicz, etal., *Trade or Conquest*? Czarnowicz, M., Ochał-Czarnowicz, A., Rosinska-Balik, K., and J. Dębowska-Ludwin, 2014. Trade or conquest? The nature of Egyptian-South Levantine relations in Early Bronze I from the perspective of Tell el-Farkha, Egypt and Tel Erani, Israel. Recherches Archeologiques NS4: 113–125.

Daoud etal., *Adaptation & Resilience*. Daoud, I., Oman, M., Alary, V., Moselhy, N., Salal, E., Naga, A., Salama, O., Duarte, L. and J. Tourrand, 2016. Adaptation and Resilience in Pastoral Management of the Mediterranean Bedouin Social–Ecological System in the Northwestern Coastal Zone of Egypt. In Dong, S., Kassam, K. A. S., Tourrand, J. F. and R. B. Boone (eds), Building Resilience of Human-Natural Systems of Pastoralism in the Developing World, 209–250. Dordrecht: Springer.

Darby etal, *Gift of Osiris*. Darby, W., Ghalioungui, P. and L. Grivetti, 1977. The gift of Osiris. London: Thames & Hudson.

David, *Pyramid Builders*. David, A., 2003. Pyramid Builders of Ancient Egypt: A Modern Investigation of Pharaoh's Workforce. London and New York: Routledge.

David, *Religion & Magic*. David, R., 2002. Religion and Magic in Ancient Egypt. London: Penguin .

Davies, *Deir el-Gebrâwi* I. Davies, N., 1902. The rock tombs of Deir el Gebrâwi, Part 1, Tomb of Aba and smaller tombs of the southern group. London–Boston: EES.

Davies, *Deir el-Gebrâwi* II. Davies, N., 1902. The rock tombs of Deir el Gebrâwi, Part 2, Tomb of Zau and tombs of the northern group. London–Boston: EES.

Davies-Colley & Nagels, *Light Penetration*. Davies-Colley, R. and J. Nagels, 2008. Predicting light penetration into river waters. Journal of Geophysical Research: Biogeosciences, 113G3: 1–9.

Davis, *Canonical Traditions*. Davis, W., 1989. The Canonical Tradition in Ancient Egyptian Art. Cambridge University Press.

Day, *Botany Meets Archaeology*. Day, J., 2013. Botany meets archaeology: people and plants in the past. Journal of Experimental Botany, 64: 18, 5805–5816.

de Jonge & Elliott, *Eutrophication*. de Jonge, V. and M. Elliott, 2008, Eutrophication. In Steele, J., Thorpe, S. and K. Turekian, (eds), Encyclopedia of Ocean Sciences, 306–323. Amsterdam: Elsevier.

de Miroschedji, *Socio-Political Dynamics*. de Miroschedji, P., 2002. The Socio-Political Dynamics of Egyptian-Canaanite Interaction in the Early Bronze Age. In van den Brink, E. and T. Levy (eds), Egypt and the Levant, Interrelations from the 4th Through to the 3rd Millennium BC, 39–57. London: Leicester University Press.

de Meyer, *Restoring the Tombs*. de Meyer, M., 2005. Restoring the Tombs of His Ancestors? Djehutinakht, Son of Teti, at Deir al-Barsha and Sheikh Said, IBAES 5: 125–135.

de Meyer, *Shadow of the Nomarchs*. de Meyer, M., 2007. In the Shadow of the Nomarchs. New Excavations in the Rock Tombs of Deir al-Barsha. In Goyon, J. and C. Cardin (eds), Proceedings of the Ninth International Congress of Egyptologists, Grenoble, 6-12 September 2004, OLA 150, 421-427. Leuven: Peeters.

de Meyer, *Two Cemeteries One Provincial Capital*. de Meyer, M., 2011. Two cemeteries for one provincial capital? Deir el-Bersha and el-Sheikh Said in the fifteenth Upper Egyptian nome during the Old Kingdom. In Strudwick, N. and H. Strudwick (eds), Old Kingdom: New Perspectives. Egyptian Art and Archaeology 2750–2150 BC. Proceedings of a Conference at the Fitzwilliam Museum Cambridge, May 2009, 42–49. Oxford University Press.

de Morgan, *Catalogue*. De Morgan, J., 1898. Catalogue des monuments et inscriptions de l'Égypte antique. Vienna: Holzhausen.

Dean, *Dynastic Warfare*. Dean, R., 2017. Warfare and Weaponry in Dynastic Egypt. Barnsley: Pen and Sword Archaeology.

Decker & Herb, *Blidlatlas zum Sport*. Decker, W. and M. Herb, 1994. Bildatlas zum Sport im alten Ägypten: Corpus der bildlichen Quellen zu Leibesübungen, Spiel, Jagd, Tanz und verwandten Themen. Leiden: Brill.

Decker etal., *Cattle Ancestry Patterns*. Decker J., McKay, S., Rolf, M., Kim, J., Molina Alcala,´ A., Sonstegard, T., Hanotte, O., Götherström, A., Seabury, C., Praharani, L. and M. Babar, 2014 Worldwide Patterns of Ancestry, Divergence, and Admixture in Domesticated Cattle. PLoS Genet 10: 3: e1004254.

Dee etal., *Chronology Reanalysis*. Dee, M., Ramsey, C., Shortland, A., Higham, T. and J. Rowland, 2009. Reanalysis of the chronological discrepancies obtained by the Old and Middle Kingdom Monuments Project. Radiocarbon 513: 1061-1070.

Dee etal., *Reservoir Offset*. Dee, M., Brock, F., Harris, S. A., Bronk-Ramsay, C., Shortland, A. and J. Rowland, 2009. Investigating the Likelihood of a Reservoir Offset in the Radiocarbon Record of Ancient Egypt. Journal of Archaeological Science 37: 2010, 687–693.

Deelder etal., *Schistome in Mummies*. Deelder, A., Miller, R., De Jonge, N. and F. Krijger, 1990. Detection of schistosome antigen in mummies. Lancet 3358691:724–725.

Delibes, *Herpestes ichneumon*. Delibes, M., 1999. Herpestes ichneumon. In: Mitchell-Jones, A., Amori, G., Bogdanowicz, W., Kryštufek, B., Reijnders, P., Spitzenberger, F., Stubbe, M., Thissen, J., Vohralík, V and J. Zima, (eds), The Atlas of European Mammals, 356–357. London: Poyser.

Deocampo, *Hippo Structures*. Deocampo, D., 2002. Sedimentary structures generated by Hippopotamus amphibius in a lake-margin wetland, Ngorongoro Crater, Tanzania. Palaios 172: 212–217.

Dhanakumar etal., *Phosphorous Fractionation*. Dhanakumar, S., Murthy, K., Mohanraj, R., Kumaraswamy, K., and S. Pattabhi, 2015. Phosphorous fractionation in surface sediments of the Cauvery Delta region, Southeast India. In Ramkumar, M., Kumaraswamy, K. and R. Mohanraj (eds), Environmental Management of River Basin Ecosystems, 477–489. Dordrecht: Springer.

Di Marco & Aello, *Cattle Walking*. Di Marco, O., and M. Aello, 1998. https://journals.uair.arizona.edu/index.php/jrm/article/viewFile/9272/8884

Di Marco & Aello, *Energy Expenditure*. Di Marco, O., and M. Aello, 2001. Energy expenditure due to forage intake and walking of grazing cattle. Arquivo Brasileiro de Medicina Veterinária e Zootecnia, 53: 1, 105–110.

Diodato & Bellocchi, *Rainfall Erosivity*. Diodato, N. and G. Bellocchi, 2009. Assessing and modelling changes in rainfall erosivity at different climate scales. Earth Surface Processes and Landforms 347: 969–980.

Diodato etal., *Erosive Rainfall Anomalies*. Diodato N, Ceccarelli M. and G. Bellocchi, 2008. Decadal and century-long changes in the reconstruction of erosive rainfall anomalies at a Mediterranean fluvial basin. Earth Surface Processes and Landforms 33: 2078–2093.

Do Linh san etal., *Herpestes ichneumon*. Do Linh San, E., Maddock, A.H., Gaubert, P. and F. Palomares, 2016. Herpestes ichneumon. The IUCN Red List of Threatened Species, e.T41613A45207211. *http://dx.doi.org/10.2305/IUCN.UK.2016-1.RLTS.T41613A45207211.en*.

Donaldson etal, *Proteins & Health*. Donaldson, A., Johnstone, A., de Roos, B. and P. Myint, 2018. Role of proteins in healthy ageing. European Journal of Integrative Medicine 23: 32–36.

Donohue & Fullerton, *Ancient Art & Historiography*. Donohue, A., and M. Fullerton, 2003. Ancient art and its historiography. Cambridge University Press.

Donovan & McCorquodale, *Principles & Themes*. Donovan, L. and K. McCorquodale, 2000. Egyptian Art, Principles and Themes in Wall Scenes Prism Archaeological Series 6. Guizeh: Foreign Cultural Information Department.

Downing, *African Climate Change*. Downing, T., Ringius, L., Hulme, M., and D. Waughray, 1997. Adapting to climate change in Africa. Mitigation and adaptation strategies for global change, 21: 19–44.

Drake, *Climatic Greek Dark Ages*. Drake, B. L., 2012. The influence of climatic change on the Late Bronze Age Collapse and the Greek Dark Ages. Journal of Archaeological Science 396: 1862–1870.

Driaux, *Water Supply*. Driaux, D., 2016. Water supply of ancient Egyptian settlements: the role of the state. Overview of a relatively equitable scheme from the Old to New Kingdom ca. 2543–1077 BC. Water History, 8: 1, 43–58.

Drysdale etal., *Old World Collapse*. Drysdale, R., Zanchetta, G., Hellstrom, J., Maas, R., Fallick, A., Pickett, M., Cartwright, I. and L. Piccini, 2005. Late Holocene drought responsible for the collapse of Old World civilizations is recorded in an Italian cave flowstone. Geology 34: 2, 101–104.

Ducassou etal., *Nile floods in Sediments*. Ducassou, E., Mulder, T., Migeon, S., Gonthier, E., Murat, A,. Revel, M., Capotondi, L., Bernasconi, S., Mascle, J., and S. Zaragosi, 2008. Nile floods recorded in deep Mediterranean sediments. Quaternary Research 70: 3, 382–391.

Dudley etal., *Hippopotamus Carnivory*. Dudley, J., Hang'Ombe, B., Leendertz, F., Dorward, L., De Castro, J., Subalusky, A., and M. Clauss, 2016. Carnivory in the common hippopotamus Hippopotamus amphibius: implications for the ecology and epidemiology of anthrax in African landscapes. Mammal Review 463: 191–203.

Duell, *Mereruka*. Duell, P., 1938. The Mastaba of Mereruka. Chicago: Oriental Institute.

Duhig, *Eating People Here*. Duhig, C., 2000. They are eating people here! Skeletal indicators of stress in the Egyptian First Intermediate Period. PhD, University of Cambridge.

Duliková & Mařík, *Complex Network Analysis*. Dulíková, V. and R. Mařík, 2017. Complex network analysis in old kingdom society: a nepotism case. In Bárta, M., Coppens, F. and J. Krejčí (eds), Abusir and Saqqara in the Year 2015, 63–83. Prague: Charles Univeristy Press.

Duliková, etal., *Ankhires*. Dulíková, V., Bárta, M., Odler, M., Peterková Hlouchová, M. and Z. Sůvová, Z., 2018. Tomb at Abusir South from a time of change belonging to Ankhires, inspector of hairdressers of the Great House. Prague Egyptological Studies XX1; 3–33.

Duliková, *Old Kingdom Viziers*. Dulíková, V., 2011. Some notes on the title of 'Vizier' during the Old Kingdom, especially on the hieroglyphic phallus-sign in the vizier's title. In Bárta, M., Coppens, F. and J. Krejčí (eds), Abusir and Saqqara in the Year 2010, 327–361. Prague: Charles Univeristy Press.

Dulíková, *Transformation*. Dulíková, V., 2018. The reign of King Nyuserre: a time of transformation. Prague Egyptological Studies, XX: 46–56.

Dumont, *Nile Basin Synopsis*. Dumont, H. J., 2009. A description of the Nile Basin, and a synopsis of its history, ecology, biogeography, hydrology, and natural resources. In Dumont, H. (ed.), The Nile: Origin, Environments, Limnology and Human Use, 1–21. Dordrecht: Springer.

Edel & Wenig, *Ne-User-Re*. Edel, E. and S. Wenig, 1974. Die Jahreszeitenreliefs aus dem Sonnenheiligtum des Königs Ne-user-re. Staatliche Museen zu Berlin, Mitteilungen aus der Ägyptischen Sammlung 7. Berlin: Akademie-Verlag.

Edwards, *Pigs*. Edwards, S., 2011. Pigs. In Webster, J., (ed.), Management and Welfare of Farm Animals: The UFAW Farm Handbook, 253–294. Ames: Wiley & Sons.

Eid etal., *Typha Growth Dynamics*. Eid, E., Shaltout, K. and T. Asaeda, 2012. Modelling growth dynamics of Typha domingensis Pers. Poir. ex Steud. in Lake Burullus, Egypt. Ecological Modelling 243: 63–72.

Ejsmond, *Nubians of Gebelein*. Ejsmond, W., 2017. The Nubian mercenaries of Gebelein during the First

Intermediate Period in light of recent field research. Journal of Ancient Egyptian Interconnections, 14: 11–13.

El-Ghani etal., *Environmental Relationships*. El-Ghani, M., El-Fiky, A., Soliman, A. and A. Khattab, 2011. Environmental relationships of aquatic vegetation in the fresh water ecosystem of the Nile Delta, Egypt. African Journal of Ecology 49: 1, 103–118.

El-Khadragy, *Asyut Soldier's Tomb*. El-Khadragy, M., 2006. The northern soldiers-tomb at Asyut. Studien zur altägyptischen Kultur 35: 147–164.

El-Khouli & Kanawati, *El-Hammimiya*. El-Khowli, A. and N. Kanawati, 1990. The Old Kingdom Tombs at El-Hammimiya. Sydney: ACE.

El-Sheekh, *Nile River Ecosystems*. El-Sheekh, M., 2016. Impact of Water Quality on Ecosystems of the Nile River. In Negm. A. M. (ed.) The Handbook of Environmental Chemistry, 1–29. Dordrecht: Springer.

El-Wakeel, *Lake Qarun Deposits*. El-Wakeel, S., 1963. A study of the bottom deposits of Lake Qarun Egypt Part I: Chemical analysis. Bulletin of the Faculty of Sciences Alexandria University 5: 51–80.

Ellery etal., *Channel Blockage*. Ellery, W., Ellery, K., Rogers, K. and T. McCarthy, 1995. The role of Cyperus papyrus L. in channel blockage and abandonment in the northeastern Okavango Delta, Botswana. African Journal of Ecology 331: 25–49.

Elsharnouby, *Linen*. Elsharnouby, R., 2014. Linen in Ancient Egypt. Journal of General Union of Arab Archaeologists 1515: 1–22.

Elster & Jensen, *Fishery Investigation*. Elster, H., and K.. Jensen, 1960. Limnological and Fishery Investigations of the Nozha Hydrodrome Near Alexandria, Egypt, 1954-1956. Cairo: Government Printing Offices.

Entz, *Lakes Nasser & Nubia*. Entz, B., 1976. Lake Nasser and Lake Nubia. In Rzóska, J. (ed.) The Nile: Biology of an Ancient Rive, 271–298. Dordrecht: Springer.

Erman, *Life in Egypt*. Erman, A., 1971. Life in ancient Egypt. London: Macmillan & Co.

Espinel, *OK Egypt & Byblos*. Espinel, A. 2002. The Role of the Temple of Ba'alat Gebal as Intermediary between Egypt and Byblos during the Old Kingdom. Studien zur altägyptischen Kultur: 103–119.

Espinel, *Desert Hunters*. Espinel, A., 2015. A neglected hunting scene from Saqqara (Pitt rivers 1926.14. 6) and the iconography of the desert hunters during the Old Kingdom. In Bárta, M., Coppens, F. and J. Krejčí (eds), 2017. Abusir and Saqqara in the Year 2015, 102–108, Prague: Charles Univeristy Press.

Estes, *Behavior Guide*. Estes, R., 1991. The behavior guide to African mammals Vol. 64. Berkeley: University of California Press.

Evans, *Animal Behaviour*. Evans, L., 2010. Animal behaviour in Egyptian art: representations of the natural world in Memphite tomb scenes. Oxford University Press.

Evans, *Domestic Animals*. Evans, L., 2000. Animals in the Domestic Environment. In Donovan, L. and K. McCorquodale (eds), Egyptian Art, Principles and Themes in Wall Scenes, 73–82. Cairo: Foreign Cultural Information Department.

Evans, *Pig Overboard*. Evans, L., 2017. Pig overboard? An enigmatic tomb scene from Beni Hassan. In Di Biase-Dyson, C. and L. Donovan (eds), The Cultural Manifestations of Religious Experience: studies in honour of Boyo G. Ockinga, 153-165. Münster: Ugarit-Verlag.

Eyre, *Cannibal Hymns*. Eyre, C., 2002. The cannibal hymn: a cultural and literary study. Liverpool University Press.

Fahmy, *Predynastic Plant Exploitation*. Fahmy, A., 2005. Missing plant macro remains as indicators of plant exploitation in Predynastic Egypt. Vegetation history and archaeobotany, 14: 4; 287–294.

Farley, *Bilharzia*. Farley, J., 2003. Bilharzia: a history of imperial tropical medicine. Cambridge University Press.

Faulkner, *Military Organization*. Faulkner, R., 1953. Egyptian Military Organization. The Journal of Egyptian Archaeology 39: 32–47.

Faust, *Pigs in Space*. Faust, A., 2018. Pigs in Space (and Time): Pork Consumption and Identity Negotiations in the Late Bronze and Iron Ages of Ancient Israel. Near Eastern Archaeology, 81: 4, 276–299.

Fernandez etal., *Degradation & Recovery*. Fernandez, R., Archer, E., Ash, A., Dowlatabadi, H., Hiernaux, P., Reynolds, J., Vogel, C., Walker, B. and T. Wiegand, 2002. Degradation and recovery in socio-ecological systems: a view from the household/farm level. In Reynolds, J. and D. Stafford Smith (eds), Global desertification: do humans cause deserts, 297–323. Berlin: Dalhelm Unniversity Press.

Fields etal., *Sex & Agricultural Transition*. Fields, M., Herschaft, E., Martin, D., and J. Watson, 2009. Sex and the agricultural transition: Dental health of early farming females. Journal of dentistry and oral hygiene, 14: 42–51.

Fields etal., *Pharaoh's Soldiers*. Field, N. Reynolds, W. and P. Bull, 2007. Soldier of the Pharaoh: Middle Kingdom Egypt. London: Osprey.

Finné, etal., *Late Bronze Age Climate Change*. Finné, M., Holmgren, K., Shen, C., Hu, H., Boyd, M. and S. Stocker, 2017. Late Bronze Age climate change and the destruction of the Mycenaean Palace of Nestor at Pylos. PloS one, 12: 12, e0189447. https://doi.org/10.1371/journal.pone.0189447.

Fiorentino etal., *Syrian Climate Change*. Fiorentino, G., Caracuta, V., Calcagnile, L., D'Elia, M., Matthiae, P., Mavelli, F. and G. Quarta, 2008. Third Millennium B.C. Climate Change in Syria Highlighted by Carbon Stable Isotopeanalysis of 14C-AMS Dated Plant Remains from Ebla. Palaeogeography, Palaeoclimatology, Palaeoecology 266; 51–58.

Fischer, *Nubian Merceneries*. Fischer H., 1961. The Nubian Mercenaries of Gebelein During the First Intermediate Period. Kush 9: 44–80.

Fischer, *Egyptian Studies*. Fischer, H., 1996. Egyptian Studies III: New Yotk: MMA.

Fisher & Sponseller, *Rivers as Ecosystems*. Fisher, S. and R. Sponseller, 2009. Streams and Rivers as Ecosystems. In Likens, G. (ed.), The Encyclopedia of Inland Waters, 491–498. Amsterdam: Elsevier.

Forshaw, *Dental Health*. Forshaw, R., 2009. Dental health and disease in ancient Egypt. British Dental Journal, 206: 9, 481–486. https://doi.org/10.1038/sj.bdj.2009.355.

Förster, *Abu Ballas Trail*. Förster, F., 2007. With donkeys, jars and water bags into the Libyan Desert: the Abu Ballas Trail in the late Old Kingdom/First Intermediate Period. International Conference of Nubian Studies, 3: 1–9.

Frank etal., *Grazing History*. Frank, A., Dickman, C., Wardle, G. and A. Greenville, 2013 Interactions of Grazing History, Cattle Removal and Time since Rain Drive Divergent Short-Term Responses by Desert Biota. PLoS ONE 87: 1–13. https://doi.org/10.1371/journal.pone.0068466.

Franke, *Arme und* Geringe *im Alten Reich*. Franke, D., 2006. Arme und Geringe im Alten Reich Altägyptens: "Ich gab Speise dem Hungernden, Kleider dem Nackten...". Zeitschrift für ägyptische Sprache und Altertumskunde 133: 104–120.

Fraser, *Early Tombs*. Fraser, G., 1903. The Early Tombs at Tehneh. Paris: Imprimerie de l'Institut français d'archéologie orientale.

Frils etal., *Fossil Water Lilies*. Friis, E., Pedersen, K. and P. Crane, 2001. Fossil evidence of water lilies Nymphaeales in the Early Cretaceous. Nature, 410: 6826, 357–360.

Fritsch, et.al., *Orthopedic Diseases*. Fritsch, K., Hamoud, H., Allam, A., Grossmann, A., Nur El-Din, A., Abdel-Maksoud, G., Al-Tohamy Soliman, M., Badr, I., Sutherland, J., Sutherland, L. and M. Akl, 2015. The Orthopedic Diseases of Ancient Egypt. The Anatomical Record, 298: 6, 1036–1046. https://doi.org/10.1002/ar.23136.

Fujita & Adachi, *Qau Palaeohealth*. Fujita, H. and H. Adachi, 2107. Paleohealth based on dental pathology and cribra orbitalia from the ancient Egyptian settlement of Qau. Anthropological Science 125: 1, 35–42.

Fuller & Lucas, *Archaeobotany*. Fuller, D. and L. Leilani, 2014. Archaeobotany. In Smith, C.(ed.), The Encyclopedia of Global Archaeology, 305–310. Dordrecht: Springer.

Gagnon Lupien, etal., *Bird Diversity*. Gagnon Lupien, N., Gauthier, G. and C. Lavoie, 2015. Effect of the invasive common reed on the abundance, richness and diversity of birds in freshwater marshes. Animal Conservation, 181: 32–43.

Galantinho & Mira, *Occurrence of Genet*. Galantinho, A. and A. Mira, 2009. The influence of human, livestock, and ecological features on the occurrence of genet Genetta genetta: a case study on Mediterranean farmland. Ecological Research, 243: 671–685.

Gallardo, *Séquito de Horus*. Gallardo, F., 2013. Séquito de Horus, Tripulación de Ra: aspectos religiosos de las luchas de poder durante la dinastía VI egipcia. ARYS: Antigüedad: Religiones y Sociedades, 11; 21–46.

García-Hernándeza etal., *Typha Wetlands Salinity Responses*. García-Hernández, J., Flessa, K., Santiago-Serrano, E., Romero-Hernández, S., Zamora-Arroyo, F. and J. Ramírez-Hernández, 2013. Salinity responses to inflow alterations in a 6500ha Typha wetland. Ecological Engineering 59: 18–29.

Gardiner, *Ancient Athletics*. Gardiner, E., 2002. Athletics in the ancient world. Chelmsford: Clarendon Press.

Gaubert, etal, *Genetta genetta*. Gaubert, P., Carvalho, F., Camps, D. and E. Do Linh San, 2015. Genetta genetta. The IUCN Red List of Threatened Species: e.T41698A45218636.

Gaudet, *Papyrus Nutrients.* Gaudet, J. J., 1977. Uptake, accumulation, and loss of nutrients by papyrus in tropical swamps. Ecology 58: 415–422.

Gautier, *Domesticated Fauna*. Gautier, A., 2005 Domesticated Fauna. In Bard, K. (ed.), Encyclopedia of the Archaeology of Ancient Egypt, 300–306. London and NewYork: Routledge.

Gee, *Old Kingdom Collapse?* Gee, J., 2015. Did the Old Kingdom Collapse? A New View of the First Intermediate Period. In Der Manuelian, P. and T. Schneider (eds), Towards a new history for the Egyptian Old Kingdom: Perspectives on the pyramid age, Vol.1, 60–75. Harvard: Brill.

Geriesh, etal., *Groundwater*. Geriesh, M., Balke, K., El-Rayes, A. and B. Mansour, 2015. Implications of climate change on the groundwater flow regime and geochemistry of the Nile Delta, Egypt. Journal of Coastal Conservation 194: 589–608.

Gerloni et.al., *Dental Status*. Gerloni, A., Cavalli, F., Costantinides, F., Costantinides, F., Bonetti, S. and C. Paganelli, 2009. Dental status of three Egyptian mummies: Radiological investigation by multislice computerized tomography. Oral Surg Oral Med Oral Pathol Oral Radiol Endod. 107: 6, 58–64.

Ghilardi etal, *Nile Evolution*. Ghilardi, M., Tristant, Y., and M. Boraik, 2012. Nile River evolution in upper Egypt during the Holocene; palaeoenvironmental implications for the Pharaonic sites of Karnak and Coptos. Geomorphologie: Relief, Processus, Environnement 18: 1, 7–22.

Ghilardi etal, *Geoarchaeology*. Ghilardi, M., Fouache, E. and R. Chiverrell, 2009. Introduction to special issue on Geoarchaeology: human-environment connectivity. Géomorphologie: Relief, Processus, Environnement, 15: 4, 227–228.

Giller & Malmqvist, *Streams & Rivers*. Giller, S. and B. Malmqvist, 1998. The Biology of Streams and Rivers. Oxford University Press.

Glenn etal., *Salinity & Growth*. Glenn, E., Thompson, T., Frye, R., Riley, J. and D. Baumgartner, 1995. Effects of salinity on growth and evapotranspiration of Typha domingensis Pers. Aquatic Botany, 52:1, 75–91.

Goedicke, *Abusir–Saqqara–Giza*. Goedicke, H., 2001. Abusir – Saqqara – Giza. In: Bárta, M. and J. Krejcí, (eds), Abusir and Saqqara in the Year 2000, 397–412. Prague: Charles Univeristy Press.

Goldberg & Macphail, *Geoarchaeology*. Goldberg, P. and R. Macphail, 2006. Practical and Theoretical Geoarchaeology. London: Blackwell.

Gorai etal., *Phragmites Germination*. Gorai, M., Vadel, A. and M. Neffati, 2006. Seed Germination Characteristics of Phragmites Communis: Effects of Temperature and Salinity. Belgian Journal of Botany, 139: 1, 78–86.

Gorman & Karr, *Habitat Structure*. Gorman, O., and J. Karr, 1978. Habitat structure and stream fish communities. Ecology 59: 507–515.

Gorny, *Environment, Archaeology & History*. Gorny, R., 1989. Environment, archaeology, and history in Hittite Anatolia. The Biblical archaeologist, 52: 2-3, 78–96.

Goulden etal., *Typha Marsh Evapo-transpiration*. Goulden, M., Litvak, M. and S. Miller, 2007. Factors that control Typha marsh evapotranspiration. Aquatic Botany 86: 2, 97–106.

Govaerts etal., *Cyperaceae*. Govaerts, R. and D. Simpson, with Bruhl, J., Egorova, T., Goetghebeur, P. and K. Wilson, 2007. World Checklist of Cyperaceae: Sedges. Kew.

Gowland & Western, *Morbidity in the Marshes*. Gowland, R., and A. Western, 2012. Morbidity in the marshes: Using spatial epidemiology to investigate skeletal evidence for malaria in Anglo-Saxon England AD 410–1050. American Journal of Physical Anthropology, 147: 2, 301-311.

Grace & Wetzel, *Typha Population Dynamics*. Grace, J. and R. Wetzel, 1998. Long-term dynamics of Typha populations. Aquatic Botany 61: 2, 137–146.

Grace, *Effects of Water Depth*. Grace, J., 1989. Effects of water depth on Typha latifolia and Typha domingensis. American Journal of Botany: 762–768.

Grace, *Nutrient Addition Effects*. Grace, J., 1988. The effects of nutrient additions on mixtures of Typha latifolia L. and Typha domingensis P. along a water-depth gradient. Aquatic Botany, 31: 1-2, 83–92.

Greenberg, *No Collapse: Transmutation*. Greenberg, R., 2017. No Collapse: Transmutations of Early Bronze Age Urbanism in the Southern Levant. In Höflmayer, F. (ed.), The Late Third Millennium in the Ancient Near East: Chronology, C14, and Climate Change, 31–58. Chicago: Oriental Institute.

Greenwood, *Kenyan Fishes*. Greenwood, P., 1966. The Fishes of Uganda. Kampala: Uganda Society.

Greenwood, *Nile Fish Fauna*. Greenwood, P. 1976. Fish Fauna of the Nile. In Rzóska, J. (ed.), The Nile: Biology of an Ancient River, 127–141. Dordrecht: Springer.

Griffith, *Beni Hasan* III. Griffith, F., 1896. Beni Hasan, Part III. London: EES.

Griffiths, *Osiris Origins*. Griffiths, J., 1980. The origins of Osiris and his cult. Leiden: Brill.

Grimal, *History.* Grimal, N., 1994. A history of ancient Egypt; translated by Ian Shaw. Oxford University Press.

Guisan & Thuiller, *Species Distribution*. Guisan, A. and W. Thuiller, 2005. Predicting species distribution: offering more than simple habitat models. EcolLett. 8: 993–1009.

Habib & Samah, *Catfish Protein Biosynthesis*. Habib, S. and A. Samah, 2013. Effect of heavy metals pollution on protein biosynthesis in catfish. Journal of Water Resource and Protection 505: 555–558.

Haldane & Henderson, *Egyptian Shadouf.* Haldane, J. and Y. Henderson, 1926. The rate of work done with an Egyptian shadouf. Nature, 118: 2965, 308–309.

Halim etal., *Last Normal Nile Flood*. Halim, Y., Guergues, S. and H. Saleh, 1967. Hydrographic conditions and plankton in the south east Mediterranean during the last normal Nile flood 1964. Internationale Revue der gesamten Hydrobiologie und Hydrographie 52: 3; 401–425.

Hall, *A Perfect Storm?* Hall, T. 2014., A 'Perfect Storm' in the Collapse of Bronze Age Civilization? Useful Insights and Roads not Taken. Cliodynamics: The Journal of Quantitative History and Cultural Evolution 5: 1, 75–86.

Haltenorth & Diller, *Field Guide*. Haltenorth, T. and H. Diller, 1986 A Field Guide to the Mammals of Africa: Including Madagascar. Charlottesville: Collins.

Hambrecht & Gielen, *Hunter-gatherer to Sedentary*. Hambrecht, R., and S. Gielen, 2005. Essay: hunter-gatherer to sedentary lifestyle. The Lancet, 366: 60–61.

Hamdan, etal., *Nile Floodplain Sediments*. Hamdan, M., Hassan, F., Flower, R., Leroy, S., Shallaly, N. and A. Flynn, 2019. Source of Nile sediments in the floodplain at Saqqara inferred from mineralogical, geochemical, and pollen data, and their palaeoclimatic and geoarchaeological significance. Quaternary International, 501: 272–288.

Hamden etal., *Faiyum Sedimentary Record*. Hamdan, M., Ibrahim, M., Shiha, M., Flower, R., Hassan, F. and S. Eltelet, 2016. An exploratory Early and Middle Holocene sedimentary record with palynoforms and diatoms from Faiyum lake, Egypt. Quaternary International. 1–13.

Hamed, *Artistic Perspectives*. Hamed, A., 2015. Sport, leisure: artistic perspectives in ancient Egyptian temples part II. Recorde: Revista de História do Esporte, 81: 1–28.

Hammer, *Freshwater Wetlands*. Hammer, D., 1997. Creating Freshwater Wetlands. Boca Raton: Lewis.

Hanneder, *Water Lilies & Lotus*. Hanneder, J., 2007. Some common errors concerning water-lilies and lotuses. Indo-Iranian Journal, 50: 2, 161–164.

Harer, *Properties of the Egyptian Lotus*. Harer, W., 1985. Pharmacological and biological properties of the Egyptian lotus. Journal of the American Research Center in Egypt 22: 49–54.

Harpur & Scremin, *Kagemni*. Harpur, Y. and P. Scremin, 2006. The Chapel of Kagemni Scene Details. Oxford University Press.

Harpur, *Decorations*. Harpur, Y., 1987. Decoration in Egyptian Tombs of the Old Kingdom: Studies in Orientation and Scene Content. London: KPI.

Harrington, *Art & Social Theory*. Harrington, A., 2004. Art and social theory: sociological arguments in aesthetics. London: Wiley & Sons.

Hartung, *Livestock Production*. Hartung, J., 2013. A short history of livestock production. In Aland, A. and T. Banhazi (eds), Livestock housing: Modern management to ensure optimal health and welfare of farm animals, 81–146. Wageningen. DOI 10.3920/978-90-8686-771-4_01.

Hartwig, *Painting & Identity*. Hartwig, M. K., 2004. Tomb painting and identity in ancient Thebes, 1419–1372 BCE, Turnhout: Brepols.

Hasgeth, *Livestock Transport*. Hagseth, M., 2015 Nilotic Livestock Transport in Ancient Egypt. MA, Texas A and M University.

Haslam, *Reed*. Haslam, S., 2009 The Reed. Updated edition, British Reed Growers Association. Norwich: Brown and Co.

Hassan & Iskander, *Neb-Kaw-Her*. Hassan, S. and Z. Iskander, 1975. Excavations at Saqqara 1937-1938. Vol.1; The Mastaba of Neb-Kaw-Her. Cairo: Antiquities Department of Egypt.

Hassan & Stucki, *Climatic Change*. Hassan, F. and B. Stucki, 1987. Nile floods and climatic change. In Rampino, R., Sanders, J., Newman, W. and L. Konigsson (eds), Climate: history, periodicity, and predictability, 37–46. New York: Springer.

Hassan etal., *Nile Flooplain Alluvium*. Hassan, F., Hamdan, M., Flower, R., Shallaly, N. and E. Ebrahem, 2017. Holocene alluvial history and archaeological significance of the Nile floodplain in the Saqqara-Memphis region, Egypt. Quaternary Science Reviews, 176: 51–70.

Hassan, *Floods & Civilisation*. Hassan, F., 1998. Climatic change, Nile floods and civilization. Nature and resources, 342: 34–40.

Hassan, *Giza* I. Hassan, S., 1932. Excavations at Giza; 1929-1930. Vol. I. Cairo: Faculty of Arts of the Egyptian University.

Hassan, *Giza* IV. Hassan, S., 1943. Excavations at Giza; Vol. IV. Cairo: Faculty of Arts of the Egyptian University.

Hassan, *Giza* V. Hassan, S., 1944. Excavations at Giza; Vol. V. Cairo: Faculty of Arts of the Egyptian University.

Hassan, *Floods & Climate Change*. Hassan, F., 1981. Historical Nile Floods and their Implications for Climate Change. Science New Series, 212: 4499, 1141–1145.

Hassan, *Holocene Lakes*. Hassan, F, 1986. Holocene lakes and prehistoric settlements of the Western Faiyum, Egypt. Journal of Archaeological Science 135: 483-501.

Hassan, *A River Runs Through*. Hassan, F., 2005. A River Runs through Egypt; Nile Floods and Civilisation. Geotimes 50: 4, 22–26.

Hassan, *Floods & Disorder*. Hassan, F., 1997. Nile floods and political disorder in early Egypt. In Dalfes, H. N., Kukla, G., and H. Weiss (eds), Third millennium BC climate change and old world collapse, 1–23. Berlin: Springer.

Hassan, *Re-reading Ipuwer*. Hassan, F., 2007. Droughts, Famine and the Collapse of the Old Kingdom: Re-Reading Ipuwer. In Hawass, Z. A. and J. Richards (eds), The Archaeology and Art of Ancient Egypt. Essays in Honor of David B. O'Connor, Vol. I, 357–377. Cairo: Supreme Council of Antiquities.

Hassan, *Saqqara* II. Hassan, S., 1975. Excavations at Saqqara 1937-1938: Vol. II. Cairo.

Hatziminaoglou & Boyazoglu, *Ancient Goats*. Hatziminaoglou, Y., and Boyazoglu, J. 2004. The goat in ancient civilisations: from the Fertile Crescent to the Aegean Sea. Small Ruminant Research 512: 123–129.

Hauschteck, *Goats in Houses*. Hauschteck, E., 2004. Goats in houses on Elephantine during the Middle Kingdom and the Second Intermediate Period. Göttinger Miszellen Beiträge zur ägyptologischen Diskussion, 202: 59–70.

Hayes, *Scepter of Egypt*, . Hayes, W., 1978. The Scepter of Egypt: A Background for the Study of the Egyptian Antiquities in The Metropolitan Museum of Art. Vol. 1, From the Earliest Times to the End of the Middle Kingdom. New York: MMA.

Hays, *Unreading the Pyramids*. Hays, H., 2009. Unreading the Pyramids. Bulletin de l'Institut Français Archéologie Orientale, 109: 26, 195–220.

Healy, *Armies of the Pharaohs*. Healy, M., 1992. Armies of the Pharaohs. London: Osprey

Hebelstrup, *Wild versus Domesticated Grains*. Hebelstrup, K., 2017. Differences in nutritional quality between wild and domesticated forms of barley and emmer wheat. Plant Science 256: 3, 1–4.

Hecker, *Pork Consumption*. Hecker, H., 1982. A zooarchaeological inquiry into pork consumption in Egypt from prehistoric to New Kingdom times. Journal of the American Research Center in Egypt 19: 59–71.

Heller & Byczyńska, *Environmental Factors & Flax*. Heller, K. and M. Byczyńska, 2015. The Impact of Environmental Factors and Applied Agronomy on Quantitative and Qualitative Traits of Flax Fiber. Journal of Natural Fibers 121: 26–38.

Hellings & Gallagher, *Flooding on Phragmites*. Hellings, S. and J. Gallagher, 1992. The Effects of Salinity and Flooding on Phragmites australis. Journal of Applied Ecology, 29: 1, 41–49.

Hély & Lézine, *Holocene Vegetation*. Hély, C. and A. M. Lézine, 2014. Holocene changes in African vegetation: tradeoff between climate and water availability. Clim. Past. 10: 681–686.

Hendrickx, *Hunting & Social Complexity*. Hendrickx, S., 2011. Hunting and Social Complexity in Predynastic Egypt. Bulletin des Séances Academie Royale des Sciences d'Outre-mer/ Koninklijke Academie voor Overzeese Wetenschappen, Mededelingen der Zittingen, 57: 2-4, 237–263.

Hendrickx & Vermeersch, *Prehistory*. Hendrickx, S., and P. Vermeersch. 2001. Prehistory. From the Palaeolithic to the Badarian Culture. In I. Shaw (ed.), The Oxford History of Ancient Egypt, 17–43. Oxford University Press.

Hendrickx etal, *Western desert Rock Art*. Hendrickx, S., Riemer, H. and F. Förster, 2011. In Riemer, H. Förster, F., Herb, M. and N. Pöllath, (eds), Desert Animals in the Eastern Sahara: Status, Economic Significance and Cultural Reflection in Antiquity. Colloquium Africanum 4, 189–244. Köln: Heinrich-Barth-Institut.

Herb & Förster, *Desert to Town*. Herb, M and F. Förster, 2009. From desert to town: The economic role of desert game in the Pyramid Ages of ancient Egypt as inferred from historical sources c. 2600 – 1800 BC. In Riemer, H., Forster, F., Herb, M. and N. Pöllath (eds), Desert Animals in the Eastern Sahara: Status, Economic Significance, and Cultural Reflection in Antiquity: Proceedings of an Interdisciplinary ACACIA Workshop Held at the University of Cologne, December 14-15, 2007, 17–45. Köln: Heinrich-Barth-Institut.

HESPOK. Smith, W., 1978. A History of Egyptian Sculpture and Painting in the Old Kingdom. Oxford University Press.

Hildebrand & Schilling, *Early Nile Agriculture*. Hildebrand, E. and T. Schilling, 2016. Storage amidst early agriculture along the Nile: Perspectives from Sai Island, Sudan. Quaternary International, 412: 81–95.

Hillman & Davies, *Domestication Rates*. Hillman, G., and M. Davies, 1990. Measured domestication rates in wild wheats and barley under primitive cultivation, and their archaeological implications. Journal of World Prehistory 4: 157–222.

Hillson, *Dental Disease*. Hillson, S. ,1979. Diet and dental disease. World Archaeology, 11: 147–162.

Hoffman, *Before the Pharaohs*. Hoffman, M.,1984. Egypt before the Pharaohs. London: Routledge.

Höflmayer, *Dating Catastrophes*. Höflmayer, F., 2014. Dating Catastrophes and Collapses in the Ancient Near East: The End of the First Urbanization in the Southern Levant and the 4.2 ka BP Event. In Nigro, L. (ed.), Overcoming Catastrophic, ROSAPAT 11, 117–40. Rome: Department of Sciences of Antiquities.

Holm etal., *World's Worst Weeds*. Holm, L., Plocknett, D., Pancho, J. and J. Herberger, 1977. The world's worst weeds: distribution and biology. Honolulu: University Press of Hawaii.

Holmgren etal., *Plant Communities*. Holmgren, M., Scheffer, M. and M. Huston, 1997. The interplay of facilitation and competition in plant communities. Ecology 78: 1966–1975.

Holz, *Man-made Landforms*. Holz, R., 1969. Man Made Landforms in the Nile Delta. Geographical Review 592: 253–269.

Hope, *Libya and the Dakhleh Oasis*. Hope, C., 2007 Egypt and 'Libya' to the end of the Old Kingdom: a view from Dakhleh Oasis. In Hawass, Z and J. Richards (eds), The Archaeology and Art of Ancient Egypt: essays in honor of David B. O'Connor, Volume 1, 349–415. Cairo: Supreme Council of Antiquities.

Hornung etal. *Chronology*. Hornung, E., Krauss, R., Warburton, D. and M. Eaton-Krause (eds), 2006. Ancient Egyptian Chronology. Boston: Brill.

Hornung, *Conceptions of God*. Hornung, E., 1983 Conceptions of God: The One and the Many. [trans. J. Baines]. Ithaca: Routledge.

Hosier, *Unwelcome Phragmites*. Hosier, P., 2014. The ecology of an unwelcome exotic, phragmites australis. http://nceppc.weebly.com/uploads/6/8/4/6/6846349/14_hosier_the_ecol_of_phragmites.pdf.

Houlihan & Goodman, *Birds*. Houlihan, P. and S. Goodman, 1986. The Birds of Ancient Egypt. Warminster: Aris & Phillips.

Houlihan, *Ti Swampland Scene*. Houlihan, P., 1996. A guide to the wildlife represented in the great swampland scene in the offering-chapel of Ti no. 60 at Saqqara. Göttinger Miszellen. 155, 19–53.

Houston, etal., *Protein Intake*. Houston, D., Nicklas, B., Ding, J.; Harris, T., Tylavsky, F., Newman, A., Lee, J., Sahyoun, N., Visser, M. and S. Kritchevsky, 2008. Health ABC Study. Dietary protein intake is associated with lean mass change in older, community-dwelling adults: The Health, Ageing, and Body Composition Health ABC Study. Am. J. Clin. Nutr. 87: 150–155.

Hughes, *Narrative Themes*. Hughes, J., 1995. Ecology and development as narrative themes of world history. Environmental History Review, 191: 1–16.

Hughes, *Deforestation Views*. Hughes, J., 1983. How the ancients viewed deforestation. Journal of Field Archaeology, 435–445.

Hughes, *Pan's Travail*. Hughes, J., 1994. Pan's Travail: Environmental Problems of the Ancient Greeks, and Romans. London: John Hopkins University Press.

Hughes, *Sustainable Agriculture*. Hughes, J., 1992. Sustainable Agriculture in Ancient Egypt. Agricultural History 66: 2, 12–22.

Huhta, *Common Reed*. Huhta, A., 2009 Decorative or Outrageous – The significance of the Common Reed Phragmites australis on water quality. Comments from Turku University of Applied Sciences 48: 1–33.

Ikram, *Choice Cuts*. Ikram, S., 1995. Choice cuts: meat production in ancient Egypt. Leeuven: Peeters.

Ikram, *Meat Processing*. Ikram, S., 2000. Meat processing. In Nicholson, P. and I. Shaw (eds), Ancient Egyptian materials and technology, 656–671. Cambridge University Press.

Iñiguez, *Goats in Dry Environments*. Iñiguez, L., 2004. Goats in resource-poor systems in the dry environments of West Asia, Central Asia and the Inter-Andean valleys. Small Ruminant Research 512: 137–144.

Iverson & Shibata, *Canon & Proportion*. Iverson, E. and Y. Shibata, 1975. Canon and Proportions in Egyptian Art. Warminster: Aris & Phillips.

Izham, *Molecular Characterisation*. Izham, N., 2014. Morphological and Molecular Characterisation of Water Lily Family: Nymphaeaceae in UNIMAS Lake East Campus. PhD, Universiti Malaysia Sarawak.

Jacomet, *Archaebotany*. Jacomet, S., 2013. Archaebotany: the potential of analyses of plant remains from waterlogged archaeological sites. In Menotti, F. and A. O'Sullivan (eds), The Oxford Handbook of Wetland Archaeology, 497–514. Oxford University Press.

Jacquet-Gordon, *Mendes* I. Jacquet-Gordon, H., 1986. Regarding Mendes I. Journal of Egyptian Archaeology 72: 207–208.

James etal., *Environmental Integrity*. James, R., Purvaja, R. and R. Ramesh, 2015. Environmental Integrity of the Tamiraparani River Basin, South India In Ramkumar, M., Kumaraswamy, K. and R. Mohanraj (eds), Environmental Management of River Basin Ecosystems, 507–523. Dordrecht: Springer.

James, *Khentika*. James, T., 1953. The Mastaba of Khentika called Ikhekhi. London: EES.

Janák etal., *Sun Temples*. Janák, J., Vymazalová, H. and F. Coppens, 2011. The Fifth Dynasty 'sun temples' in a broader context. In Bárta, M., Coppens, F. and J. Krejčí (eds), Abusir and Saqqara in the Year 2010, 430–442. Prague: Charles Univeristy Press.

Janák, *Early Ba Attestations*. Janák, J., 2011. A Question of Size. A Remark on Early Attestations of the Ba Hieroglyph. SAK 40: 143–153.

Janák, *Extinction of Gods*. Janák, J., 2018. Extinction of gods: impact of climatic change on religious concepts. Visualizing Knowledge and Creating Meaning in Ancient Writing Systems, Berliner Beiträge zum Vorderen Orient, 23: 121–131.

Janák, *Spotting the Akh*. Janák, J., 2011. Spotting the akh. The presence of the northern bald ibis in ancient Egypt and its early decline. JARCE 40: 17–31.

Janick, *Origins of Horticulture*. Janick, J., 2000. Ancient Egyptian agriculture and the origins of horticulture. In Hadid, A. A. and S. Sansavini (eds), International Symposium on Mediterranean Horticulture: Issues and Prospects, 582: 23–39.

Jansen-Winklen, *Der Untergang*. Jansen-Winkeln, K., 2010. Der Untergang des Alten Reiches. Orientalia 79: 3, 273–303.

Jansen-Winklen, *Die Rolle des Unbekannten*. Jansen-Winkeln, K., 2009. Die Rolle des Unbekannten in der ägyptischen Geschichte. In Fitzenreiter, M. (ed.), Das Ereignis: Geschichtsschreibung zwischen Vorfall und Befund, Internet-Beiträge zur Ägyptologie und Sudanarchäologie 10: 155–161. London: Golden House.

Jelinek, *Pastoralism & Social Stratification*. Jelinek, J., 2003 Pastoralism, burials and social stratification in central Sahara. Les Cahiers de L'AARS 8: 41–44.

Jirásková, *Egypt & Syria-Palestine*. Jirásková, L., 2011. Relations Between Egypt and Syria-Palestine in the Latter Part of the Old Kingdom. In Duistermaat, K., and I. Regulski (eds), Intercultural Contacts in the Ancient Mediterranean. Proceedings of the International Conference at the Netherlands-Flemish Institute in Cairo, 25th to 29th October 2008, 539–568. Leeuven: Peeters.

Johnels, *Catfish Terrestrial Locomotion*. Johnels, A., 1957. The mode of terrestrial locomotion in Clarias, Oikos 82: 122–129.

Jones & Muthuri, *Biomass in Papyrus*. Jones, M. and F. Muthuri, 1997. Standing Biomass and Carbon Distribution in a papyrus Cyperus Papyrus swamp on Lake Naivasha, Kenya. Journal of Tropical Ecology, 13: 347–356.

Jones & Muthuri, *Papyrus Canopy Structure*. Jones, M. and F. Muthuri, 1985. The Canopy Structure and Microclimate of Papyrus Cyperus Papyrus Swamps. Journal of Ecology 73: 481–491.

Jones etal., *Sediment Microbes*. Jones, J., Berner, R., Meadows, P., Durand, B. and G. Eglinton, 1985. Microbes and microbial processes in sediments [and Discussion]. Philosophical Transactions of the Royal Society of London A: Mathematical, Physical and Engineering Sciences 315: 1531, 3–17.

Jones, *Freshwater Ecosystems*. Jones, J., 2001. Freshwater ecosystems structure and response. Ecotoxicology and Environmental Safety 50: 1090–2414.

Jones, *Garden Cultivation*. Jones. G., 2005 Garden cultivation of staple crops and its implications for settlement location and continuity, World Archaeology, 37: 2, 164–176.

Jones, *Papyrus Photosynthesis*. Jones, M., 1987. The photosynthetic characteristics of papyrus in a tropical swamp. Oecologia Berlin 71: 355–359.

Kamrin, *Papyrus*. Kamrin, J., 2015. Papyrus in Ancient Egypt. In Heilbrunn Timeline of Art History, The Metropolitan Museum of Art. New York. http://www.metmuseum.org/toah/hd/papy/hd_papy.htm.

Kanawati & Abder-Raziq, *Hesi*. Kanawati, N. and M. Abder-Raziq, 1999. The Teti cemetery at Saqqara Vol. IV; The tomb of Hesi. Warminster: Aris & Phillips.

Kanawati & Abder-Raziq, *Nikauisesi*. Kanawati, N. and M. Abder-Raziq, 2000. The Teti Cemetery at Saqqara. Vol. VI; The Tomb of Nikauisesi. Warminster: Aris & Phillips.

Kanawati & Abder-Raziq, *Teti* III. Kanawati, N. and M. Abder-Raziq 1998. The Teti Cemetery at Saqqara. Vol. III; The Tombs of Neferseshemre and Seaknhuiptah. Warminster: Aris & Phillips.

Kanawati & Abder-Raziq, *Unis* II. Kanawati, N. and M. Abder-Raziq, 2003. The Unis Cemetery at Saqqara. Vol. II; The Tomb of Iynefert and Ihy re-used by Idut. Oxford University Press.

Kanawati & Abder-Raziq, *Waatetkhetor*. Kanawati, N. and M. Abder-Raziq 2008. Mereruka and his Family. Part II; The Tomb of Waatetkhetor. Oxford University Press.

Kanawati & El-Khouli, *El-Hammimiya*. Kanawati, N. and A. El-Khouli, 1990. The Old Kingdom Tombs of El-Hammimiya. Sydney: ACE.

Kanawati & Evans, *Beni Hassan* I. Kanawati, N. and L. Evans, 2014. Beni Hassan, I; The Tomb of Khnumhotep II. Oxford University Press.

Kanawati & Evans, *Beni Hassan* III. Kanawati, N. and L. Evans, 2016. Beni Hassan, III; The Tomb of Amenemhat. Oxford University Press.

Kanawati & Hassan, *Ankhmahor*. Kanawati, N. and A. Hassan 1997. The Teti Cemetery at Saqqara. Vol. II; The Tomb of Ankhmahor. Warminster: Aris & Phillips.

Kanawati & McFarlane, *Akhmin*. Kanawati, N. and A. McFarlane, 1988. The Rock Tombs of El-Hawawish. The Cemetery of Akhmim; Vol. VIII. Sydney: ACE.

Kanawati & McFarlane, *Deshasha*. Kanawati, N. and A. McFarlane, 1993. Deshasha; The Tombs of Inti, Sheshu and Others. Sydney: ACE.

Kanawati & Woods *Artists*. Kanawati, N., Woods, A. and Z. Hawass, 2009. Artists in the Old Kingdom: Techniques and achievements. Cairo: Supreme Council of Antiquities.

Kanawati & Woods, *Beni Hassan*. Kanawati, N. and A. Woods, 2010. Beni Hassan. Art and Daily Life in an Egyptian Province. Cairo: Supreme Council of Antiquities.

Kanawati, *Administration*. Kanawati, N., 1977. The Egyptian administration in the Old Kingdom: evidence on its economic decline. Warminster: Aris & Phillips.

Kanawati, *Art & Gridlines*. Kanawati, N., 2011. Art and Gridlines: The copying of Old Kingdom scenes in later periods. In Bárta, M., Coppens, F. and J. Krejčí (eds), Abusir and Saqqara in the Year 2010, 483–496, Prague: Charles Univeristy Press.

Kanawati, *Conspiracies*. Kanawati, N., 2003. Conspiracies in the Egyptian Palace: Unis to Pepy I. New York: Routledge.

Kanawati, *Deir El-Gabrawi* I. Kanawati, N., 2005. Deir El-Gabrawi. Vol I; The Northern Cliffs. London: Aris & Phillips.

Kanawati, *El-Hawawish* I. Kanawati, N., 1980. Rock Tombs of El-Hawawish Cemetery of Akmim; Vol I. Sydney: ACE.

Kanawati, *El-Hawawish* II. Kanawati, N., 1981. Rock Tombs of El-Hawawish Cemetery of Akmim; Vol II. Sydney: ACE.

Kanawati, *El-Hawawish* III. Kanawati, N., 1982. Rock Tombs of El-Hawawish Cemetery of Akmim; Vol III. Sydney :ACE.

Kanawati, *Innovations*. Kanawati, N., 2012. Innovations in the Tomb of Khnum-Hotep II at Beni Hassan. Newsletter of the Rundle Foundation of Egyptian Archaeology. 118: 1–2. Sydney: ACE.

Kanawati, *Inumin*. Kanawati, N., 2006. The Teti Cemetery at Saqqara. Vol VIII; The Tomb of Inumin. Oxford University Press.

Kanawati, *Mereruka & Teti*. Kanawati, N., 2007. Mereruka and King Teti: The Power behind the Throne. Cairo: Supreme Council of Antiquities.

Kanawati, *Meryteti*. Kanawati, N., 2004. Mereruka and his Family. Part I; The Tomb of Meryteti. Oxford University Press.

Kanawati, *Nepotism*. Kanawati, N., 2003. Nepotism in the Egyptian Sixth Dynasty. In Binder, S. (ed.),The Bulletin of the Australian Centre for Egyptology, 14: 39–60.

Kanawati, *Papyrus Thickets*. Kanawati, N., 2017. Papyrus thickets in the Old and Middle Kingdoms, with reference to the scenes in the tombs of Baqet III and Khety at Beni Hassan. In Di Biase-Dyson, C. and L. Donovan (eds), The Cultural Manifestations of Religious Experience: studies in honour of Boyo G. Ockinga, 119–132. Münster: Ugarit-Verlag.

Kanawati, *Reforms*. Kanawati, N., 1980. Governmental Reforms in Old Kingdom Egypt. Warminster: Aris & Phillips.

Kanawati, *Tomb & Beyond*. Kanawati, N., 2001. Tomb and Beyond, The: Burial Customs of Egyptian Officials. Warminster: Aris & Phillips.

Kanchan et. al., *Enamel Hypoplasia*. Kanchan, T., Machado, M., Rao, A., Krishan, K. and A. Garg, 2015. Enamel hypoplasia and its role in identification of individuals: A review of literature. Indian J Dent Revisión. 6: 2, 99–102. doi:10.4103/0975-962X.155887.

Karstad & Hudson, *Riverine Hippopotami*. Karstad, E. and R. Hudson, 1986. Social organization and communication of riverine hippopotami in southwestern Kenya. Mammalia, 50: 2, 153–164.

Katz & Voight, *Bread & Beer*. Katz, S. and M. Voigt, 1986. Bread and beer. Expedition 28: 23–34.

Kay, *Responses to Drought*. Kay, R., 1997. Responses of African livestock and wild herbivores to drought. J. Arid Environ. 37: 683–694.

Keimer, *Le Sanglier*. Keimer, L., 1935. Le sanglier égyptien. Chronique d'Egypte, 10: 19, 26–33.

Keimer, *Remarques sur le porc*. Keimer, L., 1937. Remarques sur le porc et le sanglier dans l'Égypte ancienne. Imprimerie de l'Institut français d'archéologie orientale.

Keita & Boyce, *Dental Hypoplasias*. Keita, S. and A. Boyce, 2001. Diachronic patterns of dental hypoplasias and vault porosities during the Predynastic in the Naqada Region, Upper Egypt. American Journal of Human Biology 13: 733–743.

Kemp, *Anatomy*, Kemp, B., 2019. Ancient Egypt: anatomy of a civilisation. London: Taylor & Francis.

Kenawy, *Malaria in Ancient Egypt*. Kenawy, M., 1988. Anopheline mosquitoes Diptera: Culicidae as malaria carriers in AR Egypt 'History and present status.' Journal of the Egyptian Public Health Association 63: 1-2, 67–85.

Kenward & Hall, *Urban Organic Archaeology*. Kenward, H. and A. Hall, 2014. Urban organic archaeology: an irreplaceable palaeoecological archive at risk. World Archaeology 40: 4, 584–596.

Kettenring etal., *Phragmites Invasion*. Kettenring, K., McCormick, M., Baron, H., and D. Whigham, 2011. Mechanisms of Phragmites australis invasion: feedbacks among genetic diversity, nutrients, and sexual reproduction. Journal of Applied Ecology, 485: 1305–1313.

Khan & Ansari, *Eutrophication*. Khan, F., and A. Ansari, 2005. Eutrophication: an ecological vision. The Botanical Review, 71: 4; 449–482.

Khedr & El-Demerdash, *Distribution of Aquatic Plants*. Khedr, A. and M. El-Demerdash, 1997. Distribution of aquatic plants in relation to environmental factors in the Nile Delta. Aquatic Botany 56: 75–86.

Khedr & Hegazy, *Ecology of Lotus*. Khedr, A. and A. Hegazy, 1998. Ecology of the rampant weed Nymphaea lotus L. Willdenow in natural and ricefield habitats of the Nile delta, Egypt. Hydrobiologia, 386: 1-3, 119–129.

Kim etal., *Arid Adaptations*. Kim, E., Elbeltagy, A., Aboul-Naga, A., Rischkowsky, B., Sayre, B., Mwacharo, J. and M. Rothschild, 2016. Multiple genomic signatures of selection in goats and sheep indigenous to a hot arid environment. Heredity, 116: 3, 255–264.

King et.al., *Enamel Hypoplasia*. King, T., Hillson, S. and L. Humphrey, 2002. A detailed study of enamel hypoplasia in a post-medieval adolescent of known age and sex. Arch. Oral Biol.;47: 29–39.

Kingdon, *East African Mammals*. Kingdon, J., 1988. East African Mammals: An Atlas of Evolution in Africa, Volume 3, Part A: Carnivores Vol. 4. Chicago: University Press.

Kiple & Ornelas, *Geese*. Kiple, K. and K. Ornelas, 2012. Geese. In Kiple, K. (ed.), The Cambridge World History of Food, 529–531. Cambridge University Press.

Kitchen, *Chronology*. Kitchen, K., 1991. The Chronology of Ancient Egypt. World Archaeology 23: 2, 201–208.

Klebs, *Reliefs*. Klebs, L. 1915. Die Reliefs des Alten Reiches. Heidelberg.

Knappet, *Why Networks?* Knappett, C., 2013. Introduction: Why Networks? In Knappett, C. (ed.), Network Analysis in Archaeology: New Approaches to Regional Interaction, 1–17. Oxford University Press.

Knudson, etal., *Cultural Interpretation*. Knudson, D., Cable, T., and L. Beck, 2003. In Knudsen, D. M. (ed.), Interpreting Cultural and Natural Resources. Pennsylvania.

Köbbing, etal., *Reed Utilisation*. Köbbing, J., Thevs, N., and S. Zerbe, 2013. The utilisation of reed Phragmites australis–a review. Mires Peat 13: 1–14.

Kohl, *Archaeology of Trade*. Kohl, P., 1975. The archeology of trade. Dialectical Anthropology 11: 43–50.

Kohler, *Of Pots & Myths*. Köhler, E., 2014. Of pots and myths – attempting a comparative study of funerary pottery assemblages in the egyptian Nile Valley during the late 4th millennium BC. In Mączyńska, A. (ed.) The Nile Delta as a centre of cultural interactions between Upper Egypt and the Southern Levant in the 4th millennium BC. Studies in African Archaeology 13, 155–180. Poznań: Poznań Archaeological Museum.

Kotoky etal., *Flood Management*. In Ramkumar, M., Kumaraswamy, K. and R. Mohanraj (eds), Environmental Management of River Basin Ecosystems. Dordrecht, Springer, 161–176.

Krebs, *Ecology*. Krebs, C., 2001. Ecology: The Experimental Analysis of Distribution and Abundance. San Francisco.

Krejčí, etal., *Queen Khentkaus* III. Krejčí, J. Arias Kytnarová, K. and M. Odler, 2015. Archaeological excavation of the mastaba of Queen Khentkaus III (Tomb AC 30). Prague: Charles Univeristy Press.

Krejčí, *Nyuserra revisited*. Krejčí, J., 2011. Nyuserra Revisited. In: M. Bárta, F. Coppens and J. Krejčí, (eds), Abusir and Saqqara in the year 2010, 514–524. Prague

Krom etal., *Nile Sediment Fluctuations*. Krom, M., Stanley, J., Cliff, R. and J. Woodward, 2002. Nile River sediment fluctuations over the past 7000 yrs and their key role in sapropel development. Geology 30: 1, 71–74.

Kröpelin etal., *Saharan Succession*. Kröpelin, S., Verschuren, D., Lézine, A., Eggermont, H., Cocquyt, C., Francus, P., Cazet, J., Fagot, M., Rumes, B., Russell, J., Darius, F., Conley, D., Schuster, M., Von Suchodoletz, H. and D. Engstrom, 2008: Climate-Driven Ecosystem Succession in the Sahara: The Past 6000 Years. Science 320: 765–768. doi:10.1126/science.1154913.

Kumar, *Health at Hierakonpolis*. Kumar, A., 2009. Health at Hierakonpolis, a predynastic settlement in Upper Egypt. PhD, University of Arkansas.

Kuper & Kropelin, *Saharan Holocene*. Kuper, R. and S. Kropelin, 2006. Climate-Controlled Holocene Occupation in the Sahara: Motor of Africa's Evolution. Science 313: 803–807.

Kuraszkiewicz, *Architectural Innovations*. Kuraszkiewicz, K., 2016. Architectural innovations influenced by climatic phenomena 4.2 ka event in the Late Old Kingdom Saqqara, Egypt. Studia Quaternaria, 331: 27–34.

Kushlan, *Environmental Stability*. Kushlan, G., 1976. Environmental stability and fish community diversity. Ecology 57: 821–825.

Langsjoen, *Diseases of Dentition*. Langsjoen O., 1998. Diseases of the dentition. In Aufderheide, A. C., and C. Rodriguez-Martin (eds), The Cambridge Encyclopedia of human paleopathology, 393-412. Cambridge University Press.

Langutt etal., *Late Bronze Collapse*. Langgut, D., Finkelstein, I., and T. Litt, 2013. Climate and the Late Bronze Collapse: new evidence from the Southern Levant. Tel Aviv, 40: 2, 149–175.

Larivíere & Calzada, *Genetta genetta*. Larivière, S. and J. Calzada, 2001. Genetta genetta. Mammalian species, #608, 1-6. *https://doi.org/10.1644/1545-1410200l680<0001:GG>2.0.CO;2*.

Lawler, *Collapse Revisited*. Lawler, A., 2010. Collapse? What collapse? Societal change revisited. Science 330: 907–909.

Lawler, *Die – Or Fade Away*?. Lawler, A. 2015. Did Egypt's Old Kingdom Die—or Simply Fade Away? *https://news.nationalgeographic.com/2015/12/151224-egypt-climate-change-old-kingdom-archaeology.html*.

Lawton & Wilke, *Ancient Agriculture*. Lawton, H. and P. Wilke, 1979. Ancient agricultural systems in dry regions. In Hall, A., Cannell, G. and H. Lawton (eds), Agriculture in semi-arid environments, 1–44. Berlin and Heidelberg.

Leick, *Invention of the City*. Leick, G., 2001. Mesopotamia: the Invention of the City. London.

Lepsius, *Denkmäler*. Lepsius, K., (1842-1845). Denkmäler aus Ägypten und Äthiopien nach den Zeichnungen der von Seiner Majestät dem Könige von Preußen Friedrich Wilhelm IV nach diesen Ländern gesendeten und in den Jahren 1842–1845. ausgeführten wissenschaftlichen Expedition auf Befehl Seiner Majestät herausgegeben und erläutert.

Les etal., *Classification of Water Lilies*. Les, D., Schneider, E., Padgett, D., Soltis, P., Soltis, D. and M. Zanis, 1999. Phylogeny, classification and floral evolution of water lilies Nymphaeaceae; Nymphaeales: a synthesis of non-molecular, rbcL, matK, and 18S rDNA data. Systematic Botany, 28–46.

Lichtheim, *Old & Middle Kingdom Literature*. Lichtheim, M., 2006. Ancient Egyptian Literature. The Old and

Middle Kingdoms. Volume 1. Berkeley: University of California Press.

Lind, *Ugandan Swamps*. Lind, E., 1956. Studies in Ugandan Swamps. Uganda Journal, 20: 2, 166–176.

Linseele & Van Neer, *Archaeology of Exploitation*. Linseele, V. and W. Van Neer, 2009. Exploitation of desert and other wild game in ancient Egypt: The archaeozoological evidence from the Nile Valley. In Riemer, H., Forster, F., Herb, M. and N. Pöllath (eds), 2009. Desert Animals in the Eastern Sahara: Status, Economic Significance, and Cultural Reflection in Antiquity: Proceedings of an Interdisciplinary ACACIA Workshop Held at the University of Cologne, December 14-15, 2007, 47–78. Köln: Heinrich-Barth-Institut.

Linseele & Zerboni, *Done with Fish*. Linseele, V. and A. Zerboni, 2018. Done with fish? A diachronic study of fishing in the Holocene Nile basin of Sudan. Quaternary International 471: 229–240.

Lloyd, *State & Society*. Lloyd, A., 2014. Dialogue with the Environment. In Ancient Egypt; State and Society. Oxford University Press.

Lonnie etal., *Protein for Life*. Lonnie, M., Hooker, E., Brunstrom, J., Corfe, B., Green, M., Watson, A., Williams, E., Stevenson, E., Penson, S. and A. Johnstone, 2018. Protein for life: Review of optimal protein intake, sustainable dietary sources and the effect on appetite in ageing adults. Nutrients 103: 360, 1–18.

Love, *Abusir Drill Core*. Love, S., 2001. The Abusir Drill Core Survey, Egypt. Papers from the Institute of Archaeology 12: 110–113.

Lovell & Whyte, *Enamel Defects*. Lovell, N. and I. Whyte, 1999, Patterns of dental enamel defects at ancient Mendes, Egypt. American Journal of Physical Anthropology, 110: 69–80.

Luff & Bailey, *Aquatic Basis Ancient Civilisations*. Luff, R. and G. Bailey, 2000. The aquatic basis of ancient civilisations: The case *of Synodontis schall* and the Nile Valley. In Bailey, G., Charles, R., and N. Winder (eds), Human Ecodynamics: Proceedings of the Association for Environmental Archaeology Conference, held at the University of Newcastle upon Tyne, 1998, 100–113. Oxford.

Luff & Bailey, *Catfish Growth Structures*. Luff, R. and G. Bailey, 2000. Analysis of size changes and incremental growth structures in African catfish *Synodontis schall schall* from Tell el-Amarna, Middle Egypt. Journal of Archaeological Science, 279, 821–835.

Luff, *Ducks*. Luff, R., 2000. Ducks. Cambridge World History of Food, 517–24. Cambridge University Press.

Lundqvist, *Avert Hydrocide*. Lundqvist, J., 1998. Avert looming hydrocide. Ambio 27: 428–433.

Lytle & Poff, *Natural Flow Regimes*. Lytle, D., and N. Poff, 2004. Adaptation to natural flow regimes. Trends in Ecology & Evolution 192: 94–100.

MacDonald & Edwards, *Chickens in Africa*. MacDonald, K. and D. Edwards, 1993. Chickens in Africa: The importance of Qasr Ibrim. Antiquity 67: 584–90.

Macek etal., *Typha Domingensis Spread*. Macek, P., Rejmánková, E. and J. Lepši, 2010. Dynamics of Typha domingensis spread in Eleocharis dominated oligotrophic tropical wetlands following nutrient enrichment. Evolutionary Ecology, 24: 6, 1505–1519.

Macias, *Degradación de los suelos*. Macias, M., 1994. Estimación de la degradación de los suelos por la erosion h´ıdrica en los valles de la provincia Campero Cochabamba, Bolivia [Soil degradation by erosion in the Campero Province, Cochabamba, Bolivia]. In: Iñiguez, L. and E. Tejada (eds.), roducción de Rumiantes Menores en los Valles Interandinos de Sudamérica. Memorias de un taller sobre metodolog´ıas de la investigación, Tarija, Bolivia, de Agosto 16-21, 3–18.

Macklin etal. *Nile Floodwater Farming*. Macklin, M., Woodward, J., Welsby, D., Duller, G., Williams, F. and M. Williams, 2013. Reach-scale river dynamics moderate the impact of rapid Holocene climate change on floodwater farming in the desert Nile. Geology, 416: 695–698.

Macklin etal., *River Dynamics*. Macklin, M., Toonen, W., Woodward, J., Williams, M., Flaux, C., Marriner, N., Nicoll, K., Verstraeten, G., Spencer, N. and D. Welsby, 2013. A new model of river dynamics, hydroclimatic change and human settlement in the Nile Valley derived from meta-analysis of the Holocene fluvial archive. Quaternary Science Reviews 130; 109–123.

Maclean etal. *Passerine Habitat Loss*. Maclean, I., Hassall, M., Boar, R. and I. Lake 2006. Effects of Disturbance and Habitat Loss on Papyrus-dwelling Passerines. Biological Conservation 131: 349-358.

Madella, *African Domestic Cereals*. Madella, M., García-Granero, J., Out, W., Ryan, P. and D. Usai, 2014. Microbotanical evidence of domestic cereals in Africa 7000 years ago. PloS one, 9: 10,e110177.

Mahmoud, *Vögel in Alten Reich*. Mahmoud, O., 1991. Die wirtschaftliche Bedeutung der Vögel im Alten Reich. Frankfurt am Main.

Makarewicz, *Pastoralist Manifesto*. Makarewicz, C., 2013. A pastoralist manifesto: breaking stereotypes and re-conceptualizing pastoralism in the Near Eastern Neolithic. Levant, 45: 2, 159–174.

Malby, *Waterlogged Wealth*. Malby, E., 1986. Waterlogged Wealth. London.

Málek, *Egyptian Art*. Málek, J., 1999. Egyptian Art. London.

Málek, *Old Kingdom*. Málek, J., 2003. The Old Kingdom c2686–2160 BC. In Shaw, I. (ed.), The Oxford History of Ancient Egypt. Oxford and New York.

Maltby & Acreman, *Ecosystem Services*. Maltby, E. and M. Acreman, 2011. Ecosystem services of wetlands: pathfinder for a new paradigm, Hydrological Sciences Journal, 56: 8, 1341–1359.

Manassa, *El-Mo'alla to El-Deir*. Manassa, C., 2011. El-Moalla to El-Deir. In Wendrich, W. (ed.), UCLA Encyclopedia of Egyptology. *https://escholarship.org/uc/item/4pc0w4hg*

Manlius & Gautier, *lé Sanglier*. Manlius, N and A. Gautier., 1999. Le Sanglier in Egypte. Comptes Rendus de l' Acadamie des Sciences, Serie III -Sciences de la Vie – Life Sciences 322: 7, 573–577.

Manlius, *Historical Biogeography*. Manlius, N., 2009. Historical ecology and biogeography. An example: The Barbary sheep Ammotragus lervia in Egypt. In Riemer, H., Forster, F., Herb, M. and N. Pollath (eds), Desert Animals in the Eastern Sahara; Proceedings of an Interdisciplinary ACACIA workshop, University of Cologne, December 14-15, 2007, 111–127. Köln: Heinrich-Barth-Institut.

Mannermaa, *Goose Domestication*. Mannermaa, K., 2014. Goose: Domestication. In Smith, C. (ed.), Encyclopedia of Global Archaeology, 3096–3098. New York: Springer.

Manning & Timpson, *Saharan Holocene Demographics*. Manning, K. and A. Timpson, A., 2014. The demographic response to Holocene climate change in the Sahara. Quaternary Science Reviews, 101, 28–35.

Märkle & Rösch, *Carbonization Effects*. Märkle, T. and M. Rösch, 2008. Experiments on the effects of carbonization on some cultivated plant seeds. Vegetation History and Archaeobotany 17: 1, 257–263.

Marks etal., *Phragmites Australis*. Marks M., Lapin B. and J. Randall, 1993. Element Stewardship Abstract for Phragmites australis. The Nature Conservancy.

Marriner, etal., *Nile Delta's Sinking Past*. Marriner, N., Flaux, C., Morhange, C. and D. Kaniewski, 2012. Nile Delta's sinking past: Quantifiable links with Holocene compaction and climate-driven changes in sediment supply. Geology, 401: 2, 1083–1086.

Martínez-Jauregui, etal., *People & Predators*. Martínez-Jauregui, M., Linares, O., Carranza, J. and M. Soliño, 2017. Dealing with conflicts between people and colonizing native predator species. Biological Conservation, 20: 9, 239–244.

Mascher etal., *Barley Genomics*. Mascher, M., Schuenemann, V., Davidovich, U., Marom, N., Himmelbach, A., Hübner, S., Korol, A., David, M., Reiter, E., Riehl, S. and M. Schreiber, 2016. Genomic analysis of 6,000-year-old cultivated grain illuminates the domestication history of barley. Nature genetics, 48: 9, 1089–1093.

Mashaly etal., *Vegetation Analysis*. Mashaly, I., el-Halawany, E. and G. Omar, 2001. Vegetation Analysis Along Irrigation and Drain Canals in Damietta Province, Egypt. Online Journal of Biological Sciences 1: 12, 1180–1189.

McCarthty etal., *Hippo Impact*. McCarthy, T., Ellery, W. and A. Bloem, 1998. Some observations on the geomorphological impact of hippopotamus Hippopotamus amphibius L. in the Okavango Delta, Botswana. African Journal of Ecology, 36: 1, 44–56.

McCorriston, *Barley*. McCorriston, J., 2000. Barley. In Kiple, K. and K. Ornelas (eds), The Cambridge World History of Food, 81–89. Cambridge University Press. doi:10.1017/CHOL9780521402149.012

McDowell & Woodward, *Goat Adaptation*. Dowell, R., and A. Woodward, 1982. Concepts in animal adaptation. In Adegbola, T. A. and I. Mecha (eds), Proceedings of the Third International Conference on Goat Production and Disease, Jan. 10-15, 1982, Tucson, Arizona, 387–391. Scottsdale.

McFarlane, *Irukaptah*. McFarlane, A., 2000. The Unis Cemetery at Saqqara. Vol. 1; The Tomb of Irukaptah. Warminster: Aris & Phillips.

McFarlane, *Saqqara Mastabas*. McFarlane, A., 2003. Mastabas at Saqqaara: Kaiemheset, Kaipunesut, Kaiemsenu, Sehetepu and Others. Oxford.

McGregor, *Goats Notes*. McGregor, B., 2001. Goat notes C1: What do goats really like to eat? 83–85. *http://hdl.handle.net/10536/DRO/DU:30065878*

McIlvaine, *Implications*. McIlvaine, B., 2015. Implications of reappraising the iron-deficiency anemia hypothesis. International Journal of Osteoarchaeology 25: 6, 997–1000.

McIntosh, *Ecology*. McIntosh, R., 1986. The background of ecology: concept and theory. Cambridge University Press.

McNabb & Batterson, *Phragmites Occurrence*. McNabb, C. and T. Batterson, 1991. Occurrence of the common reed, Phragmites australis, along roadsides in Lower Michigan. Michigan Academician 23: 211–220.

McNaughton, *Grazers Nutrient Cycling*. McNaughton, S., 1997. Promotion of the cycling of diet-enhancing nutrients by African grazers. Science 27: 8, 1798–1800.

McNaughton, *Typha Ecotype Function*. McNaughton, S., 1966. Ecotype function in the Typha community-type. Ecological Monographs 36: 4, 297–325.

Meadows & Saltonstall, *Phragmites Distribution*. Meadows, R. and K. Saltonstall, 2007. Distribution of native and introduced Phragmites australis in freshwater and oligohaline tidal marshes of the

Delmarva Peninsula and southern New Jersey. Journal of the Torrey Botanical Society 134: 1, 99–107.

Megahed & Vymazalová, *Djedkare Temple Circumcision.* Megahed, M. and H. Vymazalová, 2011. Ancient Egyptian royal circumcision from the pyramid complex of Djedkare. Anthropologie 49: 2, 155–164.

Mekonnen & Melesse, *Soil Erosion Mapping.* Mekonnen. M. and A. Melesse, 2011. Soil Erosion Mapping and Hotspot Area Identification Using GIS. In Melesse, A. M. (ed.), Nile River Basin Hydrology, Climate and Water Use, 207–224. Dordrecht: Springer.

Mellado, *Goat Dietary Selection.* Mellado, M., 2016. Dietary selection by goats and the implications for range management in the Chihuahuan Desert: a review. The Rangeland Journal, 38: 4, 331–341.

Menon etal., *Species Diversity.* Menon, M., Devi, P., Nandagopalan,V. and R. Mohanraj, 2015. Species Diversity and Functional Assemblages of Bird Fauna Along the Riverine Habitats of Tiruchirappalli, India. In Ramkumar, M., Kumaraswamy, K. and R. Mohanraj (eds), Environmental Management of River Basin Ecosystems, 729–748. Dordrecht: Springer.

Mercuri, *Wild Cereals.* Mercuri, A., Fornaciari, R., Gallinaro, M., Vanin, S., and S. Di Lernia, 2018. Plant behaviour from human imprints and the cultivation of wild cereals in Holocene Sahara. Nature Plants 4: 2, 71-81.

Mexicano etal., *Ecological Engineering.* Mexicano, L., Nagler, P. L., Zamora-Arrroyo, F. and E. P. Glenn, 2013. Vegetation dynamics in response to water inflow rates and fire in a brackish Typha domingensis Pers. marsh in the delta of the Colorado River, Mexico. Ecological engineering 5: 9, 167–175.

Michalowski & Guterman, *Art of Ancient Egypt.* Michalowski, K. and N. Guterman, 1969. Art of Ancient Egypt. New York.

Midant-Reynes, *Prehistory.* Midant-Reynes, B., 2000. The prehistory of Egypt: From the first Egyptians to the first Pharaohs. London.

Middleton, *Nothing Lasts Forever.* Middleton, G., 2012. Nothing lasts forever: Environmental discourses on the collapse of past societies. Journal of Archaeological Research, 20: 3, 257–307.

Miller etal., *Skulls & Dentition.* Miller, J. 2008. An appraisal of the skulls and dentition of ancient Egyptians, highlighting the pathology and speculating on the influence of diet and environment. Oxford.

Miller etal., *Predynastic Schistosomiasis.* Miller, R., De Jonge, N., Krijger, F. and A. Deelder, 1993. Predynastic schistosomiasis. In Davies, W. V. and R. Walker (eds), Biological anthropology and the study of ancient Egypt, 54–60. London.

Miller etal., *Near Eastern Archery.* Miller, R., McEwen, E. and Bergman, C., 1986. Experimental approaches to ancient Near Eastern archery. World Archaeology, 18(2), 178—195.

Miller, *Hogs & Hygiene.* Miller, R., 1990. Hogs and Hygiene. The Journal of Egyptian Archaeology, 76: 1, 125–140.

Mills, *SNA in Archaeology.* Mills, B., 2017. Social Network Analysis in Archaeology. Annual Review of Anthropology, 4: 6, 379–397.

Milotić & Hoffmann, *Dung & Competition.* Milotić, T. and M. Hoffmann, 2017. The impact of dung on inter- and intraspecific competition of temperate grassland seeds. Journal of Vegetation Science, 28: 4, 774–786.

Milotić & Hoffmann, *Reduced Germination Success.* Milotić, T. and M. Hoffmann, 2016. Reduced germination success of temperate grassland seeds sown in dung: Consequences for post-dispersal seed fate. Plant Biology, 18: 6, 1038–1047.

Minchinton & Bertness, *Phragmites Spread.* Minchinton, T. and M. Bertness, 2003. Disturbance-mediated competition and the spread of Phragmites australis in a coastal marsh. Ecological Applications 13: 5, 1400–1416.

Miranda-De La Lama, etal., *Livestock Transport.* Miranda-De La Lama, G., Villarroel, M. and G. María, 2014. Livestock transport from the perspective of the pre-slaughter logistic chain: a review. Meat Science, 981, 9–20.

Mitsch & Gosselink, *Wetlands.* Mitsch, W. and J. Gosselink, 2000. Wetlands. New York.

Moeller & Marouard, *Edfu & Dendera Development.* Marouard, G. and N. Moeller, 2018. The Development of Two Early Urban Centres in Upper Egypt During the 3rd Millennium BC. The Examples of Edfu and Dendara. In Budka, J. and J. Auenmüller (eds.), From Microcosm to Macrocosm. Individual Households and Cities in Ancient Egypt and Nubia, 29–58. Leiden: Brill.

Moeller, *FIP Famine?* Moeller, N., 2006. The First Intermediate Period: A time of Famine and Climate Change? Ägypten und Levante: Internationale Zeitschrift für ägyptische archäologie und deren nachbargebeite, 16: 153–168.

Moghaddam etal., *Salinity & Flax.* Moghaddam, M., Babaei, K. and E. Saeedi Pooya, 2018. Germination and growth response of flax *Linum usitatissimum* to salinity stress by different salt types and concentrations. Journal of Plant Nutrition, 41: 5, 563–573.

Mohamed & Serag, *Anatomy of Lotus.* Mohamed, Z. and M. Serag, 2003. Ecology and Anatomy of Nymphaea Lotus L. in the Nile Delta. Journal of Environmental Sciences, 26: 2, 1–20.

Mohammed etal., *Crayfish Development.* Mohammad, A., Awaad, A., Samir, A., Khalid, A. and M. Ghanem, 2015. Egg incubation and post-embryonic development in the red swamp crayfish *Procambarus clarkii* from the River Nile, Egypt. International Journal, 3: 8, 281–289.

Mohr, *Hetep-her-akhti*. Mohr, H., 1943. The Mastaba of Hetep-her-akhti: Study on an Egyptian Tomb Chapel in the Museum of Antiquities. Leiden.

Mona etal, *Catfish Supplementary Feed*. Mona M., Alamdeen A., Elgayar E., Heneish A. and M. El-Feky Mohamed, 2015. Evaluation on the effect of local and imported yeast as supplementary feed on African catfish *clarias gariepinus burchell,* in Egypt. International Journal of Aquaculture 5: 25, 1–8.

Montet, *Les Scenes*. Montet, P., 1923 Les scenes de la vie privee dans les tombeaux egyptiens de l'Ancien Empire. Strasbourg: Librairie Istra.

Moore, *Pollen Travel*. Moore, P., 1976. How far does pollen travel? Nature, 260; 388–389.

Morenó García, *Ancient Egyptian Administration*. Morenó García, J., 2013. The Territorial Administration of the Kingdom in the 3rd Millennium. In Morenó García, J. C. (ed.), Ancient Egyptian Administration, 85–151. Leiden/Boston: Brill.

Morenó García, *J'ai rempli les pâturages*. Morenó García, J., 1999. J'ai rempli les pâturages de vaches tachetées… bétail, économie royale et idéologie en Égypte, de l'Ancien au Moyen Empire, in: Revue d'Égyptologie 50: 241–257.

Morenó García, *Microhistory*. Morenó García, J., 2018, Microhistory. In Grajetzki, W. and W. Wendrich (eds), UCLA Encyclopedia of Egyptology. *http://digital2.library.ucla.edu/viewItem.do?ark=21198/zz002kczsg*

Morenó García, *Patronage*. Morenó García, J., 2013. The 'Other'Administration: Patronage, Factions, and Informal Networks of Power in Ancient Egypt. In Morenó García, J. C. (ed.), Ancient Egyptian Administration, 1029–1065. Leiden/Boston: Brill.

Morenó García, *Pharaonic Administration Limits*. Morenó García, J., 2013. The Limits of Pharaonic Administration: Patronage, Informal Authorities, Mobile Populations and "Invisible" Social Sectors. Diachronic Trends. In Bárta, M. and H. Küllmer (eds), Ancient Egyptian History: Studies Dedicated to the Memory of Eva Pardey, 88–101. Prague: Charles Univeristy Press.

Morenó García, *Pharaonic Landscapes*. Morenó García, J., 2007. The state and the organisation of the rural landscape in 3rd millennium BC pharaonic Egypt. In Bollig, M., Bubenzer, O., Vogelsang, R., and H.-P. Wotzka (eds), Aridity, Change and Conflict in Africa; Proceedings of an International ACACIA Conference held at Königswinter, Germany October 1-3, 2003, 313–330. Köln: Heinrich-Barth-Institut.

Morenó García, *Réinterprétation du Motif de la Famine*. Morenó García, J., 1997. Idéologie politique, culture palatine et mutation historique à la Première Période Intermédiaire: Une réinterprétation du motif de la famine. Études sur l'administration, de l'Ancien au Moyen Empire, Aegyptiaca Leodiensia 4: 1–92. Liège: C. I. P. L.

Morenó García, *Social & Economic History*. Morenó García, J., 2014. Recent Developments in the Social and Economic History of Ancient Egypt. Journal of Ancient Near Eastern History, 12, 231–261.

Morenó García, *Sociopolitocal Transformation?* Morenó García, J., 2015. Climatic change or sociopolitical transformation? Reassessing late 3rd millennium Egypt. In Meller, H., Arz, H., Jung, R. and R. Risch (eds), 2200 BC—A Climatic Breakdown as A Cause for the Collapse of the Old World? 79–94. Halle: Tagungen des Landesmuseums für Vorgeschichte.

Morenó García, *Territory, Elite & Power*. Morenó García, J., 2013. Building the Pharaonic state: Territory, elite, and power in ancient Egypt in the 3rd millennium BCE. In Hill, J., Jones, P., and A. Morales (eds), Experiencing Power, Generating Authority: Cosmos, Politics, and the Ideology of Kingship in Ancient Egypt and Mesopotamia, 185–217. Pennsylvania: University Press.

Morris, *Was It Safe to Drink?* Morris, J., 2005. A river full of water: but was it safe to drink? Ancient Egypt Life and the Hereafter 6: 1, 36–39.

Morris, *Art of Not Collapsing*. Morris, E., 2019. Ancient Egyptian exceptionalism: Fragility, flexibility, and the art of not collapsing. In Yoffee, N. (ed.), The Evolution of Fragility: Setting the Terms 61–88. Cambridge University Press.

Mortensen, *Settlement Pattern Change*. Mortensen, B., 1991. Change in the settlement pattern and population in the beginning of the historical period. Ägypten und Levante/Egypt and the Levant, 11–37.

Mourad, *OK Siege Scenes*. Mourad, A. L. 2011. Siege Scenes of the Old Kingdom. In Binder, S. (ed.), Bulletin of the Australian Centre for Egyptology, 22: 135–158. Sydney: ACE.

Moussa & Altenmüller, *Nianchchnum und Chnumhotep*. Moussa, A., Altenmüller, H., Johannes, D. and W. Rühm, 1977. Das Grab des Nianchchnum und Chnumhotep. Old Kingdom tombs at the causeway of King Unas at Saqqara. Mainz: von Zabern.

Mufarrege etal., *Typha Adaptibility*. Mufarrege, M., Di Luca, G., Hadad, H. and M. Maine, 2011. Adaptability of Typha domingensis to high pH and salinity. Ecotoxicology 20: 2, 457–465.

Müller-Wollermann, *Crime & Punishment*. Müller-Wollermann, R., 2015. Crime and punishment in pharaonic Egypt. Near Eastern Archaeology 78: 4, 228–235.

Müller-Wollermann, *End of Old Kingdom*. Müller-Wollermann, R., 2014, End of the Old Kingdom. In

Grajetzki, W. and W. Wendrich (eds), UEE: http://digital2.library.ucla.edu/viewItem.do?ark=21198/zz002hzfs18761.

Mumford, *Ras Budran*. Mumford, G., 2006. Tell Ras Budran site 345: defining Egypt's Eastern Frontier and Mining Operations in South Sinai during the Late Old Kingdom Early EBIV/ MBI. Bulletin of the American Schools of Oriental Research 342: 1, 13–67.

Murray, *Cereals*. Murray, M., 2000. Cereal production and processing. In Nicholson, P. and I. Shaw (eds), Ancient Egyptian materials and technology, 506–536. Cambridge University Press.

Murray, *Fruits*. Murray, M., 2000. Fruits, Vegetables, Pulses and Condiments. In Nicholson, P. and I. Shaw (eds), Ancient Egyptian materials and technology, 609–655. Cambridge University Press.

Muthuri & Kinyamario, *Papyrus Nutritive Value*. Muthuri, F. and J. Kinyamario, 1989. Nutritive Value of Papyrus (Cyperus Papyrus, Cyperaceae), a Tropical Emergent Macrophyte. Economic Botany 43: 1, 23–30.

Muzzolini, *Saharan Food Economy*. Muzzolini, A., 1993. The emergence of a food-producing economy in the Sahara. In Shaw, T., Sinclair, P., Andah, B. and A. Okpoko (eds), The Archaeology of Africa: Food, Metals and Towns, 227–239. London: Taylor & Francis.

Myśliwiec, etal., *Saqqara Geoarchaeology & Paleoclimate*. Myśliwiec, K., Welc, F. and J. Trzciński, 2012. Geoarchaeological and Paleoclimatic Research by the Polish Archaeological Mission in Saqqara: An Updated Overview. Études et Travaux 25: 276–296.

Myśliwiec, *Merefnebef*. Myśliwiec, K., 2004. The Tomb of Merefnebef. Warsaw: Univerity Press.

Nag etal, *Flax Genetic Diversity*. Nag, S., Jiban, M. and P. G. Karmakar, 2015. An overview on flax *Linum usitatissimum L.* and its genetic diversity. International Journal of Agriculture, Environment and Biotechnology 8: 4, 805–817.

Nana-Sinkam, *Degradation*. Nana-Sinkam, S. C., 1995. Land and Environmental Degradation and Desertification in Africa: Issues and Options for Sustainable Economic Development with Transformation. UNECA FAO Report.

Nash, *Aquatic Animals*. Nash, C., 2012. Aquatic Animals. In Kiple, K. and K. C. Ornelas (eds), The Cambridge World History of Food, 456–467. Cambridge University Press.

Negus, *A Great African Drought?* Negus, A. L., 1998. The fall of the Old Kingdom: a great African drought? PhD, University of California.

Neubauer, *Ecosystem Responses*. Neubauer, S. C., 2013. Ecosystem responses of a tidal freshwater marsh experiencing saltwater intrusion and altered hydrology. Estuaries and Coasts, 36: 3, 491–507.

Neuberger & Thornes, *Art & Climate*. Neuberger, H. and J. E. Thornes, 2005. Art and Climate. In Oliver, J. E., Fairbridge, R. W. and M. Rampino (eds), Encyclopedia of World Climatology, 94–102. Dordrecht: Springer.

Newberry, *Beni Hasan* I. Newberry, P.E., 1893. Beni Hasan, Part I. London: EES.

Newberry, *Beni Hasan* II. Newberry, P.E., 1894. Beni Hasan, Part II. London: EES.

Nikita et.al., *Saharan Dentition*. Nikita, E., Mattingly, D., and M. M Lahr, 2014. Dental indicators of adaptation in the Sahara Desert during the Late Holocene. HOMO-Journal of Comparative Human Biology, 65: 5, 381–399.

Norfolk etal., *Arid South Sinai*. Norfolk, O., Eichhorn, M. P. and F. Gilbert, 2013. Traditional agricultural gardens conserve wild plants and functional richness in arid South Sinai. Basic and Applied Ecology 14: 8, 659–669.

Nuzzolo, *Divine Legitimization*. Nuzzolo, M., 2015. Royal Authority, Divine Legitimization. Topography as an element of acquisition, confirmation and renewal of power in the Fifth Dynasty. In Coppens, F., Janák, J. and H. Vymazalová (eds), 7th Symposium on Egyptian Royal Ideology. Royal versus Divine Authority. Acquisition, Legitimization and Renewal of Power. Königtum, Staat und Gesellschaft früher Hochkulturen 4, 289–304. Wiesbaden: Harrassowitz Verlag..

O'Brien, *Boredom with Apocalypse*. O'Brien, S., 2017. Boredom with the Apocalypse: Resilience, Regeneration, and their Consequences for Archaeological Interpretation. In Cunningham, T. and J. Driessen (eds), Crisis to Collapse: The Archaeology of Social Breakdown, 295–302. Presses universitaires de Louvain.

O'Connor, *Political Systems*. O'Connor, D., 1974. Political systems and archaeological data in Egypt: 2600–1780 BC. World Archaeology, 6(1), 15–38.

Okuku etal., *Plankton Response to Damming*. Okuku, E. O., Tole, M., Kiteresi, L. I. and S. Bouillon, 2016. The response of phytoplankton and zooplankton to river damming in three cascading reservoirs of the Tana River, Kenya. Lakes & Reservoirs: Research & Management, 21: 2; 114–132.

Olea-Popelka, etal., *Human Tuberculosis*. Olea-Popelka, F., Dean, A. S., Muwonge, A., Perera, A., Raviglione, M. and P. I. Fujiwara, 2018. Mycobacterium bovis as the Causal Agent of Human Tuberculosis: Public Health Implications. In Chambers, M., Gordon, S., Olea-Popelka, F. and P. Barrow, (eds), Bovine Tuberculosis, 16–30. Wallingford: CABI.

Orozco Obando, *Sacred Lotus*. Orozco Obando, W.S., 2012, Evaluation of Sacred Lotus *Nelumbo nucifera Gaertn.* as an Alternative Crop for Phyto-remediation. PhD, Auburn University, Alabama.

Osman & Kloas. *Catfish Water Quality*. Osman, A. G. and W. Kloas, 2010. Water quality and heavy metal monitoring in water, sediments, and tissues of the African catfish *Clarias gariepinus Burchell*, from the river Nile, Egypt. Journal of Environmental Protection 1: 4, 389–400.

Östborn & Gerding, *Archaeological Data Network Analysis*. Östborn, P. and H. Gerding, 2014. Network analysis of archaeological data: a systematic approach. Journal of Archaeological Science, 4: 6, 75–88.

Oxenham & Cavill, *Porotic Hyperostosis Response*. Oxenham, M. F. and I. Cavill, 2010. Porotic hyperostosis and cribra orbitalia: the erythropoietic response to iron-deficiency anaemia. Anthropological Science, 118: 3, 199–200.

Özkurt, *Egyptian Mongoose*. Özkurt, S. Ö., 2015. Karyological and some morphological characteristics of the Egyptian mongoose, Herpestes ichneumon Mammalia: Carnivora, along with current distribution range in Turkey. Turkish Journal of Zoology 39: 3, 482–487.

Palmer etal., *Plant Archaeogenetics*. Palmer, S.A., Smith, O. and R.G. Allaby, 2012. The blossoming of plant archaeogenetics. Annals of Anatomy 194: 1, 146–156.

Parcak, *Egypt's Old Kingdom Empire*? Parcak, S., 2004. Egypt's Old Kingdom 'Empire'? A Case Study Focusing on South Sinai. In Knoppers, G. N. and A. Hirsch 9eds), Egypt, Israel, and the Ancient Mediterranean World: Studies in Honor of Donald B. Redford, 41–60. Leiden: Brill.

Parcak, *Physical Context*. Parcah, S. 2010. The Physical Context of Ancient Egypt. In Lloyd, A. (ed.) A companion to ancient Egypt. New Tork: Wiley Blackwell.

Parcak, *Remote Sensing*. Parcak, S., 2009. Satellite remote sensing for archaeology. New York: Routledge.

Park, *Class Stratification*. Park, T. K., 1992. Early Trends toward Class Stratification: Chaos, Common Property, and Flood Recession Agriculture. American Anthropologist, 94: 1, 90–117.

Partridge, *Fighting Pharaohs*. Partridge, R. B., 2002. Fighting Pharaohs. Manchester: Peartree.

Patch etal., *Dawn of Egyptian Art*. Patch, D., Eaton-Krauss, M. and S. Allen, 1947 and the Metropolitan Museum of Art New York, 2011. Dawn of Egyptian art. New York: MMA.

Paterson & Chapman, *Nile Perch*. Paterson, J. A. and L. J. Chapman, 2009. Fishing down and fishing hard: ecological change in the Nile perch of Lake Nabugabo, Uganda. Ecology of Freshwater Fish, 18: 3, 380–394.

Payne, *Moprhological Distinctions*. Payne, S., 1985. Morphological distinctions between the mandibular teeth of young sheep Ovis, and goats, Capra. Journal of Archaeological Science: 12: 139–147.

Pearce, *Hunter-Gatherer Networks*. Pearce, E., 2014. Modelling mechanisms of social network maintenance in hunter-gatherers. Journal of Archaeological Science, 50: 403–413.

Peeples & Roberts, *To binarize or Not*. Peeples, M. A. and J. M. Roberts Jr, 2013. To binarize or not to binarize: relational data and the construction of archaeological networks. Journal of Archaeological Science, 40: 7, 3001–3010.

Pethick, *Saltmarsh Geomorphology*. Pethick, J. S., 1992. Saltmarsh Geomorphology. In Allen, J. R. L. and K. Pye (eds), Saltmarshes: Morphodynamic, Conservation and Engineering Significance, 41–62. Cambridge University Press.

Pezeshki, *Cattail Photosynthetic Response*. Pezeshki, S. R., DeLaune, R. D., Kludze, H. K. and H. S. Choi, 1996. Photosynthetic and growth responses of cattail *Typha domingensis* and sawgrass *Cladium jamaicense* to soil redox conditions. Aquatic Botany, 54: 1, 25–35.

Pfeiffer & Sealy, *Foragers Body Size*. Pfeiffer, S. and J. Sealy, 2006. Body size among Holocene foragers of the Cape ecozone, southern Africa. American Journal of Physical Anthropology: The Official Publication of the American Association of Physical Anthropologists, 129: 1, 1–11.

Phillips etal., *Mid-Holocene Egypt*. Phillipps, R., Holdaway, S., Wendrich, W. and R. Cappers, 2012. Mid-Holocene occupation of Egypt and global climatic change. Quaternary International, 251: 64–76.

Pierssené, *Explaining Our World*. Pierssené, A., 1999. Explaining our world: An approach to the art of environmental interpretation. London: Routledge.

Pitman & Läuchli, *Salinity & Agriculture*. Pitman, M. G. and A. Läuchli, 2002. Global impact of salinity and agricultural ecosystems. In Läuchli, A. and U. Lüttge, (eds), Salinity: environment-plants-molecules, 3–20. Dordrecht: Springer.

Poff & Zimmerman, *Altered Flow Regimes*. Poff, N. L. and J. K. Zimmerman, 2010. Ecological responses to altered flow regimes: a literature review to inform the science and management of environmental flows. Freshwater Biology, 55: 1, 194–205.

Pollath, *Prehistoric Gamebag*. Pöllath, N., 2009. The Prehistoric Gamebag: The Archaeozoological Record from Sites in the Western Desert of Egypt. In Riemer, H., Forster, F., Herb, M. and N. Pollath (eds), Desert Animals in the Eastern Sahara; Proceedings of an Interdisciplinary ACACIA workshop, University of Cologne, December 14-15, 2007, 79–109. Köln: Heinrich-Barth-Institut.

Porter, *Wild Goats*. Porter, R., 1996. The Cretan wild goat *Capra aegagrus cretica* and the Theran "antelopes." In Reese, D. S. (ed.), Pleistocene and Holocene Fauna of Crete and Its First Settlers, 295–315. Madison.

Powel, *Sumerian Agriculture*. Powell, M. A., 1999. The Sumerian Agriculture Group: A Brief History. In Klengel, H. and J. Renger (eds), Landwirtschaft im Alten Orient, 291–299. Berlin.

Qin, *Saqqara Landscape Change*. Qin, Y., 2009. Landscape change in the Saqqara/Memphis area of Egypt from 3000 BC to the present. MSci. dissertation, Earth Sciences, University of Cambridge.

Quintana, etal., *Predation, Competition & Size*. Quintana, X. D., Arim, M., Badosa, A., Blanco, J. M., Boix, D., Brucet, S., Compte, J., Egozcue, J. J., de Eyto, E., Gaedke, U. and S. Gascón, 2015. Predation and competition effects on the size diversity of aquatic communities. Aquatic Sciences, 77: 1, 45–57.

Ramachandra etal., *Ecohydrology*. Ramachandra, T. V., Chandran, M. S., Joshi, N. V., Karthick, B. and V. D. Mukri, 2015. Ecohydrology of Lotic Systems in Uttara Kannada, Central Western Ghats, India. In Ramkumar, M., Kumaraswamy, K. and R. Mohanraj (eds), Environmental Management of River Basin Ecosystems, 621–625. Dordrecht: Springer.

Ramdani etal, *Environmental Influences*. Ramdani, M. N., Elkhiati, R. J., Flower, J. R., Thompson, L., Chouba, M. M., Kraiem, F., Ayache and M. H. Ahmed, 2009. Environmental influences on the qualitative and quantitative composition of phytoplankton and zooplankton in North African coastal lagoons. Hydrobiologia 622: 1; 113–131.

Ramkumar etal., *Land Use Dynamics*. Ramkumar, M., Kumaraswamy, K. and R. Mohanraj, 2015. Land use dynamics and environmental management of River Basins with emphasis on deltaic ecosystems: Need for integrated study-based development and nourishment programs and institutionalizing the management strategies. In Ramkumar, M., Kumaraswamy, K., and R. Mohanraj, eds., Environmental Management of River Basin Ecosystems, 1–20. Dordrecht: Springer.

Rateutschak-Pluss & Guggenheim, *Dental Plaque Accumulation*. Rateutschak-Pluss, E. M. and B. Guggenheim, 1982. Effects of a carbohydrate-free diet and sugar substitutes on dental plaque accumulation. J Clin Periodontol, 9: 239–244.

Raxter, *Egyptian Body Size*. Raxter, M. H., 2011. Egyptian body size: a regional and worldwide comparison. PhD, University of South Florida.

Redding *Pyramids & Protein*. Redding R., 2007. Pyramids and protein of cattle, sheep, goats, and pigs, http://www.aeraweb.org/spec_zoo.asp.

Redding, *Animal Use Patterns*. Redding R. W., 1992. Egyptian Old Kingdom Patterns of Animal Use and the Value of Faunal Data in Modeling Socioeconomic Systems. Paléorient, 18: 2, 99–107.

Redding, *Pig & Chicken*. Redding, R.W., 2015. The pig and the chicken in the Middle East: Modeling human subsistence behavior in the archaeological record using historical and animal husbandry data. Journal of Archaeological Research, 23: 4, 325–368.

Redding, *Pigs v Sheep/Goats*. Redding, R., 2002. Differential Consumption of Pig vs. Sheep/Goats at the Old Kingdom Site of Kom el-Hisn. Paper presented at the 53rd Annual Meeting of the American Research Center in Egypt, 2002. 39: 27–74.

Redding, *Role of the Pig*. Redding, R.W., 1991. The role of the pig in the subsistence system of ancient Egypt: A parable on the potential of faunal data. In: P. J. Crabtree and K. Ryan (eds), Animal Use and Culture Change, MASCA Research Papers in Science and Archaeology 8, 20–30. Philadelphia.

Redford, *Ancient Egypt*. Redford, D. B., 2006. A History of Ancient Egypt. Dubuque.

Reisner, *Giza* I. Reisner, G. A., 1942. A History of the Giza Necropolis, Vol. 1. Cambridge University Press.

Reisner, *Giza* II. Reisner, G. A., 1955. A History of the Giza Necropolis Vol. 2. Cambridge University Press.

Renfrew & Bahn, *Archaeology*. Renfrew, C. and P. Bahn. 2004. Archaeology: Theories, Methods and Practice. London.

Revel etal., *Nile River Dynamics*. Revel, M., Ducassou, E., Skonieczny, C., Colin, C., Bastian, L., Bosch, D., Migeon, S. and J. Mascle, 2015. 20,000 years of Nile River dynamics and environmental changes in the Nile catchment area as inferred from Nile upper continental slope sediments. Quaternary Science Reviews, 130; 200–221.

Reynolds etal, *Do humans cause deserts?* Reynolds, J., Stafford-Smith, D. and E. Lambin, 2003, July. Do humans cause deserts? An old problem through the lens of a new framework: the Dahlem desertification paradigm. In Proceedings of the 7th International Rangelands Congress (Vol. 26).

Riehl, *Archaeobotanical Evidence*. Riehl, S., 2009. Archaeobotanical evidence for the interrelationship of agricultural decision-making and climate change in the ancient Near East. Quaternary International 197, 1–2, 93–114.

Ritchie, *Dakhleh Palaeobotany*. Ritchie J., 1999. Flora, Vegetation and Palaeobotany of the Dakhleh Oasis. In: C.S. Churcher and A.J. Millis (eds), Reports from the survey of the Dakhleh Oasis, Western Desert of Egypt 1977-1987, 74–80. Oxford.

Ritchie, *Dakhla Pollen Spectra*. Ritchie, J., 1985. Modern pollen spectra from Dakhla Oasis, Western Egyptian desert. Grana Uppsala, 1–6.

Robins, *Art of Ancient Egypt*. Robins, G., 2008. The art of ancient Egypt. Harvard.

Robins, *Proportion*. Robins, G., 2010. Proportion and style in ancient Egyptian art. Austin.

Roman etal., *Salt Marsh Vegetation Change*. Roman, C., Niering, W. and R. Warren, 1984. Salt marsh vegetation change in response to tidal restriction. Environmental Management, 82: 141–149.

Romem, *Unraveling Prolonged Stability*. Romem, I., 2009. The Unraveling of Prolonged Stability: The Fall of the Old Kingdom in Ancient Egypt. Berkeley: University of California Press.

Rossel etal *Donkey Domestication*. Rossel, S., Marshall, F., Peters, J., Pilgram, T., Adams, M. and D. O'Connor, 2008. Domestication of the donkey: Timing, processes, and indicators', Proceedings of the National Academy of Sciences 105: 10, 3715-3720.

Rossel, *Food for the Dead*. Rossel, S., 2004. Food for the dead, the priest, and the mayor: looking for status and identity in the Middle Kingdom settlement at South Abydos, Egypt. In O'Day, S. J., Van Neer, W. and A. Ervynck (eds), Behaviour Behind Bones. The zooarchaeology of ritual, religion, status and identity. Proceedings of the 9th ICAZ Conference, Durham, 2002, 198–202. Oxford.

Rossel, *Tale of the Bones*. Rossel, S., 2006. A Tale of the Bones. Animal Use in the Temple and Town of Wah-Sut, http://www.penn.museum/documents/publications/expedition/PDFs/48-2/Rossel.pdf.

Roth, *Palace Attendants*. Roth, A., 1995. Giza Mastabas. Vol. 6: A cemetery of palace attendants: including G 2084-2099, G 2230+2231, and G 2240 Boston.

Rothschild, *Porotic Hyperostosis*. Rothschild, B., 2002. Porotic hyperostosis as a marker of health and nutritional conditions. American Journal of Human Biology 14: 417–420.

Rothschild, *Shift to Maize*. Rothschild, B., 2012. Extrapolation of the mythology that porotic hyperostosis is caused by iron deficiency secondary to dietary shift to maize. Advances in Anthropology 2: 157–160.

Rudolf etal., *Dissolved Oxygen*. Rudolf, A., Ahumada, R. and C. Perez, 2002. Dissolved oxygen content as an index of water quality in San Vicente Bay, Chile. Environmental monitoring and assessment 78: 89–100.

Ruffer, *Bilharzia in Mummies*. Ruffer, M., 1910. Note on the presence of Bilharzia hematobium in Egyptians mummies of the twentieth dynasty 1250-1000 BC. British Medical Journal 2, 16.

Rzóska, *Nile*. Rzoska, J., 1976. The Joint Nile in the Sudan. In Rzóska, J. (ed.), The Nile: Biology of an Ancient River, 263–270. Dordrecht: Springer.

Rzóska, *Upper Nile Swamps*. Rzóska, J., 1974. The Upper Nile Swamps, a Tropical Wetland Study. Freshwater Biology 4: 1–30.

Saad & Sami, *Nile Delta Deposits*. Saad, S. and S. Sami, 1967. Studies of pollen and spores' content of the Nile Delta deposits Berenbal Region. Pollen Spores, 9: 467–503.

Saarinen etal., *River Water Acidity*. Saarinen, T., Celebi, A. and B. Klove, 2013. Links between river water acidity, land use and hydrology. Boreal environment research, 18: 5, 359–373.

Sabri etal, *Catfish Infection*. Sabri, D. M., Eissa, I. A., Danasoury, M. and H. Khouraiba, 2010. Prevalence of Henneguya branchialis in catfish *Clarias gariepinus* in Ismailia, Egypt. Int J Agric Bio 12: 897–900.

Saffiro, *Dietary Habits*. Saffirio, L., 1972. Food and dietary habits in ancient Egypt. Journal of Human Evolution, 13, 297–305.

Sahrhage, *Fishing in Ancient Egypt*. Sahrhage, D., 2008. Fishing in Ancient Egypt. In Selin, H. (ed.), Encyclopaedia of the History of Science, Technology, and Medicine in Non-Western Cultures, 922–927. Dordrecht: Springer.

Said, *River Nile*. Said, R., 1993. River Nile: Geology, Hydrology and Utilization. Oxford and New York.

Sallam & El-Barbary, *Secondary Channels*. Sallam, G. and Z. El-Barbary, 2004. The Effect of Closing Secondary Channels on the Morphology and the Ecology of the River Nile. Proceedings of the Eight international Water Technology Conference, IWTC8, Alexandria, 489–498. Cairo.

Sampsell, *Geology of Egypt*. Sampsell, B. M., 2003. A Traveller's Guide to the Geology of Egypt. Cairo.

Samuel, *Ancient Egyptian Nutrition*. Samuel, D. W., 1997. Cereal foods and nutrition in ancient Egypt. Nutrition 13: 6, 579–580.

Santoro etal., *1000 years of Famine*. Santoro, M., Hassan, F., Wahab, M., Cerveny, R. and R. Balling, 2015. An aggregated climate teleconnection index linked to historical Egyptian famines of the last thousand years. The Holocene, 25: 5, 872–879.

Saunders etal., *Papyrus Carbon & Water Cycles*. Saunders, M., Jones, M. and F. Kanslime, 2007. Carbon and Water Cycles in Tropical Papyrus Wetlands. Wetlands Ecological Management, 15: 489–498.

Säve-Söderbergh, *Hamra Dom*. Säve-Söderbergh, T., 1994. The Old Kingdom Cemetery at Hamra Dom (El-Qasr wa es-Saiyad). Stokholm.

Säve-Soderbergh, *Hippopotamus Hunting*. Säve-Soderbergh, T., 1953. On Egyptian Representations of Hippopotamus Hunting as a Religious Motif. Lund.

Schäfer, *Principles of Egyptian Art*. Schäfer, H., Brunner-Traut, E. and J. Baines (trans.) 1986. Principles of Egyptian Art. Oxford.

Schindler, *Acid Rain & Freshwater*. Schindler, D., 1988. Effects of acid rain on freshwater ecosystems. Science, 239: 4836, 149–157.

Schliemann, *Viverrids*. Schliemann, H., 1990. Viverrids. In Grzimek, B., Illies, J., and W. Klausewitz (eds),

Grzimek's Encyclopedia of Ecology, III, 518. New York..

Schneider, *Existence Now Silenced.* Schneider, T., 2017. What Is the Past but a Once Material Existence Now Silenced?. In Höflmayer, F., (ed.) The late third millennium BC In the ancient near east and eastern mediterranean: A time of collapse and transformation. The late third millennium in the ancient Near East, 311–322. Chicago: Oreintal Institute.

Schneider, *Old Kingdom Abroad.* Schneider, T., 2015. The Old Kingdom Abroad: An Epistemological Perspective. With Remarks on the Biography of Iny and the Kingdom of Dugurasu. In Der Manuelian, P. and T. Schneider (eds), Towards a New History for the Egyptian New Kingdom, 429–455. Leiden/Boston: Brill.

Schrank & Mahmoud, *Dakhla Palynology*. Schrank, E. and M. Mahmoud, 1998. Palynology pollen, spores and dinoflagellates and Cretaceous stratigraphy of the Dakhla Oasis, central Egypt. Journal of African Earth Sciences 26: 2; 167–193.

Schultz, *Paleohistopathology*. Schultz, M., 2001 Paleohistopathology of bone: a new approach to the study of ancient diseases. American Journal of Physical Anthropology 44: 106–147.

Schurmann, *Reliefs Ii-nefret*. Schurmann, W., 1983. Die Reliefs aus dem Grab des Pyramidenoverstehers Ii-nefret. Karlsruhe: Müller.

Schwarz, *dentaire dans l'Égypte Pharaonique*. Schwarz, J., 1979. La Médicine dentaire dans l'Égypte Pharaonique. Bulletin de la société d'Égyptologie 2, 37–42.

Seidlmayer, *Beni Hassan People*. Seidlmayer, S., 2007. People at Beni Hassan: Contributions to a Model of Ancient Egyptian Rural Society. In Hawass. Z. and J. Richards (eds), The Archaeology and Art of Ancient Egypt. Essays in Honor of David B. O'Connor II, 351–368. Cairo: Supreme Council of Antiquities.

Seidlmayer, *OK Elephantine*. Seidlmayer, S., 1996. Town and State in the Early Old Kingdom: A View from Elephantine. In Spencer, J. (ed.), Aspects of Early Egypt, 108–127. London: British Museum Press.

Seidlylmeyer, *First Intermediate Period.* Seidlylmeyer, S., 2000. Ch.6; The First Intermediate Period. In Shaw, I. (ed.), The Oxford History of Ancient Egypt, 108–136. Oxford University Press.

Shafik, *Disloyalty & Punishmnet*. Shafik, S., 2010. Disloyalty and punishment: the case of Ishfu at Saqqara. In Woods, A., McFarlane, A., and S. Binder (eds), Egyptian Culture and Society. Studies in Honour of Naguib Kanawati, II, 181–191. Cairo: Supreme Council of Antiquities.

Shahin, *Hydrology.* Shahin, M., 2002 Hydrology and Water Resources of Africa. London: Springer.

Shaker etal, *Salt Tolerant Goats*. Shaker, Y., Ibrahim, N., Younis, F. and H. El Shaer, 2014. Effect of feeding some salt tolerant fodder shrubs mixture on physiological performance of Shami goats in Southern Sinai, Egypt. Journal of American Science, 10: 2s, 66–77.

Shalomi-Hen, *Rise of Osiris*. Shalomi-Hen, R., 2015. The 1Dawn of Osiris and the Dusk of the Sun-Temples: Religious History at the End of the Fifth Dynasty. In Der Manuelian, P. and T. Schneider (eds), Towards a new history for the Egyptian Old Kingdom: Perspectives on the pyramid age, I, 456–469. Harvard: Brill.

Shaltout & Azzazi, *Nile Delta Climate Change*. Shaltout, M. and M. Azzazi, 2014. Climate Change in the Nile Delta from Prehistoric to the Modern Era and Their Impact on Soil and Vegetation in Some Archaeological Sites. Journal of Earth Science and Engineering, 4, 632–642.

Shaltout & El-Sheikh, *Vegetation Environment Relations*. Shaltout, K. and M. El-Sheikh, 1993. Vegetation-environment relations along water courses in the Nile Delta region. Journal of Vegetation Science, 4: 4; 567–570.

Shaw, *Outsiders*. Shaw, I., 2000. Chapter 11: Egypt And the Outside World. In Shaw, I (ed.), The Oxford History of Ancient Egypt, 308–323. Oxford University Press.

Shaw, *Warfare & Weapons*. Shaw, I., 1991. Egyptian Warfare and Weapons. London: Shire Publications.

Shen-Miller etal., *Long-living Lotus*. Shen-Miller, J., Schopf, J. W., Harbottle, G., Cao, R.I. Ouyang, S. Zhou, K.S. Southon, J.R. and G.H. Liu, 2002. Long-living lotus: Germination and soil γ-irradiation of centuries old fruits, and cultivation, growth and phenotypic abnormalities of offspring. Amer. J. Bot. 92: 236–247.

Sherratt, *Water, Soil & Seasonality*. Sherratt, A., 1980. Water, soil and seasonality in early cereal cultivation. World Archaeology, 11: 3, 313–330.

Shimazu etal. *Dietary Patterns & Mortality*. Shimazu, T., Kuriyama, S., Hozawa, A., Ohmori, K., Sato, Y., Nakaya, N., Nishino, Y., Tsubono, Y. and I. Tsuji, 2007. Dietary patterns and cardiovascular disease mortality in Japan: A prospective cohort study. Int. J. Epidemiol. 36, 600–609.

Shirai, *Egypt's First Farmer Herders*. Shirai, N., 2010. The archaeology of the first farmer-herders in Egypt: New insights into the Fayum Epipalaeolithic and Neolithic. Leiden: Brill.

Shirai, *Ideal & Reality*. Shirai, Y., 2006. Ideal and Reality in Old Kingdom Private Funerary Cults. In Bárta, M. (ed.), The old kingdom art and archaeology: proceedings of the conference held in Prague, May 31-June 4, 2004, 325–333. Prague: Charles Univeristy Press.

Siebels, *Agricultural Scenes*. Siebels, R., 2000. Agricultural Scenes. In Donovan, L. and K. McCorquodale (eds), Egyptian Art, Principles and Themes in Wall

Scenes, 55–64. Guizeh: Foreign Cultural Information Department.

Silanikove, *Adaptation in Goats*. Silanikove, N., 2000. The physiological basis of adaptation in goats to harsh environments. Small Ruminant Research, 35: 3, 181–193.

Silliman etal., *Livestock as Control Agents*. Silliman, B., Mozdzer, T., Angelini, C., Brundage, J., Esselink, P., Bakker, J., Gedan, K., van de Koppel, J. and A. Baldwin, 2014. Livestock as a potential biological control agent for an invasive wetland plant. PeerJ, 2: 10, 1–19.

Silverman, *Anthropology & Circumcision*. Silverman, E., 2004. Anthropology and circumcision. Annual Review of Anthropology, 33: 1, 419–445.

Smith & Simpson, *Art & Architecture*. Smith, W. and W. Simpson, 1998. The Art and Architecture of Ancient Egypt. New Haven: Yale University Press.

Smith etal., *Optical Characteristics*. Smith, D., Davies-Colley, R., Knoef, J. and G. Slot, 1997. Optical characteristics of New Zealand rivers in relation to flow. Journal of the American Water Resources Association 33: 301–312.

Smith-Guzmán, *Nile Valley Malaria*. Smith-Guzmán, N., 2015. Cribra orbitalia in the ancient Nile Valley and its connection to malaria. International journal of paleopathology, 10, 1–12.

Smith, *Egyptian Imperialism*. Smith, S., 1991. A Model for Egyptian Imperialism in Nubia. GM 122: 77–102.

Smith, *Identifying Cess-pits*. Smith, D., 2013. Defining an 'indicator package' to allow identification of 'cess pits' in the archaeological record. Journal of Archaeological Science 40: 526–543.

Smith, *Origins Unidentified Reliefs*. Smith, W., 1942. The origin of some unidentified Old Kingdom reliefs. American Journal of Archaeology, 464: 509–531.

Soliman, etal., *Plant Species Nile Islands*. Soliman, A., Amer, W. and W. Hassan, 2015. Factors affecting the spatial distribution of plant species in Nile islands of mid Egypt. Current Life Sciences, 12: 70–92.

Soria-Trastoy, *Fishing Tackle Analysis*. Soria-Trastoy, M., 2012. Iconographic and Archaeological analysis of the fishing tack1e in the tomb of Niankhkhnum and Khnumhotep. Journal of Oriental and Ancient History, 1: 13–58.

Soroush & Mordechai, *Short-Term Cataclysms*. Soroush, M. and L. Mordechai, 2018. Adaptation to Short-Term Cataclysmic Events: Flooding in Pre-modern Riverine Societies. Human Ecology, 46: 3, 349–361.

Sowada, *Weather Evidence*. Sowada, K., 2013. Evidence for late third millennium weather events from a Sixth Dynasty tomb at Saqqara. Studia Quaternaria, 302: 69–74.

Sowada, *Synchronisation*, Sowada, K., 2014. Report on a project to synchronize Egyptian and Levantine chronologies of the third millennium BCE. Journal of Ancient Egyptian Interconnections, 6(4), 55–57.

Sowada, *OK Eastern Mediterranean*. Sowada, K., 2009. Egypt in the Eastern Mediterranean during the Old Kingdom: An Archaeological Perspective. Fribourg: Academic Press.

Speight & Blackith, *the Animals*. Speight, M. and R. Blackith, 1983. The Animals. In Gore, A. (ed.), Ecosystems of the World 4A: Mires, Swamps, Bog, Fen and Moor, 349–365. Amsterdam: Elsevier.

Spiegelman, *Emergency Surgery*? Spiegelman, M., 1997. The Circumcision scene in the tomb of Ankhmahor: the first record of emergency surgery? In Binder, S. (ed.), The Bulleting of the Australian Centre for Egyptology 8: 91–100. Sydney: ACE.

Stanley etal., *Nile Flow Failure*. Stanley, J., Krom, M., Cliff, R. and J. Woodward, 2003 Short contribution: Nile flow failure at the end of the Old Kingdom, Egypt: strontium isotopic and petrologic evidence. Geoarchaeology 18: 3, 395–402.

Stanley & Warne, *Nile Destruction Phase*. Stanley, J. and A. Warne., 1998. Nile Delta in its destruction phase. Journal of Coastal Research: 795–825.

Stanton, *White Nile Marshes*. Stanton, E., 1903. The Great Marshes of the White Nile. Journal of the Royal African Society, 28, 375–379.

Steenhuis etal., *Erosion & Discharge*. Steenhuis, T., Tilahun, S., Tesemma, Z., Tebebu, T., Moges, M., Zimale, F., Worqlul, A., Alemu, M., Ayana, E. and Y. Mohamed, 2014. Soil erosion and discharge in the Blue Nile Basin: trends and challenges. In Melesse, A., Abtew, W. and S. Setegn, (eds), Nile River basin: ecohydrological challenges, climate change and hydropolitics, 133–147. Dordrecht: Springer.

Steindorf, *Ti*. Steindorf, G., 1913. Das Grab des Ti. Leipzig: Hinrichs.

Sterner, *Herbivores Effect on Algae*. Sterner R., 1986 Herbivores' direct and indirect effects on algal populations. Science 231: 605–607.

Sterner, *Zooplankton*. Sterner, R. 2015. Role of Zooplankton in Aquatic Ecosystems. In Likens, G. (ed.), The Encyclopedia of Inland Waters, 678–688. Amsterdam: Elsevier.

Stevenson, *Algae of Rivers*. Stevenson, R., 2015. Algae of River Ecosystems. In Likens, G. (ed.), The Encyclopedia of Inland Waters, 114–122. Amsterdam: Elsevier.

Stevenson, *Muuseum Archaeology*. Stevenson, A., 2014. Egyptian Archaeology and the Museum. Oxford Handbooks Online.

Steward etal, *River Runs Dry*. Steward, A., von Schiller, D., Tockner, K., Marshall, J. and S. Bunn, 2012. When the river runs dry: human and ecological values of dry riverbeds. Frontiers in Ecology and the Environment, 104: 202–209.

Stewart etal, *Tracing Ancient Tracks.* Stewart, C., Lemmens, K. and M. Sala. 2015. Satellite Radar in Support to Archaeological Research in Egypt: Tracing Ancient Tracks Between Egypt and Southern Levant Across North Sinai. In G. Capriotti Vittozzi (ed.), Egyptian Curses 2. A Research on Ancient Catastrophes. Archaeological Heritage & Multidisciplinary Egyptological Studies 2: 197–221.

Stewart, *Fossil Fish.* Stewart, K., 2009. Fossil fish from the Nile River and its southern basins. In Dumont, H. (ed.), The Nile, 677–704. Dordrecht: Springer.

Stock & Gifford-Gonzalez, *Genetics & Cattle Domestication.* Stock, F. and Gifford-Gonzalez, D., 2013. Genetics and African cattle domestication. African Archaeological Review, 301: 51–72.

Stock etal., *Skeletal. Biomechanics.* Stock, J., O'Neill, M., Ruff, C., Zabecki, M., Shackelfrod, L. and J. Rose, 2011. Body Size, Skeletal Biomechanics, Mobility and Habitual Activity from the Late Palaeolithic to the Mid-Dynastic Nile Valley. In Pinhasi, R., and J. T. Stock, (eds), Human bioarchaeology of the transition to agriculture. Chichester: Wiley.

Stocker etal., *Climate Change 2013.* Stocker, T., Qin, D., Plattner, G., Alexander, L., Allen, S., Bindoff, N., Bréon, F., Church, J., Cubasch, U., Emori, S. and P. Forster, 2013. Technical summary. In Climate change 2013: the physical science basis. Contribution of Working Group I to the Fifth Assessment Report of the Intergovernmental Panel on Climate Change, 33–115. Cambridge University Press.

Stommel etal., *Hippo Vulnerability.* Stommel, C., Hofer, H. and M. East, 2016. The effect of reduced water availability in the Great Ruaha River on the vulnerable common hippopotamus in the Ruaha National Park, Tanzania. PloS one, 116, 1–18.

Strudwick, *Administration.* Strudwick, N., 1985. The Administration of Egypt in the Old Kingdom; The Highest Titles and their Holders. London: KPI.

Strudwick, *Texts*. Strudwick, N. C., 2005. Texts from the Pyramid Age. Atlanta: Society of Biblical Literature.

Stuart-Macadam, *Porotic hyperostosis*. Stuart-Macadam P., 1992 Porotic hyperostosis: a new perspective. American Journal of Physical Anthropology. 87: 1, 39–47.

Stuart, *Paleoethnobotany*. Stuart, G. S. L., 2014. Paleoethnobotany. In Smith, C. (ed.), The Encyclopedia of Global Archaeology, Springer, 5755–5760.

Suberkropp, *Seasonal Occurrence*. Suberkropp K., 1984 Effect of temperature on seasonal occurrence of aquatic hyphomycetes.Transactions of the British Mycological Society 82: 53–62.

Sukhanovski etal., *Rainfall Erosivity.* Sukhanovski, Y., Ollesch, G., Khan, K. and R. Meiβner, 2002. A new index for rainfall erosivity on a physical basis. Journal of Plant Nutritionand Soil Science 165: 51–57.

Swanson & Morrow-Tesch, *Cattle Transport*. Swanson, J. and J. Morrow-Tesch, 2001. Cattle transport: Historical, research, and future perspectives. Journal of Animal Science, 79E-Suppl, 102–109.

Swinton, *Dating Tombs*. Swinton, J., 2014. Dating the tombs of the Egyptian Old Kingdom. Oxford: Archaeopress.

Swinton, *Resources*. Swinton, J., 2012. The Management of Estates and their Resources in the Egyptian Old Kingdom. Oxford: BAR.

Tackholm, *Landscape, Flora & Agriculture*. Tackholm, V., 1976. Ancient Egypt, Landscape, Flora and Agriculture. In Rzóska, J. (ed.), The Nile: Biology of an Ancient River, 51–72. Dordrecht: Springer.

Taher & Abdel-Motelib, *Microbial Sedimentary Structures*. Taher, A. and A. Abdel-Motelib, 2015. New insights into microbially induced sedimentary structures in alkaline hypersaline El Beida Lake, Wadi El Natrun, Egypt. Geo-Marine Letters, 35: 5, 341–353.

Talling, *History of Nile Research.* Talling, J., 2009. The Nile: History of Scientific Research. In Dumont, H. (ed.), The Nile, 23–34. Dordrecht: Springer.

Tansley, *Terms & Concepts*. Tansley, A., 1935. The Use and Abuse of Vegetational Concepts and Terms. Ecology, 163: 284–307.

Tata, *Egyptian Textiles*. Tata, G., 1986. The development of the Egyptian textile industry. PhD, University of Utah.

Terrell, *SNA & History*. Terrell J., 2013. Social network analysis and the practice of history. In Knappett, C. (ed.), Network Analysis in Archaeology: New Approaches to Regional Interaction, 17–42. Oxford University Press.

Thienemann, *Grundprinzip*. Thienemann, A., 1954. Ein drittes biozonotisches Grundprinzip. Archives für Hydrobiologie 49: 421–422.

Thirukumaran & Ramkumar, *Remote Sensing*. V. Thirukumaran, V. and Mu. Ramkumar, 2015. Remote Sensing—A Fast and Reliable Tool to Map the Morphodynamics of the River Systems for Environmental Management. In Ramkumar, M., Kumaraswamy, K. and R. Mohanraj, (eds), Environmental Management of River Basin Ecosystems, 161–176. Dordrecht: Springer.

Thompson & Hamilton, *Peatlands*. Thompson, K. and A. Hamilton, 1983. Peatlands and swamps of the African continent. Ecosystems of the World. In Gore, A. (ed.),

Mires: swamp, bog, fen and moor, B: regional studies, 331–373 Amsterdam: Elsevier.

Thompson et.al., *Kerma Diet*. Thompson, A., Chaix, L. and M. Richards, 2008. Stable isotopes and diet at ancient Kerma, Upper Nubia Sudan. Journal of Archaeological Science, 352: 376–387.

Thompson, *Swamp Development*. Thompson, K., 1976. Swamp Development in the head waters of the White Nile. In Rzóska, J. (ed.) The Nile: Biology of an Ancient River, 177–196. Dordrecht: Springer.

Thompson, *Tehna* I. Thompson, E., 2014. The Old Kingdom cemetery at Tehna: Volume I: the tombs of Nikaiankh I, Nikaiankh II and Kaihep. Oxford: Aris & Phillips.

Thompson, *Tehna Report*. Thompson, E., 2012. Report for the Supreme Council of Antiquities of the Expedition of the Australian Centre for Egyptology Macquarie University at the Early Old Kingdom Cemetery at Tehna in Middle Egypt Inspectorate of Minya: November–December 2007 Season. *http://www.egyptology.mq.edu.au/08miniconf/Tehna%20Report%20for%20the%20SCA%20Nov%20Dec%202007.htm*

Thopmson etal, *Egyptian Palaeodiet*. Thompson, A., Richards, M., Shortland, A. and S. Zakrzewski, 2005. Isotopic palaeodiet studies of ancient Egyptian fauna and humans. Journal of Archaeological Science, 323: 451–463.

Tiner, *Wetland Indicators*. Tiner, R., 1999. Wetland Indicators: A Guide to Wetland Identification, Delineation, Classification and Mapping. Boca Raton: Routledge.

Tiradritti, *Art, architecture and history*. Tiradritti, F., 2002. Ancient Egypt: art, architecture and history. London: British Museum Press.

Touzeau etal., *Ancient Egyptian Diet*. Touzeau, A., Amiot, R., Blichert-Toft, J., Flandrois, J., Fourel, F., Grossi, V., Martineau, F., Richardin, P. and C. Lécuyer, 2014. Diet of ancient Egyptians inferred from stable isotope systematics. Journal of Archaeological Science, 46: 114–124.

Touzeau etal., *Nile Valley Aridity*. Touzeau, A., Blichert-Toft, J., Amiot, R., Fourel, F., Martineau, F., Cockitt, J., Keith Hall, K., Flandrois, J. and C. Lécuyer, 2013. Egyptian mummies record increasing aridity in the Nile valley from 5500 to 1500yr before present. Earth and Planetary Science Letters, 375: 92–100.

Triambelas, *Caries Prevalence*. Triambelas, K., 2014. Caries Prevalence in Ancient Egyptian and Nubians. PhD, University of Alaska Fairbanks.

Trøjelsgaard & Olesen, *Ecological Networks in Motion*. Trøjelsgaard, K. and J. Olesen, 2016. Ecological networks in motion: micro-and macroscopic variability across scales. Functional Ecology, 3012: 1926–1935.

Trzciñski, etal, *West Saqqara Geoarchaeology*. Trzciñski, J., Kuraszkiewicz, K. and F. Welc, 2010. Preliminary report on geoarchaeological research in West Saqqara, Polish Archaeology in the Mediterranean 19, 194–208.

Turner, *Grazing Effects*. Turner, M., 1987. Effects of grazing by feral horses, clipping, trampling, and burning on a Georgia salt-marsh. Estuaries 10: 54–60.

Uchytil, *Phragmites australis*. Uchytil, R., 1992. Phragmites australis. US Department of Agriculture, Forest Service, Rocky Mountain Research Station, Fire Sciences Laboratory, Fire Effects Information System. http://www. fs. fed. us/database/feis/plants/gramin oid/phraus/all. html.

Uprety etal., *Climate Change & Agriculture*. Uprety, D., Reddy, V. and J. Mura, 2019. Climate Change and Agriculture: A Historical Analysis. Singapore: Springer.

Vachala, *Ptahshepses*. Vachala, B., 2004. Abusir VIII. Die Relieffragmente aus der Mastaba des Ptahshepses in Abusir. Prague: Charles Univeristy Press.

Våge etal., *Microbial Food Webs*. Våge, S., Bratbak, G., Egge, J., Heldal, M., Larsen, A., Norland, S., Lund Paulsen, M., Pree, B., Sandaa, R., Skjoldal, E. and T. Tsagaraki, 2018. Simple models combining competition, defence and resource availability have broad implications in pelagic microbial food webs. Ecology letters, 21: 9, 1440–1452.

van De Mieroop, *Ancient Near East*. van De Mieroop, M., 2007. A History of the Ancient Near East. Oxford: Wiley.

van den Brink, *Settlement Patterns*. van den Brink, E., 1993. Settlement Patterns in the Northeastern Nile Delta in the 4th-2nd Millenium B.C. In Krzyzaniak, L., Kobusiewicz, M. and J. Alexander, (eds), Environmental Change and Human Culture in the Nile Basin and Northern Africa until the Second Millenium B.C., 301–302. Poznań: Poznań Archaeological Museum.

van der Leeuw, *Archaeology, Networks & Beyond*. van der Leeuw, S., 2013. Archaeology, networks, information processing, and beyond. In Knappett, C. (ed.), Network Analysis in Archaeology: New Approaches to Regional Interaction, 335–349. Oxford university Press.

Van der Veen, *Carbonised Plant Remains*. Van der Veen, M., 2007. Formation processes of desiccated and carbonized plant remains–the identification of routine practice. Journal of archaeological science, 346: 968–990.

van Deursen & Drost, *Cattle Defoliation*. van Deursen, E. and H. Drost, 1990. Defoliation and Treading by Cattle of Reed Phragmites Australis. Journal of Applied Ecology 27: 284–297.

Van Elsbergen, *Fischerei*. Van Elsbergen, M., 1997. Fischerei Im Alten Agypten: Untersitclnmgen zu den Fischfangdarstellungen in den Grabern der 4. bis 6. Dynastie. Berlin: Achet Verlag.

Van Lepp, *Artificial Irrigation*? Van Lepp, J., 1995. Evidence for artificial irrigation in Amratian art. Journal of the American Research Center in Egypt, 32, pp.197-209.

van Neer, *Prehistoric Fishing.* van Neer, W., 1989. Fishing along the prehistoric Nile. In Krzyzaniak, L. and M. Kobusiewicz (eds), Late Prehistory of the Nile Basin and the Sahara, 49–56. Poznań: Poznań Archaeological Museum.

Vandier, *La Famine.* Vandier, J., 1936. La Famine dans l'Egypte Ancienne. Cairo: IFAO.

Vandier, *Mo'alla.* Vandier, J., 1950. La tomhe d'Ankhtifi et la tombe de Sebekhotep. Cairo: IFAO.

Vassilika & Bourriau, *Egyptian Art.* Vassilika, E. and J. Bourriau, 1995. Egyptian Art at the Fitzwilliam Museum. Cambridge University Press.

Veiga, *Prevalent Pathologies.* daSilva Veiga, P., 2012. Some prevalent pathologies in ancient Egypt. Institut Oriental: Centro de Estudios Arqueológicos: des Universités de Genève et Porto, 63–83. Porto.

Veiga, *Health & Medicine.* daSilva Veiga, P., 2009. Health and medicine in ancient Egypt: magic and science. Oxford: BAR.

Venkatachalapathy & Karthikeyan, *Water Quality Index.* Venkatachalapathy, R. and P. Karthikeyan, 2015. Diatom Indices and Water Quality Index of the Cauvery River, India: Implications on the Suitability of Bio-Indicators for Environmental Impact Assessment. In E Ramkumar, M., Kumaraswamy, K., and R. Mohanraj, (eds), Environmental Management of River Basin Ecosystems,707–727. Dordrecht: Springer.

Venkatesan etal., *Fluvial System.* Venkatesan, A., Jothibasu, A. and S. Anbazhagan, 2015. GIS Based Quantitative Geomorphic Analysis of Fluvial System and Implications on the Effectiveness of River Basin Environmental Management. In Ramkumar, M., Kumaraswamy, K. and R. Mohanraj, (eds), Environmental Management of River Basin Ecosystems, 161–176. Dordrecht: Springer.

Vereecken etal., *Dayr al-Barsha Pottery.* Vereecken, S., De Meyer, M., Dupras, T., and L. Williams, 2009. An Old Kingdom Funerary Assemblage at Dayr al-Barsha. In Rzeuska, T. and A. Wodzińska (eds), Studies on Old Kingdom Pottery. Proceedings of the Old Kingdom Pottery Workshop, Warsaw, 20–21 August, 2007, 187–207. Warsaw: Archaeobooks.

Verner, *Dating in the Old Kingdom.* Verner, M., 2008. The System of Dating in the Old Kingdom. In Bárta, M and H, Vymazalová, (eds), Chronology and Archaeology in Ancient Egypt, 23–43. Prague: Charles Univeristy Press.

Verner, *Forgotten Pyramids.* Verner, M., 1994., Forgotten Pyramids: The Lost Pyramids of Abusir. Prague: Charles Univeristy Press.

Verner, *Ptahshepses.* Verner, M., 1986/7. Abusir I. The Mastaba of Ptahshepses I. Prague: Charles Univeristy Press.

Verner, *Sons of the Sun.* Verner, M., 2014. Sons of the Sun. Rise and Decline of the Fifth Dynasty. Prague: Charles Univeristy Press.

Verner, *Who was Shepseskara?* Verner, M., 2000. Who was Shepseskara, and when did he reign? In Bárta, M. and J. Krejčí (eds), Abusir and Saqqara in the Year 2000, 581–602. Prague: Charles Univeristy Press.

Vernus & Yoyotte, *Bestiaire.* Vernus, P. and J. Yoyotte, 2005. Bestiaire des pharaons. Paris: Perrin.

Verreth etal., *Catfish Digestive System.* Verreth, J., Torreele, E., Spazier, E., Sluiszen, A., Rombout, J., Booms, R. and H. Segner, 1992. The development of a functional digestive system in the African catfish Clarias gariepinus Burchell. Journal of the World Aquaculture Society, 234: 286–298.

Virgós & Casanovas, *Genet Habitat Selection.* Virgós E. and J. Casanovas, 1997. Habitat selection of genet Genetta genetta in the mountains of central Spain. Acta Theriol. 42: 169–177.

Vischak, *Agency.* Vischak, D., 2007. Agency in Old Kingdom Elite Tomb Programs: traditions, Locations, and Variable Meanings. Internet Beitrage zur Agyptologie und Sudanarchaologie. Vol. 6. Dekorierte Grabanlangen in Alten Reich Methodik und Interpretation: 255–276.

Vischak, *Locality & Community.*Vischak, D., 2006. Locality and community in Old Kingdom provincial tombs: The cemetery at Qubbet el Hawa. Ph.D. New York University.

Vischak, *Elephantine Identity.* Vischak, D., 2007. Identity in/of Elephantine: the Old Kingdom Tombs at Qubbet el Hawa. In Hawass, Z. and J. Richards (eds), The Archaeology and Art of Ancient Egypt: Essays in Honor of David B. O'Connor II, 443–457. Cairo: Supreme Council of Antiquities.

Vivian etal., *Wetland Plant Growth.* Vivian, L., Godfree, R., Colloff, M., Mayence, C. and D. Marshall, 2014. Wetland plant growth under contrasting water regimes associated with river regulation and drought: implications for environmental water management. Plant Ecology, 215: 9, 997–1011.

Vogelsang-Eastwood, *Textiles.* Vogelsang-Eastwood, G., 2000. Textiles. In Nicholson, P. and I. Shaw (eds), Ancient Egyptian Materials and Technology, 268–298.

von den Driesch & Peters, *Störche über Elephantine.* von den Driesch, A. and J. Peters, 2008. Störche über Elephantine. In Engel, E., Müller, V. and U. Hartung, (eds), Zeichen aus dem Sand. Streiflichter aus Ägyptens Geschichte zu Ehren von Günter Dreyer, 661–679. Wiesbaden: Harrassowitz-Verlag.

Vymazalová & Arias Kytnarova, *Sheretnebty*, Vymazalová, H. and K. Arias Kytnarová, 2017. The development of tomb AS 68c in Abusir South: Burial place of the king's daughter Sheretnebty and her family. In Bárta. M., Coppens, F. and J. Krejčí (eds), Abusir and Saqqara in the Year 2015, 435–450. Prague: Charles Univeristy Press.

Walker etal., *Anemia Hypothesis Reappraisal*. Walker, P. L., Bathurst, R. R., Richman, R., Gjerdrum, T. and V. A. Andrushko, 2009. The causes of porotic hyperostosis and cribra orbitalia: A reappraisal of the iron-deficiency-anemia hypothesis. American Journal of Physical Anthropology, 1392: 109–125.

Walker et al, *Holocene Subdivision*. Walker, M., Berkelhammer, M., Björck, S., Cwynar, L., Fisher, D., Long, A., Lowe, J., Newnham, R., Rasmussen, S. and H. Weiss, 2012. Formal subdivision of the Holocene Series/Epoch: a Discussion Paper by a Working Group of INTIMATE Integration of ice-core, marine and terrestrial records and the Subcommission on Quaternary Stratigraphy International Commission on Stratigraphy. Journal of Quaternary Science 27, 649–659.

Wapler etal., *Sudanese Cranial Pathology*. Wapler, U., Crubézy, E., and M. Schultz, Is cribra orbitalia synonymous with anemia? Analysis and interpretation of cranial pathology in Sudan. American Journal of Physical Anthropology 123: 333–339.

Warden, *Old Kingdom Economy.* Warden, L., 2014. Pottery and Economy in Old Kingdom Egypt. Leiden: Brill.

Watkins, *Hydrology, Nutrients & Salinity*. Watkins, L., 2005. The effects of hydrology, nutrients, and salinity on ten common wetland plants in Louisiana: a mesocosm study. PhD, Southern Louisiana University.

Weins, *Riverine Landscapes*. Weins J., 2002 Riverine landscapes: taking landscape ecology into the water. Freshwater Biology 47: 501–515

Weiskel, *Ecological Lessons*. Weiskel, T. 1989. The Ecological Lessons of the Past: An Anthropology of Environmental Decline. The Ecologist 193: 98–103.

Weiss, *Collapse*. Weiss, H., 2000. Beyond the Younger Dryas: Collapse as adaptation to abrupt climate change in ancient West Asia and the Eastern Mediterranean. In Bawden, G. and R. Reycraft, (eds), Environmental Disaster and the Archaeology of Human Response. Anthropological Papers, No. 7, 75–98. Albuquerque: Maxwell Museum of Anthropology.

Welc & Marks, *OK Climate.* Welc, F. and L. Marks, 2014. Climate change at the end of the Old Kingdom in Egypt around 4200 BP: New geoarchaeological evidence. Quaternary International, 324, 124–133.

Welc & Marks, *OK Climate Change*. Welc, F. and L. Marks, 2013. Climate change at the end of the Old Kingdom in Egypt around 4200 BP: New geoarchaeological evidence. Quaternary International, XXX: 1–10.

Wells, *Changes in Nutritional Diseases*. Wells, C., 1975. Prehistoric and historical changes in nutritional diseases and associated conditions. Progress in Food & Nutrition Science 1: 11–12.

Wemmer, *Comparative Ethology*. Wemmer, C., 1977. Comparative ethology of the large-spotted genet Genetta tigrina and some related viverrids. Washington: Smithsonian.

Wendrich etal, *Fayum Neolithic*. Wendrich, W., Taylor, R. and J. Southon, 2010. Dating stratified settlement sites at Kom K and Kom W: fifth millennium BCE radiocarbon ages for the Fayum Neolithic. Nuclear Instruments and Methods in Physics Research Section B: Beam Interactions with Materials and Atoms, 2687: 999–1002.

Wengrow, *Cattle Cults*. Wengrow, D., 2001. Rethinking 'cattle cults' in early Egypt: towards a prehistoric perspective on the Narmer Palette. Cambridge Archaeological Journal, 1101: 91–104.

Wenke etal., *Kom el-Hsin*. Wenke, R., Buck, P., Hamroush, H., Kobusiewicz, M., Kroeper, K. and R. Redding, 1988. Kom el-Hsin: Excavation of an Old Kingdom Settlement in the Egyptian Delta. Journal of American Research Center in Egypt 25: 5–34.

Wenke, *Egyptian Civilisation Evolution*. Wenke, R. 1991. The evolution of early Egyptian civilization: Issues and evidence. Journal of World Prehistory, 53: 279–329.

Wenke, *OK Western Delta*. Wenke. R., 1986. Old kingdom community organization in the Western Egyptian Delta. Norwegian Archaeological Review 19: 1, 15–33.

Westlake, *Light Climate for Plants*. Westlake, D., 1966, The light climate for plants in rivers. In Bainbridge, R., Evans, G. and O. Rackham (eds), Light as an Ecological Factor, 99–119. Oxford: Blackwell.

Wetterstrom, *Foraging & Farming*. Wetterstrom, W., 1993. Foraging and farming in Egypt: the transition from hunting and gathering to horticulture in the Nile Valley. In Shaw, T., Sinclair, P., Andah, B. and A. Okpoko (eds), The Archaeology of Africa. Food, metals and towns, 165–226. London: Taylor & Francis.

White & Mattingly, *Ancient Saharan Lakes*. White, K. and J. Mattingly, 2006. Ancient lakes of the Sahara. American scientist, 941: 58–65.

White, *Aswan Environmental Effects*. White, G., 1988. The environmental effects of the high dam at Aswan. Environment: Science and Policy for Sustainable Development, 307: 4–40.

WHO, *Dietary References*. World Health Organisation, 2007. Dietary Reference Intakes for Energy, Carbohydrate, Fibre, Fat, Fatty Acids, Cholesterol, Protein and Amino Acids Macronutrients; WHO Technical Report Series 935. Geneva: WHO Press.

Wild, *Ti* II. Wild, H. 1953. Le tombeau de Ti, Fasc II: La chapelle première partie. Mémoires publiés par les

Membres de l'Institut Français d'Archéologie Orientale du Caire, Le Caire.

Wild, *Ti* III. Wild, H., 1966. Le tombeau de Ti, Fasc III: La chapelle deuxième partie. Mémoires publiés par les Membres de l'Institut Français d'Archéologie Orientale du Caire, Le Caire.

Wilkinson & Stevens, *Enviromental Archaeology*. Wilkinson, K. and C. Stevens, 2010. Environmental Archaeology: Approaches, Techniques and Applications. Stroud: History Press.

Wilkinson, *Gods & Goddesses*. Wilkinson, R., 2003. The Complete Gods and Goddesses of Ancient Egypt. London: Thames & Hudson.

Wilkinson, *Marshes in the Middle East*. Wilkinson, T., 2012. Wetland Archaeology and the Role of Marshes in the Ancient Middle East. In Menotti, F. and A. O'Sullivan, (eds), The Oxford Handbook of Wetland Archaeology, 121–140. Oxford University Press

Wilkinson, *Political Unification*. Wilkinson, T., 2000. Political unification: Towards a reconstruction. Mitteilungen des Deutschen Archäologischen Instituts, Abteilung 56: 377–395, Kairo: Mitteilungen des Deutschen Archäologischen Instituts.

Wilkinson, *Reading Egyptian Art*. Wilkinson, R., 1994. Reading Egyptian art: a hieroglyphic guide to ancient Egyptian painting and sculpture. London: Thames & Hudson.

Wilkinson, *Symbolism & Design*. Wilkinson, A., 1994. Symbolism and design in ancient Egyptian gardens. Garden History: 1–17.

Willcocks, *Irrigation*. Willcocks, C., 1913. Egyptian Irrigation. London: E. & F. N. Spon.

Willems etal., *Wadi Zabayada*. Willems, H., Vereecken, S., Kuijper, L., Vanthuyne, B., Marinova, E., Linseele, V., Verstraeten, G., Hendrickx, S., Eyckerman, M., Van den Broeck, A., Van Neer, W., Bourriau, J., French, P., Peeters, C., De Laet, V., Mortier, S. and Z. De Kooning, 2009. An industrial site at Al-Shaykh Sacıd/Wadı Zabayda. Ägypten und Levante/Egypt and the Levant 19: 293–331.

Williams & Hill, *Meat & Evolution*. Williams, A., and L. Hill, 2017. Meat and nicotinamide: a causal role in human evolution, history, and demographics. International Journal of Tryptophan Research, 10: 1, 1–23.

Williams & Nottage, *Extreme Rainfall Impacts*. Williams, M., and J. Nottage, 2006. Impact of extreme rainfall in the central Sudan during 1999 as a partial analogue for reconstructing early Holocene prehistoric environments. Quaternary international, 1501: 82–94.

Williams etal., *Late Quaternary Floods & Drought*. Williams, M., Williams, F, Duller, G., Munro, R, El Tom, O., Barrows, T., Macklin, M., Woodward, J., Talbot, M., Haberlah, D and J. Fluin, 2010. Late Quaternary floods and droughts in the Nile valley, Sudan: new evidence from optically stimulated luminescence and AMS radiocarbon dating. Quaternary Science Reviews, 299, 1116–1137.

Williams, *Ancient Nile Environments*. Williams, M., 2009. Late Pleistocene and Holocene environments in the Nile basin. Global and Planetary Change, 691: 1–15.

Williams, *Human Impact Nile Basin*. Williams, M., 2009. Human Impact on the Nile Basin: Past, Present, Future. In Dumont H. (ed.), The Nile, 771–779. Dordrecht: Springer.

Williamson & Payne, *Animal Husbandry*. Williamson, G. and W. Payne., 1978. Animal Husbandry in the Tropics. London: Longman.

Wilson, *Mendes*. Wilson, K. L., 1982. Cities of the Delta Volume II: Mendes. Malibu: ARCE Reports.

Winckler & Wilfing, *Anthropologische un den Skelettresten*. Winkler, E. and H. Wilfling. 1991 Tell el-Dab'a VI: Anthropologische Untersuchungen an der Skelettresten der Kampagnen 1966–1969, 1975–1980, 1985. VI. Vol. 10. Wien: Österreichische Akademie der Wissenschaften.

Wing, *Animals as Food*. Wing, E., 2000. Animals Used for Food in the Past: As Seen by Their Remains Excavated from Archaeological Sites. In Kiple, K. and K. Ornelas (eds), The Cambridge World History of Food, 51–58. Cambridge University Press..

Wiskens & Rzóska, *Vegetation*. Wickens, G. and J. Rzóska, 1976. Vegetation of the Nile Basin. In Rzóska, J. (ed.), The Nile: Biology of an Ancient River, 79–86. Dordrecht: Springer.

Wittfogel, *Oriental Despotism*. Wittfogel, K., 1957 Oriental Despotism: A Comparative Study of Total Power. New York: John Wiley & Sons.

Woods, *Five Features*. Woods, A., 2015. Five significant features in Old Kingdom spear-fishing and fowling scenes. In Kousoulis, P. and N. Lazaridis (eds), Proceedings of the Tenth International Congress of Egyptologists, University of the Aegean, Rhodes, 22-29 May 2008, 1897–1910. Leuven: Peeters.

Woods, *Marshes*. Woods, A., 2007. A Day in the Marshes. PhD, Macquarie University, Sydney.

Woodward etal., *Nile Evolution*. Woodward, J., Macklin, M., Krom, M., and M. Williams 2007. The Nile: evolution, Quaternary river environments and material fluxes. Large Rivers, Geomorphology and Management, 261–289. New York: John Wiley & Sons

Wörster, *Rivers of Empire*. Wörster, D., 1985. Rivers of Empire: Water Aridity and the Growth of the American West. New York: Pantheon.

Wright, *Humans as Agents*. Wright, D., 2017. Humans as Agents in the Termination of the African Humid Period. Frontiers in Earth Science. 5: 4, 1–14.

Yang et.al, *Phragmites Stress*. Yang, Z., Xie, T. and Q. Liu, 2014. Physiological responses of Phragmites australis to the combined effects of water and salinity stress. Ecohydrology, 72: 420–426.

Yilmaz, etal., *Domesticated Donkey*. Yılmaz, O., Saim, B., and E. Mehmet, 2012. The domesticated donkey: I–species characteristics. Can J App Sci, 42: 339–353.

Yitayew & Melesse, *Water Resource Issues*. Yitayew, M. and A. Melesse, 2011. Critical Water Resources Issues in the Nile River Basin. In Melesse, A. (ed.) Nile River Basin Hydrology, Climate and Water Use, 401–416. Dordrecht: Springer.

Zahran & Willis, *Vegetation of Egypt*. Zahran, M. and A. Willis, 2008. The Vegetation of Egypt Vol. 2. Dordrecht: Springer.

Zakrzewski, *Life Expectancy*. Zakrzewski, S., 2015. Life Expectancy. UCLA Encyclopedia of Egyptology, 11. *http://escholarship.org/uc/item/7zb2f62c*.

Zakrzewski, *Continuity or Change*. Zakrzewski, S., 2007. Population continuity or population change: formation of the ancient Egyptian state. American Journal of Physical Anthropology 132: 501–509.

Zakrzewski, *Variation in Ancient Egyptians*. Zakrzewski, S., 2003. Variation in ancient Egyptian stature and body proportions. American Journal of Physical Anthropology, 1213: 219–229.

Zanchetta etal., *Tephrostratigraphy.* Zanchetta, G., Sulpizio, R., Roberts, N., Cioni, R., Eastwood, W., Siani, G., Caron, B., Paterne, M. and R. Santacroce, 2011. Tephrostratigraphy, chronology and climatic events of the Mediterranean basin during the Holocene: an overview. Holocene, 211: 33–52.

Zeder, *Origins of Agriculture*. Zeder, M., 2011. The origins of agriculture in the Near East. Curr. Anthropol. 52: 221–235.

Zeuner, *Domesticated Animals.* Zeuner F., 1963. A history of domesticated animals. London: Hutchinson.

Zhao etal., *Nutrient Accumulation*. Zhao, Y., Xia, X. and Z. Yang, 2013. Growth and nutrient accumulation of Phragmites australis in relation to water level variation and nutrient loadings in a shallow lake. Journal of Environmental Sciences, 25: 1, 16–25.

Ziada etal., *Economic Potentialities*. Ziada, M., Mashaly, I., El-Monem, M. and M. Torky, 2008. Economic potentialities of some aquatic plants growing in north east Nile delta, Egypt. Journal of Applied Sciences, 8: 8, 1395–1405.

Zink etal., *Tuberculosis from Mummies*. Zink, A., Haas, C., Reischl, U., Szeimies, U. and A. Nerlich, 2001. Molecular analysis of skeletal tuberculosis in an ancient Egyptian population. Journal of Medical Microbiology 50: 355–366.

Zohary & Hopf, *Old world Plant Origins*. Zohary, D. and M. Hopf, 2000. Domestication of plants in the old world: the origin and spread of cultivated plants in West Asia, Europe and the Nile Valley. Oxford University Press.

Zorn & Komac, *Erosivity*. Zorn M. and B. Komac, 2013 Erosivity. In Bobrowsky P.T., (ed.), Encyclopedia of Natural Hazards (Vol. 1135). Dordrecht: Springer.

www.ingramcontent.com/pod-product-compliance
Lightning Source LLC
LaVergne TN
LVHW071211210926
848482LV00015B/782